GOD

AND THE

Multiverse

ALSO BY VICTOR J. STENGER

GOD

AND THE

Multiverse

HUMANITY'S
EXPANDING VIEW
OF THE COSMOS

Victor J. Stenger

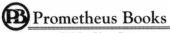 Prometheus Books

59 John Glenn Drive
Amherst, New York 14228

Published 2014 by Prometheus Books

Cover image © Ken Crawford / Media Bakery
Cover design by Jacqueline Nasso Cooke

Inquiries should be addressed to
Prometheus Books
59 John Glenn Drive
Amherst, New York 14228
VOICE: 716–691–0133 • FAX: 716–691–0137
WWW.PROMETHEUSBOOKS.COM

18 17 16 15 14   5 4 3 2 1

Library of Congress Cataloging-in-Publication Data

Stenger, Victor J., 1935-
    God and the multiverse : humanity's expanding view of the cosmos / by
Victor J. Stenger.
        pages cm
    Includes bibliographical references and index.
    ISBN 978-1-61614-970-3 (hardcover) — ISBN 978-1-61614-971-0 (ebook)
    1. Cosmology. I. Title.
BD511.S74 2014
523.1 — dc23

2014013909

Printed in the United States of America

Dedicated to the memory of David Norman Schramm (1945–1997)

CONTENTS

CONTENTS   **13**

ACKNOWLEDGMENTS

The author is indebted to many who have provided feedback on this manuscript. In particular, I must single out Bob Zannelli, Don McGee, and Brent Meeker for their meticulous multiple readings and their many critical corrections and recommended rewordings. They and other members of the Google discussion group atvoid ("Atoms and the Void") have been a continual help to me with my previous books and the numerous essays and blogs I generate. This group includes Greg Bart, Lawrence Crowell, Anne O'Reilly, Kerr Regier, Christopher Savage, Brian Stilson, Pete Stewart, Phil Thrift, Jim Wyman, and Roahn Wynar. I am also grateful for comments from author Kim Clark, physicist Taner Edis, and mathematician James Lindsay.

Thanks also go to several professional astronomers, physicists, and cosmologists who have provided invaluable comments. I am especially grateful to astronomer Jay Pasachoff for his meticulous reading of the manuscript and for providing many corrections. Suggested changes to specific portions of the manuscript where they had deep expertise were also made by Sean Carroll, Alan Guth, Lawrence Krauss, Andrei Linde, Robert Nemiroff, Brent Tully, and Alex Vilenkin.

Finally, I have once again benefitted greatly from the support of my work given by Jonathan Kurtz, Steven L. Mitchell, and their dedicated, talented staff at Prometheus Books.

PROLOGUE

In July 2012, after I sent the first-draft manuscript for a book titled *God and the Atom: From Democritus to the Higgs Boson* to the publisher, a dramatic news conference was held at the European Organization for Nuclear Research (CERN) in Geneva that made headlines around the world. It was announced that two experiments, each costing about a billion dollars and involving thousands of scientists from dozens of nations, had independently confirmed, with high empirical significance, the existence of an elementary particle called the Higgs boson.

This Higgs boson had been proposed forty-eight years earlier to provide the mechanism by which other elementary particles obtain mass. Its discovery corroborated the standard model of elementary particles, which was developed in the 1970s. Since then, the standard model has successfully described the basic constituents of subatomic matter and the forces by which they interact to construct the physical universe. No empirical violation of the standard model has yet been reported.

Fortunately, I was able to work a full account of the Higgs discovery into *God and the Atom*, which was published in 2013.

In March 2014, I had submitted the first draft of the manuscript for the present book to the publisher when this story was repeated. A news conference at Harvard attracted similar worldwide attention, making front pages everywhere. An international collaboration working at the South Pole announced that it had observed, with high significance, a type of polarization in the cosmic microwave background called "B-mode" that was interpreted as a signal for gravitational waves produced by quantum fluctuations in space-time when the universe first came into existence 13.8 billion years ago.

Most media reports failed to mention that the result was pro-visional, that all other explanations have not been ruled out, and that the result remains to be independently verified. However, if the observation passes these tests, the implication is momentous.

B-mode polarization in the early universe had been predicted by a cosmological model, first proposed in 1980, called *inflation*. According to inflation, during the first tiny fraction of a second after it came into being, the universe expanded exponentially by many orders of magnitude. The model solved a number of outstanding problems in cosmology and has since passed several rigorous tests that could have falsified it.

Inflation strongly implies that our universe is not alone but is just one of an unlimited number of other universes in what has been termed the *multiverse*. This multiverse exists endlessly in space and time. It had no beginning, no creation. It always existed and always will.

Once again I was able to include this discovery in the present book, which contains a detailed discussion of the cosmic micro-wave background (CMB). We will see how precise measurements of the CMB by extraordinary instruments in space and on Earth have provided us with a deep knowledge of the earliest moments of our universe and its evolution to the present day.

The aim of *God and the Multiverse* is to show how the current picture of our vast universe and the real possibility of multiple uni-verses developed over the millennia since humans first looked at the sky and asked what was out there. We will examine how the ancient notion of a supernatural creation arose to explain what eventually became explicable by purely natural processes.

The story is necessarily a long one, and the reader is asked to be patient and follow step-by-step the chronological progression as humanity's view of the cosmos expanded from a flat Earth with the heavens above and waters below to a hundred billion galaxies, each with a hundred billion stars, and countless planets capable of sus-taining life, and then on to an infinite and eternal multiverse.

PREFACE

In my 2013 book *God and the Atom*, I traced the history of the ancient notion that the universe is nothing more than atoms and the void, with no role for gods, spirits, or the supernatural of any kind. I began the story 2,500 years ago with the Greek philosophers Leucippus (fifth century BCE) and Democritus (ca. 460–370 BCE), and wound up with the observation of the Higgs boson in 2012, which serendipitously happened just as I was finishing the manuscript. This provided the final verification for the standard model of elementary particles that had been in place since the 1970s and has agreed with all the data since.

That is not to say it's the final word. With the Large Hadron Collider now being doubled in energy, we can look forward to moving to the next level in our understanding of the nature of matter—what may lie beyond the standard model.

In this book I move from the very small to the very large and examine how humanity's view of the cosmos has dramatically changed over the last ten thousand years, from the primitive picture of a flat Earth with heaven above and an underworld below to the current majestic array of a hundred billion galaxies of stars lighting the night sky.

We will explore how the disciplines of particle physics and cosmology have collaborated to give a common picture of a boundless and eternal *multiverse* in which our universe is just one among countless others and has neither a beginning nor an end.

Of course, the existence of other universes besides our own has not been empirically established—at least not yet. We will see that the verification of the multiverse is not beyond the realm of possibility.

Given the current scientific respectability of the multiverse hypothesis, philosophers and theologians cannot simply sit back and ignore the implications of such a profound notion. All major religious traditions are likely to have great difficulty reconciling their concept of a creator god with a multiverse that is eternal and uncreated.

Even without the multiverse, there exist in our own, visible universe quadrillions of planets capable of supporting life. How will theists square that with their believe that humans are a special creation of God?

As we will learn in the early chapters, observations by ancient Greek astronomers led them, with one or two exceptions, to develop a picture of the sun, moon, planets, and stars circling a spherical Earth. In a remarkable feat of mathematics, in the second century of the Common Era, Alexandrian astronomer Claudius Ptolemy developed a complex geocentric model of the solar system. His model enabled the accurate prediction of the motions of the planets, which even the ancient saw did not follow exact circles but wandered around the sky.

This model stood until the sixteenth century when the Polish astronomer and Church canon Nicolaus Copernicus (1473–1543) proposed that Earth is just another planet that revolves around the sun.[1] This revived the scheme that was originally put forward by Aristarchus of Samos (ca. 310–230 BCE) in the third century before the Common Era and was finally confirmed in the early seventeenth century when the Florentine physicist Galileo Galilei (1564–1642) turned the newly invented telescope on the heavens. We will cover these events in detail and show that their true implications are not exactly as commonly thought.

Galileo also founded the new science of particle mechanics that would be fully developed in England a generation later by perhaps the greatest scientific mind of all time, Isaac Newton (1642–1727).

Central to the success of the new science was the replacement of sacred authority by observation as the final arbiter of what was to be regarded as physical reality, at least among those scholars who were not bound by Church dogma. This doctrine of empiricism was first proposed by Francis Bacon (1561–1626) and seconded by John Locke (1632–1628), becoming the operating principle of the institution that set the standards for the scientific age that followed: the Royal Society of London for Improving Natural Knowledge, granted a charter by Charles II in 1660, shortly after he was restored to the throne.

Once the scientific revolution was underway, telescopes became steadily more powerful and it was realized that the sun is just another star in a galaxy of stars called the Milky Way. With large reflecting telescopes, originally invented by Newton, placed on mountaintops, the twentieth century saw incredible advances in our knowledge of the cosmos. The Milky Way in which we reside was discovered to be just one galaxy among others in the observable universe, now known to number on the order of 100 billion, each containing on the order of a 100 billion stars.

Equally startling was the discovery that the universe is expanding, with most galaxies receding from one another at speeds that increase with their distances from each other. This suggested that the universe is the remnant of an explosion called the *big bang* that occurred 13.8 billion years ago, by our current estimate.

The size of the observable universe calculated by astronomers went from around 1,000 light-years in 1900 to 46.5 billion light-years in 2000.[2] Any signal traveling at the speed of light that we might receive from that distance would have come from the origin of the big bang. A signal from a lesser distance would have been emitted later. Any signal from a greater distance would not have reached us in the age of the universe. Such a signal would be beyond our *light horizon*. But note that this does not mean nothing exists beyond that distance.

As the twentieth century neared its close, cosmologists had accumulated convincing evidence that a tiny fraction of a second after the universe appeared, it briefly expanded exponentially by many orders of magnitude. As a result, the total universe that resulted from that original explosion is now far larger than that the portion within our horizon, perhaps hundreds of orders of magnitude larger.

As if that isn't sufficiently mind-boggling, as mentioned we now have strong indications that this vast universe in which we live is just one of an endless number of other universes within a multiverse that possibly extends indefinitely in space and endlessly in both the past and future. If it exists, it would have no beginning and no end.

These conclusions are not wild speculations but are based on vast amounts of ever-increasing and increasingly precise quantitative observations on mountaintops, in space, and underground. The main sources of information have been photons, the particles of electromagnetic radiation that are emitted from astronomical objects. They are observed not only in the spectrum to which our eyes are sensitive but also over a range of twenty or more orders of magnitude in energy that vastly exceeds the visible range.

And now, the budding field of neutrino astronomy is joining photon astronomy to produce yet another window on the universe. Neutrinos interact very weakly and are able to pass through a great thickness of matter as if it were not there. They promise to reveal processes and details about the cosmos that are inaccessible to photons.

While all these observational methods have contributed to a spectacular enrichment of our knowledge of the universe, perhaps the most significant has been the incredible wealth of information gained by measurements of small anisotropies across the sky in the cosmic microwave background (CMB), the radiation left over from the big bang. These reveal the spectrum of sound waves produced in the very early universe by the quantum fluctuations in space that ultimately led to the complex structure of matter that formed the galaxies we see today. The big bang really made a bang. The CMB data, which exhibit the fundamental tones and harmonics of a crude musical instrument being played in a hall with poor acoustics, have enabled cosmologists to determine not only characteristics of that instrument but also something of the structure of the hall itself.

From these and other observations, it has been determined that luminous matter—the stars and hot gas we see in the sky by eye and instrument—constitutes a mere 0.5 percent of the total mass of our universe. Another 4.5 percent is nonluminous matter, such as planets and dead stars, made of the same familiar atoms. In addition, 26 percent is composed of something different than atoms or their constituent elementary particles. Dubbed *dark matter*, it remains unidentified. The remaining, dominant 69 percent of the universe is an even more mysterious *dark energy* that has resulted in an accelerating expansion of the universe that will continue indefinitely into the future, making the universe increasingly dilute.

Today the term *atom* conventionally refers to the elements of the chemical periodic table. However, if we take the term as defined by the ancient Greeks to refer to the basic irreducible ("uncuttable") material stuff of the universe, whatever it is, then we can retain their description of the natural world as "atoms and the void." Void, then, identifies the empty space between atoms.

Note that the existence of dark "energy" does not imply anything immaterial, since it still exhibits the qualities of inertia and gravitation that characterize matter. Energy is one of the properties of matter and is not separate stuff.

But this book is more than a particle physicist's presentation of the history of cosmology. From the earliest days when humans contemplated the world around them, they have sought explanations for what they saw. Until very recently, they lacked the tools, both physical and mental, that were needed to provide a reliable picture free of magic and superstition. Not seeing the actual forces behind many events, they mythologized those events in terms that were familiar but simply invisible and more powerful than those forces that they did see or, such as the wind, that they could feel.

When it came to the cosmos, the sky was the "heavens" and the precise motions of the heavenly bodies contrasted sharply with the unpredictability of events on Earth, suggesting they were gods or under the control of gods. For most of history, astronomy and astrology were related activities. Astronomy provided an accurate clock by which events such as the Nile flood could be usefully predicted. Astrology provided useless predictions such as when to go to war.

Copernicus, Galileo, and Newton gave us a less human-centered model of the world. But for the centuries that followed until very recently, most scientists still saw the need for a divine providence behind everything.

In this book I describe the events that led to the remarkable success of current cosmology and the picture it now paints of our universe, including the marvelous prospect that many other universes exist as well. The vast majority of humanity does not appreciate most of these findings. I hope I can make an incremental addition to their accessibility.

I will show how the universe that presents itself to our senses and instruments can be described without any need for the introduction of forces other than those that are purely natural. This conclusion is disputed by those who declare that they cannot understand how all the complexity we see around us, in life on Earth and in the vast galactic structures in the sky, can have happened except by the hand of some supreme power. So they conclude that a creator God must exist.

As we will see in many examples throughout the book, this argument from complexity is nothing more than the God-of-the-gaps argument—often less politely called the "argument from ignorance." Just because a particular author cannot personally understand how some phenomenon can be explained naturally, it does not follow that no explanation is possible other than a supernatural one.

On the other hand, science is in the business of finding natural explanations for events because those explanations often prove useful to human life. Where would today's teenager be without the smartphone, which is an application of electromagnetic theory?

Indeed, imagine life without electricity. In fact, you don't have to imagine it—just look at history.

Perhaps more important than practical applications, the more science is able to demonstrate the bankruptcy of the magical thinking inherent in all religion, moderate as well as extreme, the less humans will rely on this worthless and dangerous way to rule their lives. The future survival of humanity depends on it ridding itself of magical thinking.

Our current picture of the physical world is well described by the standard model of elementary particle physics and various cosmological models . Of course, not every phenomenon has a proven explanation within that framework. We still don't know everything, and we never will. And you can be sure that much that is in this book eventually will be outdated in the future. All we can do is provide plausible models for the observations made to date. Because these natural models are more parsimonious, that is, require fewer assumptions than supernatural alternatives, their very viability suffices to refute any claims that science provides support for gods or spirits of any kind.

Throughout this book, the primary emphasis is on science and, within science, more on observation and experiment than theory. My own background is that of a retired professor of physics who spent forty years teaching and doing research in experimental elementary particle physics and astrophysics.

In this regard, I differ from the preponderance of professional physicists and cosmologists who write popular books, most of them quite excellent and to which I often refer, who are mostly theorists. I generally try to avoid the numerous theoretical speculations you will find in their books, entertaining as they are, and stick as much as possible to what the data say.

Of course, I can't avoid some theoretical interpreting, since there is a close tie between theory and experiment. But where I do this, I try to choose the simplest proposals that are consistent with all observations and make the fewest hypotheses not required by the data. In particular, I do not worry myself when some theory has a logical or mathematical problem requiring its proponents to generate paper after paper proposing solutions. That's their business, not mine. If one of their equations goes to infinity, then the equation is wrong since, as we will see, there are no infinities in the empirical world. And this includes the so-called singularity that so many people still think spawned our universe.

Theoretical physicists and mathematicians understandably marvel at the wonders of mathematics and how surprisingly, even "unreasonably," successful it is when put to practical use. This leads them — with several notable exceptions I will mention — to view their equations and the other mathematical objects in their theories as the actual elements of ultimate reality. They adopt the concept, first proposed by Plato (438–427 BCE), in which our observations are simply shadows or distortions of the true reality that exists in ideal "forms."

I disagree. The approach I take throughout is that observation is our only source of information about the world. The models we then formulate are attempts to rationalize our observations and put them to practical use. These models may contain mathematical abstractions, but it is a mistake to assume that these abstractions relate in any direct way to whatever may be the elements of an ultimate reality that lies beyond what we detect with our senses and scientific instruments. Of course, for the models to be successful they must have some relation to reality. But we have no way of knowing what that relation may be. Furthermore, they are constantly being replaced by new and better models, so how can they possibly represent absolute reality?

We must make a careful distinction, which unfortunately the majority of scientists do not make, between the objects in their models and the quantities they measure. This includes our most basic notions, such as space, time, mass, and energy. In actual scientific practice, these are all defined *operationally*, that is, in terms of carefully prescribed, repeatable measuring procedures. And so time is what you read on a clock. Temperature is what you read on a thermometer. All these notions are human contrivances introduced to quantify our observations. An alien species might make different definitions. And neither theirs nor ours are likely to bear a one-to-one correspondence with the elements of reality.

Short of divine revelation, for which no evidence exists, I know of no method by which we can determine what is ultimately real. The best we can do is make ever-improving observations and describe them with ever more accurate models.

Although they are human inventions, the models of physics are not subjective. They are discarded if they do not agree with objective observations. In this regard, physics models are not socially constructed, as the now largely defunct philosophy of social-constructivist postmodernism once tried to tell us.[3] I am often misquoted, without any citation, as having said, "There is no connection between models and reality." Allow me to make it as clear as I can: If a model agrees with the data, then it has something to do with reality. We simply have no way of knowing if the elements of that model correspond to any elements of reality.

Many of the arguments you will hear for the existence of a world beyond the senses, whether from theologians or secular academics, rest on the notion that pure reason, absent of any empirical data, is capable of gaining knowledge of reality. Reason often is associated with deductive logic, but it also includes other methods such as induction.

However, nothing new can be learned by logical deduction that is not already embedded in its original premises. At best, the process provides only a check on whether a statement is consistent with its premises.

As for other forms of "pure thought," over the centuries not a single verifiable fact has ever been discovered by either pure reason or by mystical experience. This is sufficient to rule it out, along with divine revelation, which has been equally unproductive.

In this book I will be treading on the territories of many experts who know more than I do about the precise terrain. Their domains include history, philosophy, theology, theoretical physics, astronomy, and cosmology. These experts are bound to complain, as experts always do, that the actual facts are more complicated and I have oversimplified their subjects. Of course, no one can be expert in everything. At the same time, I think all of these subjects are simpler, or at least less arcane, than the experts often claim when we just limit ourselves to observational implications. What today's theorists call "deep problems" are usually associated with their own current theories, not with observable facts. Quantum mechanics is the prime example, where we have a theory that has agreed with all observations for almost a century and people still argue over "what it means" without ever coming to a consensus.

Let me say a few words about the technical level of the book. The early chapters should provide no problem for the general reader, but as we get into modern physics and cosmology I will necessarily have to use some technical language that I will try to define as I go along and that can be easily searched on the web. However, the mathematically challenged need not freak out. When I write an equation, it will be no more than shorthand for a statement that can be also put in words. Why write, "Energy equals mass times the speed of light squared" when $E = mc^2$ says the same thing far more compactly? Also, why write "a hundred thousand million million" (as Stephen Hawking was forced to do by his publisher in order to sell more books) when 10^{17} will do?

While I will not use any higher mathematics, I do present many graphs that are crucial for demonstrating quantitative results. I suspect that anyone picking this book up in the first place will have no trouble understanding these illustrations, nor the meaning of 10^{17}.

Finally, to fend off the often-repeated charge that my ideas and those of other scientists are just as "dogmatic" as those of any religion, allow me to state unequivocally: if in the future convincing empirical evidence shows that any result or conclusion presented here is wrong, I will be perfectly happy to make the necessary corrections.

1.

FROM MYTH TO SCIENCE

THE WORD

Upper Paleolithic cave art in Europe demonstrates that humans of forty thousand years ago were already thinking abstractly and expressing creativity by projecting their mental images of animals onto the rock environment. During the Neolithic era, stone megaliths sprang up across Europe, most famously Stonehenge, which was erected in 2,300 BCE and was used for royal burials as well as charting the course of the sun and hosting solstice festivals for pilgrims. This suggests that humans had begun relating to their outside environment in an entirely new way—a way that has been exclusive to *Homo sapiens* ever since.

Earlier, around 3,500 BCE, writing, called cuneiform, had been invented in Mesopotamia in the form of wedge-shaped characters carved into clay, and for the first time information could be accurately preserved and transferred without depending on the memory of a human messenger. When the ancient Sumerians first took wedge to clay tablet, they couldn't have known that they were writing the first chapter of humankind's journal. But they were. And that story begins roughly six thousand years ago, which is why some still cling to the belief that planet Earth is only six thousand years old—despite conclusive scientific evidence that it is 4.5 billion years old and that our universe goes back 13.8 billion years.

Indeed, *our* world is six thousand years old. *Our Story* began when we started to write about it, 3000–4000 BCE. Everything that happened before then we call "prehistory." In a way, it is a prologue because we don't know the names of any Neolithic people. They speak to us only in images, structures, and artifacts. In the early Bronze Age, humans start using words and so that's when individuals emerge and Our Story begins. "In the beginning was the Word."

With the development of civilized culture and writing, we have on record the fact that Bronze Age humans thought in terms of myths (from Greek *muthos*, meaning "story"), sacred tales that served as explanations for what they observed in the world around them. And the words that they used to describe these myths had power, which gave them a magical quality. Written cosmologies were more systematic than prehistoric, animistic attempts to describe the world, but the supernatural was still a major component.

In Egypt, cosmological myths were woven into the religion with the heavenly bodies being associated with gods. Creation stories with differing details depending on city have been deciphered from various texts and tomb-wall decorations; I will just mention their common elements.

In most Egyptian creation myths, the world arose out of a watery chaos called Nu, an obvious allusion to the river Nile so central to Egyptian life. A pyramid-shaped mound emerged from the waters. The sun god Ra, or in some cases Khephri, appeared out of the mound—a self-created god who then created other gods and eventually humans.

Since it was imperative that priests could anticipate the appearance of gods in the sky and, more important, predict the Nile flood, they developed a calendar with twelve thirty-day months within a year of 365 days based on astronomical observations. And so ancient astronomy made the first step toward applied science.

Science involves two activities, which are today largely carried out by separate groups of professionals—observers or experimentalists who gather the data and theorists who develop models, usually mathematical, to describe what is observed. While the observational side of ancient astronomy was remarkably accomplished, the theory side remained heavily magical and mythological.[1]

The Egyptian theory of creation is an example. According to the most common Egyptian creation myth, the sky goddess Nut gives birth to the sun god Ra once a year, thus describing a universe that is self-creating and eternal.

The Egyptian cosmos was composed of a flat, rectangular Earth with the Nile at its center. In the south, in the sky supported by mountains, there was a river, and on this river the sun god made his daily trip. The sky was a roof placed over the world, supported by columns placed at the four cardinal points. The stars were suspended from the heavens by strong cables, but no apparent explanation was given for their movements.[2]

The civilizations of Mesopotamia and Canaan all had similar concepts of the cosmos and humanity's place in it.[3] Earth was a disk with the firmament, or vault of heaven, resting on its edge. Beneath Earth were the waters of the abyss.[4] Once again, we note the role of water, obviously derived from the fact that the civilization developed along the Tigris and Euphrates Rivers.

Babylonian astronomy was remarkable, with records going back to 800 BCE that are the oldest scientific documents in existence. Babylonian astronomers discovered the 18.6-year cycle of the rising and setting of the moon. They made the earliest recorded sighting of Halley's Comet in 164 BCE, making possible accurate calculations of its orbit by Halley and other astronomers centuries later. Babylonians produced the first almanacs of the movements of the sun, moon, and planets. Although primarily intended for astrological forecasting, the techniques they developed enabled accurate predictions of planetary motion and eclipses.

Curiously, Babylonian scientific astronomy did not lead them to a scientific cosmology; their cosmology remained highly mythological. However, the Greeks would utilize Babylonian astronomy to produce the first cosmology that could be clearly identified as natural rather than magical or mythological.[5]

As in Egypt, Mesopotamian cosmology was integrated into religious belief and a cosmogony, a myth of creation. That myth is described in a poem known as the *Enûma Elish* dating from a little before 1000 BCE. It describes a clash among the gods in which Marduk (or Assur) defeats the ocean goddess Tiamat and rips her body into two halves that become the heavens and earth. He then organizes the sun, moon, stars, planets, and the weather. He creates humans from the blood of Tiamat's husband, Kingu, so they can do the work of the gods.

The Christian faith, the biggest in the world today, emerged two thousand years ago out of the older faith of a tiny tribe of desert dwellers in Canaan called the Hebrews. The Hebrews divided the universe up into heaven, earth, sea, and the underworld, as shown in figure 1.1. Earth is a more-or-less flat disk floating on water with the vault of heaven resting on foundations on the edge of the sea. Beneath earth were the waters of the abyss. God sits at the apex of a series of heavens above the vault of the sky.[6]

Of course Christians today, even those who insist that the Bible is infallible, know that there is more to the universe than what is depicted in figure 1.1. Furthermore, God is now generally imagined to be either everywhere in space and time or outside both (depending on theologian), rather than at one special place. Nevertheless, it must be kept in mind that Christianity was built on the superstitions of very simple people for whom the model in figure 1.1 represented the best knowledge available at the time—based on their own observations of the land, sea, and sky. It was easy for them to conceive of God as a great king watching everything that happens and making sure events proceed according to his divine plan.

There are actually two contradictory creation myths in the Hebrew Bible. For example, in Genesis 1, vegetation is made before animals and animals are made before Adam and Eve. In Genesis 2 Adam is made first, then vegetation, then animals, then Eve.[7] Chapter 1 was composed in the sixth century BCE when the Jews were in exile in Babylon, and it is clearly descended from the ancient

Babylonian *Enûma Elish*. The creation myth in chapter 2 of Genesis came from Canaan a few centuries earlier.

Figure 1.1. The ancient Hebrew conception of the universe.
Image © Michael Paukner.

The biblical creation myth familiar to us today, based on Genesis 1, needs little elaboration: God creates the world in six days, resting on the seventh. On the first day, he starts by creating heaven and earth. But they are in darkness and he (using the power of the Word) says, "Let there be light." On the second day, he makes a firmament to divide the waters above, which he called heaven, from the waters below. On the third day, God demands dry land to appear in the waters below, and he calls it earth and the waters seas. Then he tells earth to bring forth vegetation yielding seed and trees bearing fruit.

On day four, God adds the sun, moon, and stars to the firmament of the heavens. On the fifth day, he orders the waters to bring forth fish and birds and blesses them, saying, "Be fruitful and multiply." On the sixth day, God has earth bring forth cattle, beasts, and creepy things. And finally, God makes man in his image and gives him dominion over all Earth and its living things. And then he saw it was good and took a day off to recover from his labors.

Genesis, of course, also tells the story of the first humans that God kicked out of Eden because they ate from the tree of knowledge of good and evil. In Christianity, this became the original sin that Christ died for.

Since the late nineteenth century, most scholars have concluded that Moses did not write the book of Genesis, as tradition claims, but it most likely arose during the Jewish captivity in Babylon. This is primarily based on the discovery in 1872 of Babylonian cuneiform tablets containing a deluge story closely resembling the biblical flood.[8] It has even been argued by some that the Hebrew Bible and Judaism itself derives directly from Mesopotamian mythology, although how much remains in dispute.[9]

CREATION *EX NIHILO*

Today's prominent Christian apologists, such as William Lane Craig and Dinesh D'Souza, argue that Genesis is the only ancient creation story that is consistent with modern cosmology in featuring a transcendent deity that created the universe out of nothing, the doctrine known as creation *ex nihilo*.[10] However, it is to be noted that Genesis does not speak of any creation *ex nihilo*, which was a doctrine that developed much later.

The first words of the Bible are, "In the beginning, God created the heavens and the earth." But, as biblical scholar Tim Callahan points out, this is not the only translation of the original Hebrew.[11] The word *bara* can be translated not only as "create" but also as "chose," "divide," and other meanings depending on context. And Hebrew translations depend very much on context.

Old Testament scholar Ellen van Wolde says that in the context of Genesis, the first sentence should be translated, "In the beginning, God separated the heavens and the earth."[12] This is more consistent with the other "separations" that are then described: light from darkness, waters above from waters below, and land from water. Furthermore, this interpretation is more in line with the Babylonian creation myth, on which the Hebrew myth was surely based. In the *Enûma Elish* mentioned above, Marduk slices Tiamat's body into the heavens above and earth below.

But even if creation *ex nihilo* is the correct interpretation, it is not unique to the Bible. Callahan observes that, before being killed by Marduk, Tiamat ruled over a chaotic, formless void.[13] As we will see, formless chaos is about as *nihilo* as *nihilo* can be. Neither have any information or structure that you can use to define them as something other than nothing.

The late keeper of the Egyptian and Assyrian antiquities at the British Museum, Sir Wallis Budge, lists a series of what he oddly calls "epithets" collected by the nineteenth-century German Egyptologist Heinrich Brugsch (1827–1894) from hieroglyphs. Here is a sample:

> God is from the beginning, and He hath been from the beginning;
> He hath existed from of old and was when nothing else had being.
> He existed when nothing else existed, and what existeth he created after He had come into being. . . .
> God is the eternal One, He is eternal and infinite; and endureth forever and aye; He hath endureth for countless ages, and He shall endure to all eternity. . . .

God hath made the universe, and He hath created all that therein is: He is the Creator of what is in this world, and it was He Who fashioned it with His hands before there was any beginning.[14]

Budge says these epithets, which "are applied to the gods, from texts of all periods," show that "the ideas and beliefs of the Egyptians concerning God were almost identical to the Hebrews and Mohammadans at later periods."[15]

However, no actual texts are cited. It makes one wonder if we have a bit of apologetics going on here, designed to show that Judeo-Christian-Islamic beliefs are universal. Ironically, these claims contradict those of the apologists mentioned above who want to make us think that the Christian creation myth is unique and original.

In Budge's own translation of the *Papyrus of Nesi Amsu* in the museum (No. 10,188), the god Neb-er-tcher identifies himself as Ausares (Osiris) and says, "I developed myself out of the primeval matter which has evolved multitudes of evolutions from the beginning of time." This sounds more like creation from already-existing matter. Incidentally, the god does this by word alone: "I uttered my own name, as a word of power, from my own mouth, and I straightway evolved myself."[16] Again we see how much stock the ancients put in the power of words by themselves.

Moving to India, in his *Creation Myths of the World*, David Leeming gives the following quotation from the Hindu Rig Veda, which goes back at least to 1000 BCE:

In the beginning there was neither Being nor Non-Being, neither air nor sky. What was there? Who or what oversaw it? What was it when there was no darkness, light, life or death? We can only say that there was the One that which breathed of itself deep on the void, that which was heat and became desire and the germ of the spirit.[17]

In any case, according to Leeming, creation *ex nihilo* myths are as common as creation from a preexisting watery chaos. And, as I will often remark, complete chaos is indistinguishable from nothing.

OTHER CULTURES

It is not my purpose here to give a comprehensive survey of the many rich creation myths and cosmologies that can be found in the oral and written traditions of ancient cultures. While differing greatly in details, most early cosmogonies, as mentioned, attempted to explain the origin of the world in human terms with anthropomorphic gods bringing order out of chaos or out of nothing (*ex nihilo*).

Of course, many complex traditions can be found in ancient India and China. It would take me far afield to do justice to them. In any case, they have little bearing on my topic, which is the development of scientific cosmology to the present day and its clash with the religions of yesterday and today. Modern science emerged exclusively in Europe, with its origins in the great river civilizations of the Middle East and Greece where the religions it most clashes with also arose.

Still, it is worth mentioning that Hindu thinking envisaged a world that passed endlessly between phases of creation and destruction. The concept of a cyclical universe, which differs markedly from the Judeo-Christian-Islamic view of a created universe finite in time, has raised its head in some scientific cosmologies of today.[18]

In astronomy, the Hindu astronomical work called the *Surys Siddhanta*, written about 400 CE by an unknown author, gave the average length of the sidereal year that is only 1.4 seconds longer than the modern value, and it stood for over a thousand years as the most accurate measurement of that quantity made anywhere in the world.

A common picture of the cosmos is found across most ancient cultures, probably because it described what people observed: a flat, central Earth surrounded by water, celestial bodies rotating above, and a dark, forbidding underground associated with death.

Ancient astronomers did remarkable work in measuring the motions of the sun, moon, planets, and stars. This enabled them to produce calendars from which they could predict the seasons and repetitive events, such as the annual Nile flood. Indeed, these motions behaved so predictably compared to events on Earth, such as storms, earthquakes, and floods (other than the Nile flood), that the heavens were viewed as a divine province that controlled the lives of humans and the world around them. And so, in addition to predicting the seasons, astronomers took on a second role as astrologers to be consulted on virtually every decision.

GREEK PHYSICAL COSMOLOGY

Starting in the sixth century BCE, a group of thinkers residing in the Greek cities of Ionia on the western coast of Asia Minor began to remove gods and the supernatural from any major role in the world. Together with several philosophical schools in Greek colonies in Italy, they are called the Presocratics, although some of the later ones were contemporaneous with Socrates (ca. 469–399 BCE). The designation was introduced in the nineteenth century to distinguish this group of philosophers from Socrates, who was more concerned with human problems while the Ionian Presocratics, at least, dealt with cosmology and physics where the role of humanity is minimal.[19]

It should be kept in mind that few of the writings of the Presocratics have survived and most of what we know about them comes from later sources, in particular, Aristotle (384 –322 BCE), who was not all that sympathetic to their teachings and was in strong disagreement with some. Aristotle named the Ionians *physikoi* or *physiologoi*, from their concern with *physis* or nature.[20] Today we call them physicists. As we will see, by contrast the Italian Presocratics were mostly mystical so they really should not be grouped with the Ionians.

The Ionians began a movement described by the term *Kosmos*, which denoted an ordered world in contrast to *Chaos*, the unpredictable, capricious world of divine intervention described in the epic poems of Homer (ca. 700–800 BCE) and Hesiod (ca. 650–750 BCE).

Thales

Thales of Miletus (ca. 624–546 BCE) is generally regarded as the first physicist. He sought natural explanations for phenomena that made no reference to mythology. For example, he explained earthquakes as a result of Earth resting on water and being rocked by waves. Thales is famous for supposedly predicting the eclipse of the sun that modern astronomers calculate occurred over Asia Minor on May 28, 585 BCE (current calendar). However, most historians today doubt the truth of this tale.

Thales's most significant contribution was to propose that all material substances are composed of a single elementary constituent, namely, water. While he was (not unreasonably) wrong about water being elementary, Thales's proposal represents the first recorded attempt to explain the nature of matter without the invocation of invisible spirits.

Thales and the other Ionian philosophers that followed espoused a view of reality now called *material monism* in which everything is matter and nothing else.

Anaximander

Anaximander (ca. 610–546 BCE) was the second master of Thales's school in Miletus. He proposed that the universe is boundless and has no origin. Furthermore, boundlessness or *apeiron* was the source of all things:

> The first principle of existing things is the Boundless for from this all things come into being and into it all perishes. Wherefore innumerable worlds are brought to birth and again dissolved into that out of which they came.[21]

Despite an unlimited number of worlds, Earth still floats at its center. Anaximander's cosmos is shown in Figure 1.2.[22]

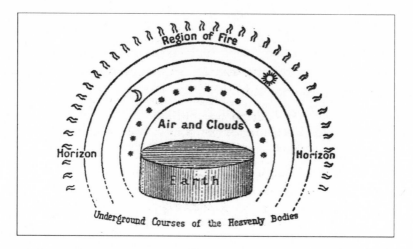

Figure 1.2. Anaximander's cosmology. Image from Mary Ackworth Orr, *Dante and the Early Astronomers* (Port Washington, NY: Kennikat, 1969).

Anaximenes

Anaximenes (585–528 BCE) was a student or at least a younger colleague of Anaximander. He proposed that everything is made of air, which in ancient times was widely regarded as the breath of life. In Anaximenes's cosmology, Earth is formed by a condensing or felting of air and rests on a cushion of air. The heavens are like a felt cap with the heavenly bodies fixed to its surface that turns around the head. Interestingly, the sun does not circle Earth as in most geocentric cosmologies but is hidden behind mountains at night. (In chapter 13 we will see how this theory is falsified by a photograph of the sun taken at night with neutrinos passing through Earth).

The Atomists

Leucippus and Democritus were the founders of atomism. In my previous book *God and the Atom*, I told the story of how their notion that everything is composed of elementary particles of matter has been triumphantly corroborated by modern physics.[23]

According to Orr, Leucippus and Democritus, along with the Pythagoreans we will discuss below, made a great step forward in cosmology when they realized that the sky is not a hemisphere ending at the horizon but surrounds Earth like a sphere.[24] She attributes the diagram shown in figure 1.3 to Democritus. In a similar diagram she ascribes to Leucippus, Earth covers the whole lower hemisphere, which is air in the Democritus version.

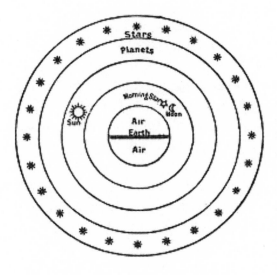

Figure 1.3. Cosmology of Democritus. The cosmology of Leucippus is the same except Earth extends down in the lower hemisphere shown here as air. Image from Mary Ackworth Orr, *Dante and the Early Astronomers* (Port Washington, NY: Kennikat, 1969).

Epicurus (341–270 BCE) transformed the atomism and cosmology of Leucippus and Democritus into a major philosophical school.[25] Three hundred and fifty years later, the Roman poet Titus Lucretius Carus (ca. 99–55 BCE) immortalized the teachings of Epicurus and the earlier atomists in his epic work *De rerum natura* (*The Nature of Things*). What follows is Lucretius's presentation of atomist cosmology:

> Since empty space is limitless on all sides and the amount
> Of atoms meandering in the measureless universe, past count,
> All flitting about in many different ways, endlessly hurled
> In restless motion, it is *most unlikely* that the world,

This sky and rondure of the earth, was made *the only one,*
And all those atoms *out*side of our world get nothing done;
Especially since this world is the product of Nature,
    the happenstance
Of the seeds of things colliding into each other by pure chance
In every possible way, no aim in view, at random, blind,
Till sooner or later certain atoms suddenly combined
So that they lay the warp to weave the cloth of mighty things:
Of earth, of sea, of sky, of all the species of living beings,
That's why I say you must admit that there are other cases
Of congregations of matter that exist in other places
Like this one here of ours the aether ardently embraces.[26]

In short, the universe of the atomists, like that of Anaximander, is unbounded, eternal, uncreated, and composed of many worlds.

Furthermore, according to atomism, the universe is in large measure the product of random chance. As with atoms, the atomists got this one right, too–long before anyone else did.

Some historians dispute that the atomists' universe was eternal.[27] While it is true that the stuff made of atoms decayed, the atoms themselves are eternal, as Lucretius explains:

If atoms are solid, therefore, and without void, as I've taught,
Then they must be eternal. And what's more, if they were not,
Everything would have met with its demise,
And any objects that we see here, now, before our eyes
Would have been born again from nothingness.
    But since I've taught
That nothing's made of nothing, and the nothing can be brought
To nothingness once it is made, then there must be first bodies
Made of stuff that last forever — atoms — and it is these
That everything is broken down to in its final hour
So there is a supply of matter on hand to re-power
The world. These atoms thus are pure and solid
    through an through;
How else could they survive infinite time to make things new?[28]

Pythagoras and Philolaus

Pythagoras (ca. 570–495 BCE) is probably the best known of the Presocratics and so unlike the Ionians that it is incongruous to lump them together as historians do. He was the founder of a religious and political community in Croton, Sicily, where he achieved wide fame as a mystic, a wonder-worker, and the purveyor of the doctrine of an immortal soul able to transmigrate into animals. It is to be noted that the notion of an immortal soul, so widespread today, was not prevalent in Greece or Italy at that time.

Pythagoras's modern image as a master mathematician, geometer, and scientist is less secure. Neither Plato nor Aristotle attributes the "Pythagorean theorem" to Pythagoras.

According to historian John North, Pythagoras taught that the universe was produced by heaven inhaling infinity to produce numbers.[29] Orr presents the diagram in figure 1.4 as the "earliest form" of the universe of the Pythagoreans, although there is no record of Pythagoras himself having held this picture.[30] It may be the earliest cosmology in which Earth is a sphere. The region labeled "Cosmos" contains the five planets, the sun, and the moon. "Olympos" contains the stars. "Ouranos" refers to the sky. The whole is surrounded by the celestial fires and beyond that by Apeiron, the infinite space from which the world draws its breath.

This model was modified by Philolaus (ca. 470–385 BCE) of the Pythagorean school in the dramatic way illustrated in figure 1.5.[31] Philolaus proposed that Earth is not the center of the universe. But neither is the sun. Rather, Earth, the sun, and the seven planets visible to the naked eye revolve around a central fire. The stars are fixed and at a great distance. Earth rotates every twenty-four hours about the central fire, the sun once a year. In addition, another planet, a counter-Earth called Antichthon exists that also revolves every twenty-four hours and is always on the other side of the central fire from Earth. The reason we do not see either the central fire or Antichthon is that only the side of Earth turned away from each is inhabited.

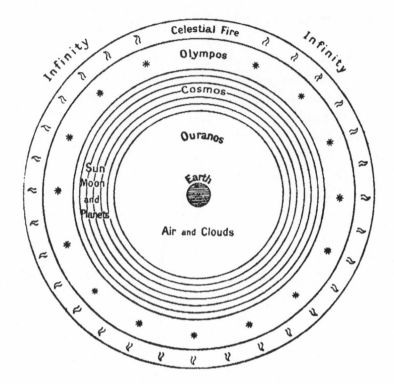

Figure 1.4. The early Pythagorean cosmos. Image from Mary Ackworth Orr, *Dante and the Early Astronomers* (Port Washington, NY: Kennikat, 1969).

Compared to the Ionians, I think it can be safely said that the Pythagoreans, even if they were great mathematicians, were not scientists. Their ideas were based on mystical arguments rather than observations. The five planets visible to the naked eye plus the sun and moon numbered seven, a sacred number since it was the number of notes on the musical scale that the Pythagoreans had discovered. Adding Earth and the central fire made nine, so Philolaus introduced Antichthon to bring the total to ten, another sacred number (presumably because we have ten fingers and toes). According to Orr, the Pythagoreans regarded the central fire as the Watch Tower of Zeus, the Hearth of the Universe, where was placed the purest element, fire.[32]

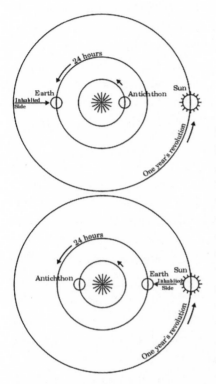

Figure 1.5. The system of Philolaus. Earth, the sun, and the counter-Earth, Antichthon, revolve around a central fire. Only the side of Earth opposite the central fire and Antichthon is occupied, so both the central fire and Antichthon are unseen. The top diagram illustrates night on Earth and the lower diagram illustrates day. Not shown, the moon's orbit is inside that of the sun, and the planets' orbits are all outside. Image from Mary Ackworth Orr, *Dante and the Early Astronomers* (Port Washington, NY: Kennikat, 1969).

Curiously, despite their obsession with numbers, it never occurred to the Pythagoreans to make quantitative measurements of the heavenly bodies, which would have quickly disproved Philolaus's cosmology. But it was a bold guess nonetheless and the earliest record of a cosmology that was not centered on Earth. As Orr writes,

It was the braver of the Pythagoreans to shake the steady earth from her centre, and set her whirling in the depths of space, that they realized, as no one had done before, how large she must be; for Greece and the surrounding lands, the Middle and the other seas, instead of making the whole of the earth, were now understood to be only a portion of a great globe.

Here, then, is a conception of the universe widely different from Homer's. The little flat disc has become a round ball swiftly moving through space; the crystal dome that tenderly covered it like a bell-glass over some fragile flower, has lifted and the vast sphere is seen, infinitely distant and studded with enormous stars. Man himself is now a tiny creature on a great earth, and his world but one of many, but if he is humiliated by his insignificance, is he not elevated by the vastness of his outlook?[33]

Empedocles

Empedocles (490–430 BCE) was a citizen of Agrigentum in Sicily and was closer to the Ionian physicists than he was to the Pythagorean mystics. Empedocles was the first to propose that matter is composed of the four elements: earth, water, air, and fire. Since Aristotle adopted this picture, it became the "standard model" for millennia until it was replaced in the nineteenth century by the chemical elements of the periodic table, and then in the late twentieth century by the quarks, leptons, and bosons of what we now call the standard model of elementary particles.

Empedocles's cosmos contained a number of insights, such as the moon shines by reflected light from the sun and a solar eclipse is the result of the moon blocking off the sun.

More mystically, Empedocles proposed that two opposing forces called "Love" and "Strife" (or, perhaps, attraction and repulsion) governed the elements, resulting in cyclical patterns as the two competed with one another. In Empedocles we find a thinker deeply suspicious that matter has many different fundamental forms. Also, Empedocles believed the behavior of matter was controlled by intangible but real natural forces that came in multiple forms as well. This philosophical breakthrough laid the foundations of all future study of matter, and survives to this day.

Plato

Eventually, the Presocratics in Ionia and Italy gave way to the Socratics in Athens. We know about Socrates mainly from the writings of Plato (429–347 BCE), which indicate Plato was no scientist, believing that true knowledge of reality can only be obtained by pure reason without the help of the senses.

According to Plato, Socrates tried to move philosophical interest from physics and cosmology to human concerns such as ethics and politics. But he did not totally succeed. Greek philosophers in Athens and elsewhere continued to think about the universe, although most were not quite willing to adopt the naturalist views of the Ionians, especially the atomic theory.

In Plato's *Timaeus*, the character by that name is introduced as an astronomer and student of the universe. He describes a creation out of chaos, "where there is no order, and no matter which can be distinguished by name, but all is confused and seething with random restless motions."[34] That was a good guess, but the rest was not. The heavenly bodies, according to Timaeus, are divine, intelligent beings. They all are perfect spheres composed of fire. Earth is also a perfect sphere at the center.

In *Timaeus*, Plato clearly rejects the notion of multiple universes: "The Maker made neither two, nor yet an infinite number of worlds. On the contrary, our universe came to be as the one and only thing of its kind, is so now and will continue to be so in the future."[35]

Plato was certainly not a material monist—someone who believes the world is matter and nothing more. He argued that what we humans detect with our eyes and other senses are but shadows of a true, more perfect reality. Plato taught that a divine craftsman called the Demiurge constructed the cosmos according to a sacred plan. There are two realms of reality: a realm of *forms* or ideas that is perfect and a material realm in which these forms or ideas are imperfectly replicated.[36]

Plato's creation is to be contrasted, however, with the notion that the universe was created *ex nihilo*—from nothing—while the Demiurge used already-existing materials.

For example, the heavenly bodies are perfect spheres rotating in perfect circles around the sun. The fact that the planets appear to wander about the sky (*planet* means "wandering star" in Greek) is explained as a distortion of the senses, the way a lens distorts what we see.

Since far more writings of Plato and Aristotle have survived than those of their predecessors, or immediate successors for that matter, they have tended to have the greatest influence of any ancients on human thinking throughout the ages. Not to take anything away from their enormous contributions, but the often-unquestioned authority of Plato and Aristotle has not always been to the benefit of human progress.

Plato's importance to history may have been his reintroduction of divinity into the scheme of things after it had been dismissed by the Ionians. Not only did the Demiurge create the universe, but the planets and stars hosted the celestial gods. According to historian David Lindberg, "Plato restored the gods in order to account for precisely those features of the cosmos that, in the view of the *physikoi*, required the banishing of the gods."[37]

Aristotle

Unlike his mentor, Plato, Aristotle was a legitimate scientist who made observations and performed experiments, although he still trusted his reason over his data. It was not until Galileo that the priority of reason over observation was reversed and observation became the final arbiter of knowledge—at which point, coincidentally and not accidentally, science began its great climb to where it resides today, joining other advances at the pinnacle of human accomplishments.

Aristotle, like almost all the other post-Pythagoreans, put Earth back at the center of the cosmos. He said the form of the universe had to be a sphere because it is the most perfect solid. For Aristotle, the universe had no beginning and will have no end (a teaching

Christianity would ignore). Here the reason he gives is that the gods are immortal, and since they live in the highest heaven of the universe, so too must the universe be immortal.

As mentioned, Aristotle adopted Empedocles's model in which earth, water, air, and fire are the basic elements from which all of matter is composed. This led to its adoption as the "standard model" of matter until the rise of modern chemistry and physics. Aristotle added a fifth element, a quintessence called *aether*, as the stuff of the heavens.

Aristotle declared that there are three forms of natural movement, exemplified by the motions of five elements: (1) earth and water move in a straight line toward the center of Earth; (2) air and fire move in a straight line away from the center; (3) celestial bodies move in circles around the center. The last provided an explanation for the ancient question of why the heavenly bodies never fall to the ground.

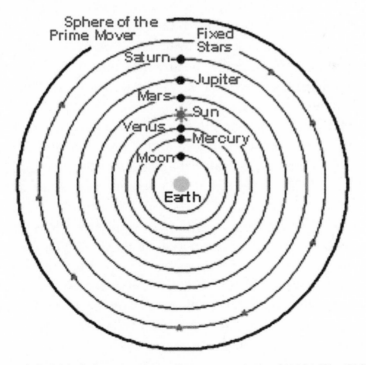

Figure 1.6. Aristotle's cosmology. Image courtesy of NASA/StarChild.

Aristotle's cosmology is illustrated in figure 1.6.[38] He made a major advance in proposing that the planets and stars were actual physical bodies, although still perfect spheres. This contrasted with Plato's image of the celestial bodies as divine beings. Moreover, Aristotle disagreed with Plato's claim the universe was created. He said it was eternal, but not infinite.

Note the "Sphere of the Prime Mover" in the figure. Aristotle's physics required that all motion have a cause, and the ultimate cause, the Prime Mover, was the "First Cause Uncaused." This entity was interpreted by Christian theologians in the Middle Ages, notably Thomas Aquinas, to be the creator God. However, Aristotle introduced the unmoved mover in *Physics* not as the creator but as a spiritual something located at the outermost part of the universe that is the source of all celestial motion.[39]

Aristotle disagreed with the atomists. In their view, earth, water, air, and fire are not elementary but are composed of atoms that are more elementary. Aristotle claimed to prove that the atomists' void was impossible, while the key principle of atomism is that everything is atoms and the void. The atomists agreed that the universe was eternal but disagreed that it was finite. They also claimed that there were multiple worlds.

Aristotle's teachings on physics and cosmology were not totally accepted in pre-Christian Greece and Rome. Not only the atomists, led by Epicurus, but also the Stoics, led by Zeno of Citium (ca. 334–262 BCE), claimed that the universe was infinite as well as eternal.

Nevertheless, Christian theology adopted Aristotle as an ancient authority on many subjects, especially the idea of First Cause. As a result, the great European universities built by the Catholic Church in the Middle Ages became so deeply entrenched in what is called Aristotelian scholasticism that the scientific revolution, which repudiated much of Aristotle's science—especially his physics—occurred outside these institutions.

Aristarchus

Aristarchus of Samos (310–230 BCE) is the first recorded astronomer to place the sun at the center of the universe. The Babylonian astronomer Seleucus (ca. 190 BCE) is said to have made the same proposal a bit later. Aristarchus also located the planets in their correct distance-order from the sun. Although he used proper geometrical arguments, his distance estimates were too low because of faulty data.[40] He did recognize, however, that the stars had to be far away because of their absence of measurable parallax.

Hipparchus

Hipparchus (ca. 190–120 BCE) was born in Nicaea in Bithynia but lived most of his life in Rhodes. He made the first accurate models of motions of sun and moon, based on Babylonian tables. Hipparchus also discovered the precession of the equinoxes, calculated the length of the year to within six and a half minutes, compiled the first known star catalog, and made an early formulation of trigonometry. As such, he is regarded as the first to apply numerical data from observations to geometrical models. Hipparchus laid the foundation for the work of Ptolemy three centuries later, and he has been called "the greatest astronomer in antiquity."[41]

Ptolemy

Claudius Ptolemy (ca. 90–168 CE) resided in Alexandria, where he had available to him the immense store of the writings of the Greeks and Romans that had been gathered in the Library of Alexandria, the greatest library of the ancient world. In 48 BCE, Julius Caesar (100–44 BCE) accidentally burned a large number of books stored near the docks when he set fire to the ship in the harbor. However, two other collections in the city were spared. These were largely destroyed during Christian riots in 390 CE, but Alexandria continued to be a center of Greek learning.

Ptolemy also was able to utilize excellent astronomical instruments in the observatory associated with the library. With these facilities, he combined astronomical data accumulated over centuries, including those of Hipparchus, with his own observations to frame a system that mathematically described all celestial motions observed to that day. He published his system in a set of thirteen books originally named the "Mathematical System of Astronomy," which was renamed by others as *Magiste Syntaxis* (*Great System*), and later by the Arabs as the *Almagest* (*The Great System* or just *The Majestic*), by which it has since been known. This work remained the definitive treatise on astronomy until Copernicus's *De revolutionibus orbium celestium* (*On the Revolutions of the Heavenly Spheres*) appeared in 1543.[42]

Ptolemy maintained the Earth-centered system of almost all his predecessors, although Earth is slightly off-center. He described his physical premises as follows:

> The heaven is spherical in shape, and moves as a sphere; the Earth too is sensibly spherical in shape, when taken as a whole; in position it lies in the middle of the heavens very much like its centre; in size and distance it has the ratio of a point to the sphere of the fixed stars; and it has no motion from place to place.[43]

In order to maintain Earth near the center of his system and still accurately describe planetary motions, which are far from circular about Earth, Ptolemy constructed a model of enormous complexity, illustrated in simplified form in figure 1.7.[44]

In Ptolemy's system, the planets move in circles called *epicycles* whose centers themselves move in circles about a point called the *equant* that is not exactly centered on Earth. In some cases, the epicycles themselves move in epicycles.

The *Almagest* is primarily a technical work that enabled trained astronomers to predict planetary motions. Ptolemy's cosmology is presented in another work, *Planetary Hypotheses*, which is known mainly from Arabic sources. It is essentially the same as Aristotle's cosmology. Earth is composed of the four elements earth, water, fire, and air. It does not rotate about its axis. The heavens are composed of ten concentric spheres composed of the transparent fifth element, quintessence or aether, proposed by Aristotle, which rotate in circles around Earth. There is no void, no empty space, between the spheres, in agreement with Aristotle's insistence that a void cannot exist. They are filled with quintessence.

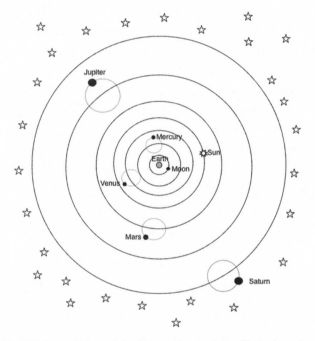

Figure 1.7. The Ptolemaic model of the solar system. The planets move in circles on the surfaces of concentric spheres that revolve in circles around a point not precisely at the center of Earth. The spheres are not empty but are composed of Aristotle's quintessence or aether. Image by the author.

In order of distance from Earth are the spheres containing (1) the moon, (2) Mercury, (3) Venus, (4) the sun, (5) Mars, (6) Jupiter, (7) Saturn, and (8)–(10) the "fixed stars," which are fixed in the sense that they do not move with respect to one another as their spheres rotate. Of course, everything rotates with a period of twenty-four hours. Ptolemy's model does not provide the distance to the objects.

Ptolemy counted 1,022 stars in forty-eight constellations and listed the fifteen brightest stars.[45] As well as solar, lunar, and planetary motion, his system enabled one to predict the rising and setting of stars.

The Ptolemaic system marked the end of the advance of Greek astronomy. As the Greek language gave way to Latin in the Roman Empire, few could read Ptolemy's books. The Romans themselves made no significant contributions to the science of astronomy but typically found value in its practical use for measuring time. By the time of Julius Caesar, the calendar had gotten all out of kilter, so he called upon the Alexandrian astronomer Sosigenes (first century BCE) to provide a new one. The so-called Julian calendar that Sosigenes provided is still in use with only small corrections made over the centuries since. Note, however, that it was still Greek astronomy that provided the mechanism.[46]

2.

TOWARD THE
NEW COSMOS

COSMOLOGY IN CHRISTENDOM

Some early leaders of the Church objected to the conception of the world of the Greek scientists. Tertullian (ca. 160–225) asked, "What indeed has Athens to do with Jerusalem?" The influential bishop Lactantius (ca. 240–320), advisor to Emperor Constantine (272–337), rejected the sphericity of Earth as a heretical belief, as well as the ridiculous notion that people on the other side "have footprints above their heads" and that "rain and snow and hail fall upwards."[1]

However, several Church fathers seemed to accept the notion that Earth is spherical, notably Clement of Alexandria (ca. 150–215), Origen (ca. 184–253), and Ambrose (ca. 340–397).[2]

The great theologian Augustine of Hippo (354–430) respected science as a means for learning about God's creation, although he still insisted that reason and revelation must prevail over observation. Galileo would find much to agree with in Augustine's philosophy and mentions the theologian's statement that the Holy Spirit behind revelation "did not desire that men should learn things that are useful to no one for salvation."[3] This helped him to argue, ultimately unsuccessfully, that his own sun-centered cosmology based on the Copernican model did not conflict with Church teachings at the time.

Much of Greek and Roman scholarship was suppressed by the early Church. However, by the seventh century, scientific literature began to reappear in Europe, though it was mainly from Roman authors and rather primitive.[4] Still, despite the destruction of its library by Christian fanatics in 390, Alexandria remained a bastion of Greek scholarship. There, John Philoponus (490–570) sought to bring Aristotle's natural philosophy in line with monotheism. He disputed Aristotle's theory of light as incorporeal but said they were rays caused by the fire of the sun. He also defended the existence of the void.

In cosmology, Philoponus rejected the notion of an eternal universe, obviously in conflict with Christian teaching, in a work called *On the Eternity of the World against Aristotle.* One of his arguments was that a temporally infinite universe is impossible because it implies an infinite chain of causes is needed to reach the present. That is, if the universe is eternal, it would never have reached the present. But Philoponus was assuming that an eternal universe still had a beginning, one that occurred an infinite time ago. This argument continues to be used by modern-day theologians, as we will see in a later chapter. To put things simply for now, it is wrong. The universe need not have had a beginning. The time from the present to any moment in the past no matter how distant — last year, a thousand years ago, ten billion years ago — is still finite.

By the twelfth century, Greek learning had seeped back into Europe as some texts were translated directly from Greek to Latin, although most came by way of Arabic books and commentaries (see the next section). Gerard of Cremona (1114–1187) translated Euclid's *Elements*, Aristotle's works on natural philosophy, and Ptolemy's *Almagest* from Arabic to Latin. By the mid-thirteenth century, Aristotelian-Ptolemaic cosmology became the paradigm in the new universities founded by the Church.[5]

However, as mentioned, Christian scholars made a very important modification to Aristotle's cosmos. Instead of being eternal in time, having no beginning or end, the world was created by God a finite time ago and will end a finite time in the future when Christ returns to establish God's Kingdom. As early as the late second century, Theophilus of Antioch (ca. 180) estimated that the universe came into existence in 5529 BCE. Since then to the present, those refusing to accept the much longer times estimated by science have settled on an age, inferred from the Bible, of about six thousand years.[6]

Even before Aristotle's teachings became widespread, European scholars had begun to think in terms of viewing nature as an autonomous entity with its own laws, written by God of course but following causal principles. And, since Aristotle provided no natural theory of creation, they would have to do so.

In the 1220s, the first chancellor of the University of Oxford, Robert Grosseteste, provided a naturalist explanation of the origin of the universe that did not rely on miracles or other divine intervention once God got it started. His model superficially resembles our modern scenario of the radiation-dominated early stage of the inflationary universe. Here's how historian Helge Kragh describes Grosseteste's model:

> The universe, he said, was originally created by God as a point of light instantaneously propagating itself into an expanding sphere, thereby giving rise to spatial dimensions and eventually, by means of light emanating inwards from the expanding light sphere, to the celestial spheres of Aristotelian cosmology.[7]

Some modifications to Aristotelian cosmology were made for theological reasons, placing additional spheres beyond the sphere of the fixed stars and including an immobile "empyrean heaven" that was the abode of angels.

Ptolemy's highly mathematical system was gradually absorbed into medieval scholarship and used for making astronomical predictions. Many medieval astronomers did not indulge in philosophical speculation about the nature of the cosmos and viewed the Ptolemaic system as simply a model. Natural philosophers, on the other hand, thought astronomy was more than model making, that it was telling us something about reality.

COSMOLOGY IN THE ARABIC EMPIRE

Medieval philosophers included not only those in Christendom but also scholars in the Arabic empire that at the time was going through its remarkable golden age. In addition to translating ancient Greek and Roman texts into Arabic, these scholars made many advances on ancient thinking. And, it is important to note, they included many Jews, other non-Muslims, and non-Arabs who were able to thrive in the Arabic-speaking world that was so different then than it is now.

The prominent Anglo-Iraqi physicist Jim Al-Khalili tells the story of Arabic scholarship in his very fine book, *The House of Wisdom: How Arabic Science Saved Ancient Knowledge and Gave Us the Renaissance.*[8] As was the case for Christian scholars, those in Islamic countries developed a high regard for the teachings of Aristotle but found the need to modify those teachings that conflicted with their religious beliefs.

In cosmology, this again meant proving that the universe could not be eternal since the notion of a creation is as deeply embedded in the Qur'an as it is in the Bible. Al-Khalili identifies Ya'qūb ibn Ishāq al-Kindī (ca. 800–873) as the very first philosopher of Islam. Al-Kindī adopted the argument from Philoponus mentioned earlier that if the universe is eternal, it would have taken an infinite time to reach the present.[9] I showed earlier why this argument, which is still heard today, fails.

While the bulk of medieval Arabic scholars, like those in Christendom, accepted the geocentric model of the solar system, a heliocentric system was proposed by the Baghdadi astronomer Abu Sa'id Ahmed ibn Mohammed ibn And al-Jalili al-Sijzi (ca. 950–1020).[10] Earlier, the Persian astronomer Abū Ma'shar al-Balkhi (a.k.a. Albumasar) (ca. 787–886) had proposed a unique system in which all the planets except Earth revolve around the sun, while the sun revolves around Earth, which sits at the center of the universe.[11]

As the golden age of Islam began to wane with the invasion by the Mongols in the thirteenth century, one remarkable figure arose who would have a prominent influence on Copernicus and the astronomical revolution that followed.[12]

Nasr al-Dīn al-Tūsi (1201–1274) was a Persian scholar from northern Iran. When in around 1220 the Mongols ravaged the cities and towns in the area, killing hundreds of thousands, al-Tūsi joined a secretive religious sect called the Hashashim. Operating out of a mountain stronghold Alamūt, the Hashashim would come down out of the mountains to conduct raids and assassinations, often for hire. So they are well known to history as "the Assassins."[13] Indeed, the word *assassin* derives from Hashashim, as does *hashish*, which originally just meant "dry weed" or "grass" but now refers to a form of cannabis.

Al-Tūsi conducted research for almost thirty years at Alamūt, building an observatory and attracting scholars from all over to work with him. Somehow he survived when in 1256 the Mongols, led by Hūlāgū Khan, the grandson of Genghis Khan, were able to attack the mountain fortress and wipe out the sect. Al-Tūsi convinced Hūlāgū that the Khan needed a science advisor. He then built another observatory at Marāgha, east of Teheran, that soon became the world's greatest center for astronomy. Scholars from as far away as China joined al-Tūsi there.

Al-Tūsi extended and advanced the work of earlier mathematicians in trigonometry and number theory, including one named Omar Khayyām, and led a temporary revival of the science after the Mongols had destroyed libraries in Baghdad and elsewhere in the Arabic empire.

Al-Tūsi's *Memoir on Astronomy* (1261) is considered by many historians to be the most important book on astronomy written in the medieval period. Most significantly, it seems to have influenced

Copernicus. Al-Tūsi improved on the Ptolemaic model using a geometrical construction now known as the *Tūsi-couple*. This is basically a circle revolving around the rim of another circle of twice the diameter. Essentially the same figure would appear in Copernicus's *De revolutionibus orbium celestium* published in 1543.

While neither al-Tūsi nor the other astronomers at Marāgha proposed the heliocentric model, Copernicus seems to have used their mathematical techniques to develop that physical picture.

COPERNICUS AND THE SUN-CENTERED UNIVERSE

In chapter 1, we saw that in the third century BCE Aristarchus of Samos laid out a sun-centered solar system with the known planets in their correct order of distance from the sun. However, this model met with such strong philosophical and theological objections, especially from the highly influential Aristotle and later the Catholic Church, that it was largely ignored for two millennia.

As dawn was slowly breaking on the Dark Ages in the fifteenth century, European scholars began to have new thoughts about the cosmos. Nicolas de Cusa (1401–1464) was a German scholar and Roman Catholic cardinal of wide accomplishment in philosophy, theology, church politics, and astronomy. He considered the possibility that Earth is not at rest at the center of the universe. He also suggested that the celestial bodies are not perfect spheres and their orbits are not exactly circular.

However, de Cusa's arguments were based not on observations but on theological reasoning. He conjectured that God is everywhere and nowhere, both the center and the circumference of the universe. Nevertheless, de Cusa set the stage for what became known as the Copernican revolution.

The person responsible for that revolution was Nicolaus Copernicus (1473–1543). However, for whatever reason, the details of his life are rarely presented in the literature, which tends to focus more on Galileo and his troubles with the Church.

I will attempt to remedy this neglect somewhat, relying heavily on an outstanding recent biography of Copernicus called *A More Perfect Heaven* by Dava Sobel.[14] Sobel relates how Copernicus somehow managed to find time to observe and ponder the heavens while holding important positions in the Polish Catholic Church as well as practicing medicine.

Copernicus was born in Torun, Poland. He was raised under the tutelage of his uncle, Lukasz Watzenrod, a canon of the Catholic Church who rose to become Bishop of Varmia. In 1491, Copernicus enrolled in Jagiellonian University in Krakow where he studied logic, poetry, rhetoric, natural philosophy, and mathematical astronomy, with the last as his greatest interest. He enthused, "What could be more beautiful than the heavens, which contain all beautiful things."[15]

In 1496, at his uncle's command, he traveled to Italy, where he studied canon law at the University of Bologna. The next year he was appointed by his uncle as a canon in Varmia, and in the jubilee year 1500 he visited Rome, where he also lectured briefly on mathematics.

After a short visit home, in 1501 Copernicus returned to Italy, where he studied medicine in Padua, learning all about bloodletting, corrupt humors, and the other medical wisdom of the day. He also studied the applications of astrology to diagnosis and treatment, although he ever took astrology seriously.

In 1503, Copernicus received a doctorate in canon law from the University of Ferrara, after which he returned home to Poland and became "healing physician" to the bishop and canons of Varmia. He also treated peasants free of charge.

In 1510 Copernicus took up permanent residence in Frauenberg (Frombork), a small town in northeastern Poland within the Varmia diocese. There he engaged in astronomical research while carrying out many high-level ecclesiastical duties, although he never became an ordained priest.

Sobel tells us that, by 1510, Copernicus had "leapt to his Sun-centered conclusion via intuition and mathematics. No astronomical observations were required."[16] Perhaps not directly, but it would be incorrect to assume that the idea was the results of "pure thought." After all, Copernicus was an observer and was familiar with the motions of planets.

Copernicus was apparently unaware of Aristarchus's proposal and wrote up a forty-page outline of his ideas, which included no proofs, called "Commentariolis," which Sobel refers to as *Brief Sketch*. He sent it around to friends who, in turn, made copies and distributed in further. There he makes his sun-centered position unequivocal:

> All spheres surround the Sun as though it were in the middle of all of them, and therefore the center of the universe is near the Sun. What appears to us as motions of the Sun arise not from its motion but from the motion of the Earth and our sphere with which we revolve about the Sun like any other planet.[17]

The equivocating came later.

In 1511, Copernicus was named chancellor of his chapter, which made him responsible for all its financial accounts, which were extensive. This required him to administer 150,000 acres of farmland controlled by the Church. He had to deal with the most mundane transactions between the peasants who worked the land, all of which needed his approval.

Despite these heavy duties, Copernicus was able to make a series of observations that enabled him to determine the length of a year within seconds, far more accurate than any clock at the time could measure.

Being by this time a reputable astronomer, in 1512 Copernicus was invited by Rome to consult on calendar reform, since the Julian calendar was now way out of whack. Of particular importance to the Church, Easter had drifted far from the Sunday after the first full moon of the vernal equinox, where it had been set by the First Council of Nicaea in 325. Although a record of Copernicus's contribution has not survived, the new Gregorian calendar, which we still use today, was officially adopted by the Church in 1582. It improved over the Julian calendar in use since the days of Julius Caesar by reducing the number of leap years from 100 per four centuries to 97 and making 3 out of 4 years beginning a century normal rather than leap years.

Copernicus also fretted about debasing the currency, publishing in 1517 *Meditata*, a meditation on how to solve the money problem by properly minting coins. Interestingly, late in his life Newton would serve as warden of the Royal Mint in England.

In 1517, Martin Luther (1483–1546) triggered the Protestant Reformation with his *Ninety-Five Theses*. One of his early disciples was a former Catholic priest Andreas Osiander (1498–1552), who would later play a significant role in the publication of Copernicus's great work. Copernicus, however, never deviated from his dedication to the Mother Church.

Sobel's book contains a delightful play in two acts dramatizing the visit made to Copernicus in 1539, when he was sixty-five years old, by twenty-five-year-old German Georg Joachim Rheticus (1514–1574). Rheticus was a Lutheran, and, in fact, a professor of mathematics and poetry at the University of Wittenberg, where Luther resided. At the time, Lutherans were barred from that region of Poland, but Rheticus somehow sneaked in.

Although Luther scorned astrology as "framed by the devil," Rheticus believed it carried information about the deepest matters of the cosmos and sought astronomical knowledge that he thought would enhance the discovery of that information. He had cast many horoscopes of himself and was convinced he would have a short life and so had to move fast to accomplish anything. He lived to age sixty, a ripe old age in those days.

While in Nuremberg, Rheticus heard about this Polish canon who had described the motion of the planets by centering them on the sun. He traveled to northern Poland to find out more.

Sobel's play covers the period from May 1539 when Rheticus appears on Copernicus's doorstep and gradually becomes his assistant in producing his masterwork *De revolutionibus orbium celestium* (*On the Revolutions of the Heavenly Spheres*) to Copernicus's death on May 24, 1543.

Copernicus was reluctant to publish his findings, but Rheticus urged him on. So did the friendly Tiedemann Giese (1480–1550),[18] Bishop of Kulm, who was pleased that Rheticus knew a highly respected printer of scientific texts in Nuremberg, Johannes Petreius (1497–1550).

Late in 1539, Rheticus wrote a summary of Copernicus's model in the form of a letter to a mentor in Nuremberg, Johann Schöner (1477–1547) called *A First Account*, which was published the following year by Petreius.

Shortly thereafter, Andreas Osiander enters the picture. As mentioned, Osiander was an early disciple of Luther, also a theologian and amateur mathematician. Osiander was a friend of Petreius who may have asked him to consult on publishing Copernicus's book. Osiander wrote to Copernicus suggesting that they include an introduction making the point that mathematical hypotheses "are not articles of faith but the basis of computation; so that even if they are false it does not matter, provided that they reproduce exactly the phenomena of the motions."[19] He made a similar suggestion to Rheticus, saying it will placate the "peripatetics and theologians if the hypotheses were not claimed to be in reality true."

Manuscript in hand, Rheticus left Frauenberg in September 1541, arriving in Wittenberg in October, where he was reappointed to the faculty and made dean of the Faculty of Arts. He was not totally accepted. Reportedly, Luther stopped by for lunch one day and was reported to have remarked, "So it goes now. Whoever wants to be clever must agree with nothing that others esteem. He must do something of his own [Luther himself included?]. This is what that fellow does who wishes to turn the whole of astronomy upside down. Even in those things that are thrown into disorder, I believe the Holy Scriptures, for Joshua commanded the Sun to stand still and not the Earth."[20]

Rheticus and Giese had attempted heroically to reconcile Joshua

(Joshua 10:12–13) to the Copernican model. They also tried to deal with Psalm 93, which declares that the foundation of Earth remain forever unmoved, and other biblical contradictions. Rheticus wrote a tract attempting to rectify Copernicus with holy scripture, but it was never published.

Rheticus was finally able to get to Nuremberg with most of the manuscript in hand in May 1542 and deliver it to Petreius, who began printing it immediately. Rheticus proofread every page as it came off the press, but in October with only half done, he left Nuremberg for a job as professor of higher mathematics at the University of Leipzig.

At that point, Osiander took over and completed the work. In November 1542, at age sixty-nine, Copernicus, who had been checking progress of the manuscript, suffered a stroke that left him unable to continue. On May 24, 1542, Copernicus's friend Jerzy Donner took the final pages, just arrived from Nuremberg, to the invalid's bed and put them in his hands. At the next moment, Donner saw the life go out of him.[21] Copernicus's work was complete.

At the time, the Church did not deem *De revolutionibus* heretical. Copernicus had dedicated the work to Pope Paul III. First, the work was highly technical so only the most mathematically sophisticated reader could follow it. Second, it did not ask to be taken as literal reality. Osiander had included an unsigned preface that presented the model only as a tool for astronomical calculations with no philosophical or theological implications:

> You may be troubled by the ideas in this book, fearing that all of liberal arts are about to be thrown into confusion. But don't worry. An astronomer should make careful observations, and then frame hypotheses so that planetary positions can be established for any time. This our author has done well. But such hypotheses need not be true nor even probable. Perhaps a philosopher will seek truth, but an astronomer will just take what is simplest, and neither will find anything certain unless it has been divinely revealed to him. So if you expect to find truth here, beware, lest you leave a greater fool than when you entered.

The Copernican model is shown in figure 2.1.

De revolutionibus is actually six books. It presents a unified system with the planets in their correct order from the sun and their orbital periods estimated with what turned out to be remarkable accuracy. However, Copernicus still had to introduce epicycles to maintain circular orbits, so the published model was not as simple as figure 2.1 suggests, although it was still simpler than Ptolemy's system.

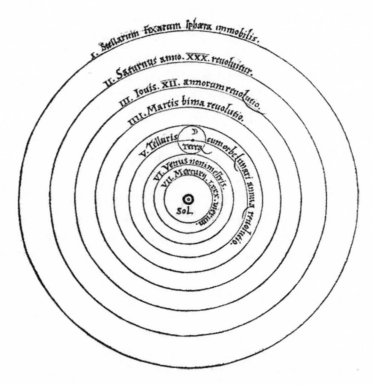

Figure 2.1. The Copernican heliocentric model of the solar system. The picture Galileo (1564–1652) would later promote is the same, except he added the orbits for the four moons about Jupiter discovered with his telescope. Image courtesy of NASA's Earth Observatory, adapted from Nicolaus Copernicus, 1543, *De revolutionibus orbium celestium* (*On the Revolutions of the Heavenly Spheres*).

INITIAL REACTION

While the Church did not immediately reject the Copernican model, forces were coming into play at the time that would eventually lead to its condemnation in theological circles. The Protestant Reformation produced a deep split with the Roman Church on the location of the source of authority in Christendom. Although the Church revered the Bible, it did not regard it as the final authority on theological matters. That authority rested ultimately with the pope, based on the claim of an unbroken line of succession from Peter, who was given rule over all earthly matters directly from Christ. The reformation had to find a replacement for papal authority, and the only alternative was the Bible, which meant that it was to be taken literally as the word of God.

Even today, we find Bible inerrancy to be a basic dogma in many Protestant sects, which leads them to reject those scientific results such as evolution and the age of Earth, while the Catholic Church has little problem with them. Popes have declared most scientific discoveries acceptable, as long as they do not deny divine creation and the immaterial nature of soul. (In this regard, it is debatable whether the Catholic Church or moderate Christians really believe in biological evolution as it is understood by science, which is not God-guided).

Luther and other reformers who preached that the Bible was literal truth objected to the Copernican picture since it disagreed with scriptures. However, it should be noted that Luther died just three years after the publication of *De revolutionibus* when the model was far from being established on scientific grounds. Nevertheless, under the pressure of the reformation, over the next century the Roman Catholic Church, led by the Society of Jesus, found it necessary to take more conservative stands on many issues and that included backing off from Copernicus and objecting to the mathematical method of infinitesimals that would eventually lead to calculus.[22]

Another thorn in the Church's side at the time was the Italian Dominican friar Giordano Bruno (1548–1600), whom they could only shut up by burning him at the stake. Bruno committed many heresies to earn his toasting, but his cosmology is particularly relevant to our story. He seems to have picked up and expanded on some of the ideas of Nicolas de Cusa mentioned earlier.[23] Bruno proposed that the sun was just another star in a universe containing an infinite number of worlds and having no center. Furthermore, these worlds were populated by other intelligent beings.

Next in our chronological tale, in 1572 Danish astronomer Tycho Brahe (1546–1601) observed a supernova, a bright flash in the sky that quickly disappeared. This provided the first evidence that the heavens can change in an unpredictable fashion, contrary to the traditional belief that they are perfect and unchanging. English astronomer Thomas Digges (1546–1595) tried and failed to measure the parallax of Tycho's supernova and concluded it must be beyond the orbit of the moon.

Digges also published the first account of the Copernican model in English, in 1576. He proposed a major change of cosmological import when he discarded the notion of a finite sphere of fixed stars beyond the solar system and replaced it with an infinite cosmos of stars. The lack of observed parallax convinced him they were at great distances, as was suggested by Copernicus.

However, while Tycho agreed that the Copernican model "expertly and completely circumvents all that is superfluous or discordant in the system of Ptolemy," he objected that it "ascribes to the Earth, that hulking, lazy body, unfit for motion, a motion so quick as that of the aethereal torches, and a triple motion at that."[24]

So Tycho published a model, which had been suggested earlier by others, in which Earth remains at the center with the sun and moon revolving about it, while the other planets revolve around the sun. It was actually more accurate than the Copernican system in fitting the observations of the day. And it was more in keeping with Church teaching, which made it quite popular for a while.

KEPLER AND THE LAWS OF PLANETARY MOTION

While the Copernican model is conceptually simpler than the Ptolemaic model, especially for picturing the motions of the planets, it was not immediately superior as a calculational tool because the input data were flawed. This would change with the improved observations of Brahe and then Johannes Kepler (1571–1630). Furthermore, Kepler made a huge advance, proposing three laws of planetary motion that described the new observations with great precision.

Kepler's laws of planetary motion

1. The orbit of every planet is an ellipse with the sun at one of the two foci.
2. A line joining a planet and the sun sweeps out equal areas during equal intervals of time.
3. The square of the orbital period of a planet is directly proportional to the cube of the semimajor axis of its orbit.

TURNING THE TELESCOPE ON THE HEAVENS

The story of Galileo is familiar but still widely misunderstood. I have discussed his gigantic contributions to physics in several previous books, most recently in *God and the Atom*,[25] and will focus here on his astronomy. Unlike Osiander, Galileo was not satisfied to simply assent that the heliocentric model was merely a useful tool for predicting celestial events. He insisted it was a fact of nature. Kepler also was of the same mind.

The refracting telescope had been invented in 1608 by the Dutch eyeglass maker Hans Lipperhey (or Lippershey) (1570–1619). The earliest versions had a magnification of only a few times, and Galileo was able to improve that to a factor of thirty and turned his superior instrument on the heavens.

The initial publication of his observations appeared in 1610 in *Sidereus nuncius* (known popularly today as *Starry Messenger*).[26] He reported seeing mountains and craters on the moon. He viewed ten times as many stars than are visible to the naked eye and fuzzy nebulae that he interpreted, along with the Milky Way, as collections of stars too far away to resolve.

See figure 2.2 for Galileo's sketches of the moon from *Sidereus nuncius*. This, along with sunspots observed later, provided the direct empirical falsification of the common belief, as taught by Aristotle, that the heavenly bodies are perfect spheres.

In 1610 Galileo also made the important discovery that, similar to the moon, Venus exhibited phases that result from the partial illumination it gets from the sun as it circles around inside Earth's orbit. This observation cannot be explained in the Ptolemaic system, although it did not rule out other geocentric models such as Tycho's.

But perhaps the most dramatic set of observations by Galileo was the four moons of Jupiter. From January 7, 1610, through March 1, he sketched the positions of four bodies near Jupiter, except when clouds obscured his view. Some sixty-four such sketches can be found in *Sidereus nuncius*, extending from page 65 to page 83 in the cited edition. A sample set is shown in figure 2.3.

It must be mentioned that Galileo also made a number of claims that were not supported by observations, notably that the tides were caused by, and evidence for, the movement of Earth around the sun. Here the great observer failed to consider the fact that two tides occur daily, while his model predicted only one. Kepler had earlier given the correct explanation: the tides are caused by the moon.

In 1616, the Church ordered Galileo not to claim that it is a fact that the sun was the center of the universe and that Earth moves, which violated several biblical references where it is stated that Earth "cannot be moved" (Psalms 93:1, 96:10, 104:5, Chronicles 16:30). Furthermore, the Church regarded Aristotle's physics, which included the concept of absolute motion, as authoritative. Note that this comports well with the doctrine that Church teachings are absolute. However, Galileo seemed to go along and was not prohibited at this time from continuing his work.

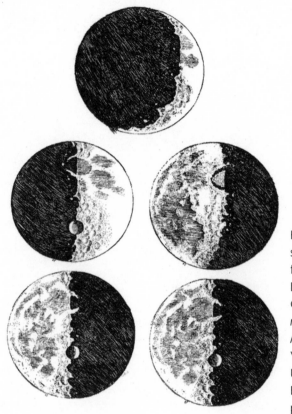

Figure 2.2. Galileo's sketches of the moon from *Sidereus nuncius*. Images from Galileo Galilei, *Sidereus nuncius, Or, the Sidereal Messenger*, trans. Albert Van Helden (Chicago: University of Chicago Press, 1989); first published in 1610.

And so Galileo continued his telescopic observations. Then he got himself in deep trouble. In 1632, Galileo published *Dialogues on the Two Chief World Systems*, which argued forcefully for the Copernican model. In *Dialogues*, the character Simplicio, who represents the Aristotelian position, makes arguments that the pope himself had previously expressed. Pope Urban VIII, who was previously a friend and supporter of Galileo, was not pleased. Galileo was tried by the Inquisition for disobedience, forced to recant on his knees, and sentenced to permanent but quite comfortable house arrest.

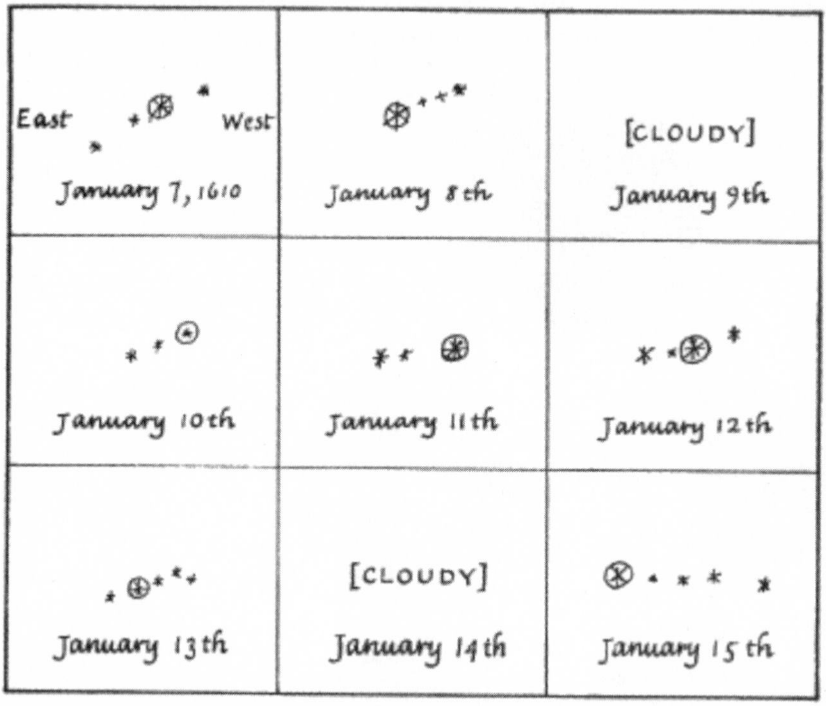

East + ⊕ West January 7, 1610	⊕ * * * January 8th	[CLOUDY] January 9th
* * ⊙ January 10th	⧣ * ⊕ January 11th	✗ *⊕ * January 12th
* ⊕* * * January 13th	[CLOUDY] January 14th	⊗ ▴ * * * January 15th

Figure 2.3. A sample of Galileo's sketches of the positions of the moons of Jupiter. These observations left no doubt that these bodies were moons revolving around Jupiter and not Earth. But, once again, one could conceive of a geocentric model, such as Tycho's, in which not every body circles Earth. Images from Galileo Galilei, *Sidereus nuncius, Or, the Sidereal Messenger*, trans. Albert Van Helden (Chicago: University of Chicago Press, 1989); first published in 1610.

When the celebrated French philosopher René Descartes (1596–1650) heard of Galileo's fate, he withheld publication of his *Le Mond*, which was based on Copernican principles. He wrote to a friend, "I wouldn't want to publish a discourse which had a single word that the Church disapproved of; so I prefer to suppress it rather than publish it in a mutilated form."[27]

Although technically forbidden from writing any more about physics and astronomy, Galileo continued work in physics that laid the foundation for Newtonian mechanics that appeared a generation later. Newton was born the year Galileo died, 1642.

GALILEAN RELATIVITY

One of Galileo's most important discoveries is rarely mentioned outside physics classrooms and is grossly misunderstood elsewhere. Yet it represents the major deviation of Galileo from the common understanding of the day that was based on both scripture and Aristotle.

It is common sense to most people that we can tell when we are moving and when we are at rest. There are "obviously" two different states of motion. So, as Galileo was asked, if Earth is moving and we are sitting on Earth, how come we don't notice it? It was a good question.

Common sense is not the same thing as careful scientific observation. It's everyday common sense that the world is flat. Galileo was one of the many scientists in history who have shown that you can't always trust common sense. From his careful observations, Galileo was sure Earth moved about the sun. But he had to provide an explanation for why we don't experience that motion. He proposed what is now known as the *principle of Galilean relativity*. Allow me to express it in the following modern, operational form:

Principle of Galilean relativity

> There is no experiment that can be performed inside a closed capsule to measure the velocity of that capsule.

The *velocity* of a body is the time rate of change of position of the body. It is a three-dimensional vector whose magnitude is called *speed* and whose direction is the direction of motion of the body. That is one measure of motion. Another measure of motion is *acceleration*, which is the time rate of change of velocity.

Applied to Earth's motion, if we are sitting in a closed room on Earth, we have no sense of motion. We can only determine the velocity of Earth by looking outside and making some astronomical measurement. Recall that Copernicus had made reasonable estimates of the orbital radii of the planets, which were based on observations. Using today's best number, Earth is 150 million kilometers from the sun, which means that it travels 942 million kilometers in a year. Therefore, Earth moves around the sun at a speed of 30 kilometers per second.

The implication of the principle of relativity is that the velocity of an object is not absolute. Velocities can be measured only *relative* to other objects. The speed of Earth relative to the sun is 30 kilometers per second. Earth's speed relative to me sitting here at my desk is zero.

And this is really where the conflict between Galileo and the Church, and Aristotle, arises. According to Aristotle, motion is absolute. A body at rest is absolutely at rest. A body in motion is absolutely in motion. And scripture is unequivocal in insisting that Earth cannot be moved—even relatively.

Although Galileo supposedly said, "*Eppur si muove*" ("And yet it moves"), this does not accurately describe his discovery, as we now fully understand it. The principle of relativity says that Earth moves in some reference frames and is at rest in others. Motion is relative. And that was a new and revolutionary idea.

And Galileo did not prove that empirically. It really wasn't until the eighteenth century that Earth's (relative) motion about the sun was confirmed by the observation by James Bradley (1693–1762) of *stellar aberration*. This is an apparent shift in the position of stars and other astronomical objects that results from the (relative) motion of the observer on Earth.

As we will see, when the principle of relativity was challenged in the early twentieth century, Albert Einstein (1879–1955) preserved it in the *special theory of relativity*, leading to a profound restructuring of our ideas of space, time, and motion.

THE MECHANICAL UNIVERSE

Galileo and Newton drew a picture of the cosmos that was very similar to the one of Democritus and the other ancient atomists. The universe is composed of particles moving around in empty space, colliding with one another or otherwise interacting by means of a long-distance force called gravity. A particle can be thought of as a body that appears to the observer as an infinitesimal point. This appearance could be the result of the body being very small so that any structure it might have is not detectable by eye or with whatever magnifying instruments the observer has at her disposal. Or the body could be as large as a galaxy, such as a quasar, but so far away that, using the most powerful telescope, it looks like a particle.

Newton's laws of motion can be most easily expressed in terms of particles. Rather than discuss them in their original form, I will put them in a modern context. A particle of mass m and speed v has a momentum that is a vector whose magnitude $p = mv$ and direction is equal to the direction of the velocity vector. (The actual magnitude formula as we now know is more complicated, but this is only significant for speeds near the speed of light and we need not consider that here). Newton defined p as the "quantity of motion."

The basic principle of Newtonian mechanics is then:

Newton's second law of motion

> The total force on a particle is equal to the time rate of change of the momentum of the particle.

If no force is applied, the particle's momentum will be unchanged. This is called the *law of conservation of momentum* and applies not to just a single particle but to any system of particles where the net force on the system is zero.

Newton's *first law of motion* is simply the special case when the force is zero and so the momentum is constant. When the body's mass is constant, its velocity will be constant.

The *third law of motion*, "for every action there is an equal and opposite reaction," is just another way of saying momentum is conserved.

It is interesting that, although it can be derived from Newton's laws of motion, the *law of conservation of energy* was not formulated until well into the nineteenth century.

When the mass of a body is constant, Newton's second law of motion can be written $F = ma$, where F is the total force on the body, m is its mass, and a is the acceleration or time rate of change of velocity of the body. From this equation it is possible to predict how far a body acted on by a force will go in a given time.

If the force is not constant, you can break the motion down into infinitesimal time intervals and use calculus, invented independently by Newton and Gottfried Wilhelm Leibniz (1646–1716), to sum up the intervals to get the net effect. You can also use calculus to calculate the motion of larger bodies by breaking the body up into infinitesimal parts and treating these parts as particles. They don't have to be elementary particles. This works for solids, liquids, and gases. It's all remarkably simple, once you learn how it works.

In his universal law of gravitational attraction, Newton proposed that the gravitational force F between two particles with masses m_1 and m_2 is proportional to the product $m_1 m_2$ and inversely proportional to the square of the distance r between them. The value of the proportionality constant G, called *Newton's constant*, was unknown to Newton and was first measured in the laboratory in 1798 by the British physicist and chemist Henry Cavendish (1731–1810).

Using calculus, Newton proved that two large spherical bodies can be treated as particles of the same mass located at the centers of the spheres. In this manner, the planets can be described as particles moving in the void of empty space.

Newton's laws of motion and gravity provided the final confirmation of the validity of the sun-centered model of the solar system. As mentioned, Kepler had introduced the idea that the orbits of the planets are not circles but ellipses. Newton was able to prove this mathematically. When he showed his proof to astronomer Edmund Halley (1658–1742), Halley convinced Newton to publish (at Halley's expense) what is considered the greatest scientific work in history, *Philosophiae naturalis principia mathematica.* Referred to simply as *Principia*, it presented the three laws of motion and the law of universal gravitation, from which Kepler's laws of planetary motion were then derived. Today this is a simple exercise in freshman physics.

Using Newton's laws, Halley calculated that a comet seen in 1682 was the same as one that had been recorded by astronomers as far back as 240 BCE, circling the sun in a highly elliptical orbit every 75–76 years. Halley predicted it would return in 1758. The success of Halley's prediction, after his and Newton's deaths, may have been the single most important event in scientific history. It established the power of the new science in the minds of scholars and laypeople alike.

OPTICKS

In every field of physical science today, the workload is generally divvied up between observers/experimenters who build the instruments and gather the data, and theorists who develop the mathematical models used to describe that data and attempt to make predictions from those models. As we have seen, this was not true in the olden days. Galileo was an observer, experimenter, and theorist. Newton was a great theorist and experimenter. While *Principia* was Newton's theoretical masterpiece, *Opticks*, published in 1704, was a masterwork of experimentation.

In *Opticks*, Newton presents the results of his laboratory experiments with light and the conclusions about the nature of light that he drew from them. Of course, light is our primary source of information about the world and, until the twentieth century, the only source humans had to learn about the universe beyond Earth. You can't touch, hear, or smell a star. Well, I suppose we feel the heat from the sun, but that's about it. (And, as we will see later, we can hear the big bang).

Once again, Newton overthrew an erroneous concept of Aristotle that had been entrenched in European thought for millennia. In *De anima* (*On the Soul*), Aristotle presents an incorporeal theory of perception based on Plato's forms. According to Aristotle, when you look at an object, your eye somehow becomes the form of the object. No need to waste your time trying to make any sense out of that.

The atomists, on the other hand, were closer to the modern understanding of perception. Democritus proposed that visual perception results from atomic emanations from the body colliding with atoms in the eye. Today we know these emanations are particles called *photons*.

Newton's great achievement in *Opticks* was to show that white light is composed of all the colors of the rainbow. He recognized that color is not an inherent property of an object but results from the way the object emits or reflects the various colors. Many of our perceived colors, such as maroon or brown, are not present in the spectrum of light but result as a mixture of light from different parts of the spectrum.

In *Opticks*, Newton also made several conjectures about the nature of light that he could not demonstrate empirically, such as that it is made up of particles ("corpuscles"). He had earlier presented this idea to the Royal Society and had been challenged by the curator of experiments and prominent physicist in his own right, Robert Hooke (1635–1703). Hooke had his own pet theory that light is a wave and had clashed with Newton in the past on other matters. Because of this bitter disagreement, Newton held off publishing *Opticks* until after Hooke's death.

NEWTON AND GOD

Descartes had proposed an alternative to the atomic model in which the universe is a continuum of matter whirling in vortices about the sun. Gravity was somehow the product of this whirlpool-like motion but, despite his mathematical genius, Descartes did not provide a quantitative model.

However, Descartes made a major philosophical breakthrough when he proposed that God created the universe as a clockwork of perfect motion that required no further intervention. Thus, although he remained a committed Christian, Descartes was the first to envisage the alternative to the Judeo-Christian-Islamic God called *deism*.

During the period in the eighteenth century called the Enlightenment, the clockwork universe would become associated with Newtonian particle physics and form the basis of the belief in a deist god who creates the universe and then leaves it alone to carry on according to the laws it established at the creation.[28] Since god is perfect, so must be his laws and thus it has no need to step in to make changes. The deist god clearly contradicts Christian belief, although this did not seem to get Descartes into trouble when he made his original proposal.

Although the Newtonian clockwork universe was perhaps the primary motivating factor for deism, Newton did not profess deistic views himself and remained a Christian who saw the need for God to step in from time to time to keep the universe running properly.

Newton was an unconventional Christian who rejected the Trinity. But he still was very much a believer not only in the Christian God but also in the occult. He spent more time and wrote more on alchemy and biblical interpretation than he did on physics. And God entered his physics in important ways. Unlike Galileo, who did not mix his religion and science but still claimed belief, Newton turned to God to provide explanations for what he could not explain himself. He may have been the first to use what we now call the "God-of-the-gaps" argument, also known as the "argument from ignorance." When you lack a natural explanation for a phenomenon you, conclude God made it happen.

Newton's gravitational theory admitted only attractive forces, which implied that the stars, which everybody at the time assumed to be fixed, should collapse upon one another as a result of their mutual attraction. He conjectured that God had placed them in just the right positions where their attractions balanced.

In 1718, Halley discovered that three bright stars were no longer in the same positions as reported in ancient observations, thus showing that stars were not fixed after all; it was another blow against biblical cosmology.

Newton also invoked God to provide for the stability of planetary motion. He was fully aware that his derivation of Kepler's laws assumed each planet moved independently of the others while, in fact, the planets also exert gravitational pulls on one another. Newton reasoned that the orbits of planets would not maintain their regularity by blind chance. Thus, he concluded, God had to intervene from time to time to keep things in order.

Newton's archrival Leibniz scoffed at this:

> Sir Isaac Newton and his followers have also a very odd opinion concerning the work of God. According to their doctrine, God Almighty wants to wind up his watch from time to time: otherwise it would cease to move. He had not, it seems, sufficient foresight to make it a perpetual motion.[29]

Newton and Leibniz also quarreled over the priority of the invention of calculus, which both developed independently. We still use Leibniz's superior notation today.

3.

BEYOND UNAIDED HUMAN VISION

EIGHTEENTH-CENTURY ASTRONOMY

While Newton did not make any significant astronomical observations, in 1668 he built the first *reflecting* (as opposed to refracting) telescope. By using a curved mirror rather than a lens to provide the magnification, the reflecting telescope avoids the chromatic aberration that occurs in a lens because of the dependence of the refractive index of glass on the color of light. The reflecting telescope is the primary type of astronomical optical telescope in use today, in space and on Earth, although it wasn't until the late nineteenth century that reflectors were used at altitude.

With the telescope, astronomy was now able to move beyond the limitation of unaided human vision, and a never-imagined cosmos began to reveal itself. In 1655, the Dutch physicist and astronomer Christiaan Huygens (1629–1695) built a refracting telescope with a magnification of fifty and discovered Saturn's moon Titan, the largest moon in the solar system. From his observations of Saturn, he concluded that the planet is surrounded by a solid ring. Later Huygens was able to resolve the Orion nebula into separate stars and observed the transit of Mercury across the sun.

Huygens was also one of the great physicists of all time. He invented the pendulum clock and the internal combustion engine, derived the formula for centrifugal force, and wrote a book on probability theory.

Among his many achievements, in 1678 Huygens quantitatively described the wave theory of light, which Robert Hooke had proposed in 1672. Newton's corpuscular theory actually came later. Huygens demonstrated how waves are able to bend ("diffract") around corners. Newtonian's corpuscular theory of light was ostensibly refuted when in 1800 Thomas Young (1772–1829) demonstrated conclusively that it exhibits the interference and diffraction effects associated with waves. As we will see, in the twentieth century, light was shown to be composed of particles called photons and the waves associated with light refer not to individual photons but to the statistical behavior of beams of light containing many photons.

KANT'S COSMOS

In 1755, a young privatdocent at the University of Königsberg named Immanuel Kant (1724–1804) published a book on the structure of the universe titled *Allgemeine naturgeschichte und theorie des himmels* (*Universal Natural History and Theory of the Heavens*).[1] Unlike Newton but like Leibniz, Kant saw no need for any divine miracles to explain the universe. He wrote, "A constitution of the world which did not maintain itself without a miracle, has not the character of that stability which is the mark of the choice of God."[2]

Kant's universe starts with a divinely created chaos of particles in an infinite void. The particles attract one another by Newtonian gravitation and form condensations, which then evolve into orderly structures such as the solar system. The Milky Way is disk-shaped conglomeration of stars, and the observed nebulous bodies are not individual stars but conglomerations like the Milky Way wending indefinitely throughout infinite space.[3]

The book remained largely unknown until 1854 when the German physicist Hermann von Helmholtz (1821–1894) mentioned it in a lecture. Kant is still better known for his philosophy than his astrophysics, but the latter was not too shabby either.

HERSCHEL'S HEAVENS

Perhaps the most productive astronomer in the eighteenth century was Frederick William Herschel (1738–1822). During the latter part of the century, Herschel built his own reflecting telescopes and made a series of important discoveries. He helped establish that Uranus, which had previously been thought to be a star, was in fact a planet. He discovered two moons of Saturn and two of Uranus. He determined that the Milky Way was in the form of a disk. Herschel observed binary and multiple stars. He showed that the sun emits infrared light. And, by the way, using a microscope he proved that coral is an animal, not a plant.

Between 1782 and 1802, Herschel made a systematic study of nonstellar objects, that is, diffuse bodies called nebulae. He produced a catalogue of over a thousand nebulae, classifying them according to brightness, shape, size, and other features.

Starting in 1785, Herschel wrote a series of papers under the common title "The Construction of the Heavens" that suggested the nebulae were far away. Because of the finite speed of light, by observing them the astronomer should have been able to follow their movements through the heavens. However, no such movement was evident, indicating the nebulae were very distant.[4]

In 2009, the European Space Agency launched the Herschel Space Observatory, a large infrared telescope that operated until 2013.[5]

OLBERS'S PARADOX

Halley, along with Kepler and the Swiss astronomer Jean-Philippe de Cheseaux (1718–1751), recognized a problem with the idea proposed by de Cusa and Digges that the universe contains an infinite number of stars. After it was formulated by the German astronomer Heinrich Wilhelm Matthias Olbers (1758–1840), the problem became known as *Olbers's paradox*: if the universe is infinite in size and eternal in age, then the sky should be bright rather than dark at night because of the light from all those stars.

To see this, picture a thin spherical shell of a certain thickness at a distance r from Earth. Its volume will be equal to that thickness multiplied by the surface area of the shell, $4\pi r^2$. Assuming a uniform density of stars, the luminosity from the shell will be proportional to that volume. However, the intensity of light (power per unit area) reaching Earth falls off as $1/r^2$. Thus, if the average luminosity of stars is independent of their distances from Earth, the contribution from each shell of the same thickness throughout the universe will be the same and add to give the total light observed on Earth from all the stars in the universe. For an infinite universe, that intensity would appear brighter than the sun—indeed have infinite intensity. Clearly this is not what we observe.

There are a number of possible explanations for why the sky is, in fact, dark at night. In an 1848 essay, "Eureka," Edgar Allan Poe (1809–1849) suggested that light has not yet reached us from the most distant stars. That is, the universe has a finite lifetime so we are only observing the stars out to the distance that light can travel in that time.

This is true, but another factor that contributes is the expansion of the universe, which we will discuss later. The energy of light coming from great distances as light is reduced because it is *redshifted* to longer wavelengths or lower energies.

SUFFICIENT REASON

In 1710, Leibniz wrote *Theodicy: Essays on the Goodness of God, the Freedom of Man, and the Origin of Evil*.[6] This term *theodicy* became associated with the still-unsuccessful attempts to justify the undeniable evil and suffering in a world supposedly under the complete control of an omnibenevolent, omnipotent, omniscient God. Leibniz proposed that God had created the "best of all possible worlds." The evil that exists is just part of that optimization. Without it, the world would be even worse.

Leibniz also proposed the *principle of sufficient reason*, also known as the *cosmological argument*, as a proof of the existence of God. Basically, the argument says that everything that is true requires a complete explanation, a sufficient reason. Since the universe does not explain itself, God must exist as its sufficient reason for the existence of the universe.

These arguments fails for a simple, sufficient reason, which I have already emphasized. Like all purely logical arguments that contain no empirical input, it tells you nothing not already embedded it its premises. Applied here, God must exist because he must exist.

THE CENTER OF THE UNIVERSE

In our historical survey so far, we have seen that a conflict has long persisted over the location of what is termed "the center of the universe." For the most part, we humans have found it easy to position ourselves at the center. All living species are self-centered, if not always the individual organism but what Richard Dawkins termed their "selfish genes."[7] Species would not have survived if they weren't selfish at some level.

There is also a good empirical reason for us to think we are the center of the cosmos. When we look at the sky, everything seems to revolve around us. The planets sometimes turn around and go back the other way, but they soon turn back again and resume circling Earth.

Even if we now know that a sun-centered system is the simplest for visualizing the motions of the planets, how often in our everyday lives do we need to worry about where a planet will be tomorrow, a month from now, or where it was on March 28, 585 BCE?

In fact, for most of our purposes, an Earth-centered model of motion does just fine. It would be silly to calculate the route an airliner should take in going from Tokyo to London in a sun-centered coordinate system. And we could still use a geocentric model for predicting planetary motion if we wanted to.

Of course, we know today that the sun is not the center of the universe, as it was thought to be when Copernicus's picture of the cosmos was one of a solar system of seven planets surrounded by a shell of fixed stars. As telescopes improved, astronomers discovered that our sun is just another star. As I have mentioned, the ancient atomists proposed a picture of a cosmos that is limitless in time and space and that no point in space can be designated as that special place we can call the center of the universe. Similarly, no moment in time can be designated as that special moment when the universe began (or will end). As we will see, scientific consensus today is converging on precisely this cosmological model. But, again, I must emphasis it is a human-constructed model.

4.

GLIMPSES OF THE UNIMAGINED

THE ADVANCE OF CELESTIAL MECHANICS

T he telescope, combined with Newtonian mechanics, made it possible for humanity to glimpse a universe never before imagined and to describe what they saw with mathematical precision. Kepler's laws of planetary motion, which Newton derived from his laws of mechanics and gravity, provided a huge improvement over previous attempts to describe planetary motion. But they still were not perfect. They treated each planet as interacting gravitationally solely with the sun, neglecting the planet's interactions with other planets or with any other bodies, such as comets, asteroids, and moons.

Fortunately, this is a good approximation for our solar system because, as we saw in the last chapter, the gravitational force between two bodies is proportional to the product of their masses, and the masses of planets are very much less than the mass of the sun. Furthermore, the force falls off as $1/r^2$, and the planets are very distant from one another.

Still, the planets in our solar system are sufficiently massive and they are sufficiently close so that their interactions with one another produce nonelliptical orbits. However, these deviate only very slightly from those given by Kepler's laws. These deviations became measurable as telescopes improved, which is just another example of how technology drives scientific advancement.

In this case, the effects are small enough that they can be treated as approximately as "perturbations" to the two-body Keplerian orbits. In 1747, the Swiss mathematician Leonhard Euler (1707–1783) (pronounced "Oiler") developed a technique that won for him a prize the Academy of Sciences of Paris had offered for developing analytical methods for calculating the motions of Jupiter and Saturn. Euler laid the groundwork for *perturbation theory*, which is still a major tool used in physics for solving by a sequence of approximations problems that cannot be solved exactly. And this method is not limited to celestial mechanics. For example, the highly successful theory of quantum electrodynamics, developed by physicists in the late 1940s, is based on calculating a series of increasingly accurate perturbation approximations.

One can imagine that other star systems might exist in the universe where the interactions of other planets cannot be treated as small perturbations. In that case, perturbation theory would produce such inaccurate predictions of planetary motion as to be worthless. And only the two-body problem can be solved exactly. Astronomers in those planetary systems would be forced to use numerical techniques to calculate orbits, but they could do this if they had computers at least as advanced as ours are today.

While his calculations were only partially successful, Euler set the stage for the methods developed by the French mathematician, astronomer, and physicist Pierre-Simon Laplace (1749–1827) in his five-volume tome *Mécanique céleste* (*Celestial Mechanics*) published between 1799 and 1805.

The equations of celestial mechanics were also worked out by another great French mathematician, astronomer, and physicist, Joseph-Louis Lagrange (1736–1813), who was born Giuseppi Luigi Lagrangcia of Italian parents in Torino, Italy. His treatise *Mécanique analytique* (*Analytic Mechanics*), first published in 1788, placed Newtonian mechanics on a comprehensive mathematical footing.

Lagrange's equations still provide the means for students to solve problems in Newtonian mechanics in the most general way, independent of the choice of coordinate system. Furthermore, today many physics models, including those in relativistic quantum field theory, start by writing down a mathematical function called the *Lagrangian*.

Recall, Newton recognized that his derivation of Kepler's laws was based on the two-body assumption, which made it mathematically tractable. From the limited data available to him at the time, he concluded that because of all the random interactions between planets, the solar system could not retain its stable, predictable behavior by gravitational forces alone. And so, Newton conjectured, God had to step in occasionally to correct their motions.

A century later, Laplace and Lagrange independently calculated the long-term variations in a planet's semimajor axis resulting from the effects of perturbations by the other planets. Their calculations showed that, up to the first order in the masses of the planets, these variations vanish. Later still, the French mathematicians Siméon Denis Poisson (1781–1840) and Henri Poincaré (1854–1912) would show that this result remains true through the second order in the masses of the planets, but not through the third order.[1] In short, the solar system is pretty stable but not ultimately so.

Laplace was able to account for all of Ptolemy's observations within a minute of arc, including the motions of Jupiter and Saturn that were inconsistent with previous calculations. Thus, Laplace showed that Newton's laws were, in themselves, sufficient to explain the movement of the planets throughout previous history.[2] This led him to propose a radical notion that Newton had rejected: *nothing besides physics is needed to understand the physical universe.*

Like Lagrange, Laplace's name also appears frequently in physics, engineering, and mathematics classrooms, where students use the *Laplacian* in calculus problems and perform Laplace transformations in electrical-engineering classes. Poisson and Poincaré are names that are also prominently represented in these classes.

However, physics students usually do not hear, at least in the physics classroom, the story of Laplace's encounter with the emperor Napoleon in or about 1802. Physics professors generally avoid lecturing about anything that can't be calculated. Here's one version, but it is still disputed and the whole story is probably apocryphal:

Laplace had an audience with Napoleon and presented him a copy of *Mécanique céleste*. Someone had told Napoleon that the book contained no mention of God. Napoleon received it with the remark, "M. Laplace, they tell me you have written this large book on the system of the universe, and have never even mentioned its Creator." Laplace, answered, "*Je n'avais pas besoin de cette hypothèse-là.*" ("I had no need of that hypothesis"). Napoleon, greatly amused, told this reply to Lagrange, who exclaimed, "*Ah! c'est une belle hypothèse; ça explique beaucoup de choses*" ("Ah, that's a beautiful hypothesis; it explains many things").

Laplace never wrote anything denying the existence of God, and he may have been a deist. As discussed in the previous chapter, deism—as opposed to theism—supposes a creator god who set the universe in motion and then left it alone to carry out the instructions coded into the natural laws by that god. The quotation above, if true, may simply say that Laplace did not need to hypothesize anything beyond the laws of mechanics and gravity to describe the motions of the heavenly bodies.

In his 1796 book *Exposition du système du monde*, Laplace quotes Newton as saying, "The wondrous disposition of the Sun, the planets and the comets, can only be the work of an all-powerful and intelligent Being." Laplace expresses the deist position when he comments that Newton "would be even more confirmed, if he had known what we have shown, namely that the conditions of the arrangement of the planets and their satellites are precisely those which ensure its stability."[3]

Laplace agreed with Leibniz's criticism of Newton: "This is to have very narrow ideas about the wisdom and the power of

God . . . that God has made his machine so badly that unless he affects it by some extraordinary means, the watch will very soon cease to go." We will see this mistake being made today by those who say that the universe created by God is so imperfect that he had to fine-tune it so that life could evolve. Laplace and Leibniz would counter, "God is smarter than that." I would put it another way: the universe is smarter than that.

LAPLACE'S DEMON

Whatever his religious proclivities, Laplace laid out the principle that became known as the *clockwork universe* or, alternatively, the *Newtonian world machine*: the universe is like a giant machine or clockwork, operating according to the laws of physics so that everything that happens is predetermined by what happened before.

Here's how Laplace expressed it in *A Philosophical Essay on Probabilities*:

> We may regard the present state of the universe as the effect of its past and the cause of its future. An intellect which at a certain moment would know all forces that set nature in motion, and all positions of all items of which nature is composed, if this intellect were also vast enough to submit these data to analysis, it would embrace in a single formula the movements of the greatest bodies of the universe and those of the tiniest atom; for such an intellect nothing would be uncertain and the future just like the past would be present before its eyes.[4]

This "intellect" is usually referred to as *Laplace's demon*, although he did not use the term. Laplace just called it "an intellect" and did not associate it with any god. While not at odds with a creator deist god, the clockwork universe is clearly incompatible with the Judeo-Christian-Islamic God, or any other supreme being that is believed to not only create the universe but also play a major role in the operation of the universe after it is created, stepping in constantly to change the course of events — as Newton believed.

Deism became prominent during the period called the Enlightenment in the eighteenth century when science and reason began to hold sway over theology and revelation. Deists, following the previously cited opinions of Leibniz and Laplace, regarded it as illogical that a perfect god needed to step in at any time after the creation to fix things that may have gone awry.

Many prominent people of the time either were avowed deists or were regarded as deists based on their published views. In Europe these included Adam Smith (1723–1790), Frederick the Great (1712–1786), James Watt (1736–1819), and Voltaire (1694–1778). In America, Benjamin Franklin (1706–1790), Thomas Paine (1737–1809), and at least the first four presidents: George Washington (1732–1799), John Adams (1735–1826), Thomas Jefferson (1743–1826), and James Madison (1751–1836) were deists.

However, the Enlightenment and its version of deism did not survive into the nineteenth century for a lot of reasons that had little to do with Laplace's demon. The impersonal deist god did not provide the comfort of religion sought by the average person. Christian revivalism, which appealed to rich and poor alike with emotion in place of reason, spread throughout Europe and America. At the same time, the intellectual world of art and literature reacted against scientific rationalization and replaced it with an emphasis on intuition and emotion that was termed *Romanticism.*

Unlike the American Revolution, the French Revolution, which also grew out of the Enlightenment, had been a disaster. Furthermore, the Industrial Revolution had greatly increased the wealth of the upper and middle classes, who then proceeded to make the lives of the working class miserable as they were forced to labor for long hours and meager pay in "dark Satanic mills," as the poet William Blake (1757–1827) called the mines and factories. At least in the fields where they worked from dawn to dusk at the mercy of landowners, peasants breathed fresh air, ate fresher food, and drank clean water.

NINETEENTH-CENTURY ASTRONOMY

The nineteenth century saw many advances in astronomy as both technology and mathematical calculations continued to improve. Here I will briefly review some of the more-important developments.

Let's begin late in the eighteenth century to add one more contribution to cosmology by Laplace. In *Exposition du système du monde* mentioned earlier, Laplace provided a model of the formation of the solar system that explained a number of facts that had previously puzzled astronomers, in particular, why all the planets revolve around the sun in the same direction and in pretty much the same plane. Swedish philosopher Emanuel Swedenborg (1688–1772) had proposed the same model back in 1734 and Immanuel Kant followed up on Swedenborg's idea in 1755. Laplace put it on a mathematical footing. Called the *nebular hypothesis*, the model assumes that the solar system evolved from a rotating globular mass of incandescent gas. As it cooled, the mass contracted and successive rings broke off from its outer edge. These rings cooled and condensed into the planets, with the remaining central core becoming the sun.

While Laplace's model was widely accepted in the nineteenth century, it was abandoned in the twentieth century when it could not explain the fact that the planets contain 99 percent of the angular momentum of the solar system. However, the notion of a rotating ball of gas is basically correct. Today's astronomers can see discs of diffuse material orbiting around young stars and protostars—objects in the process of becoming stars. Planets are believed to form in these disks as matter is pulled together into lumps by gravity. However, the theory still has problems with the formation of gas giants such as Jupiter and Saturn.

In 1801, the French astronomer and popularizer Joseph Jérôme Lefrançois de Lalande (1732 –1807) published a catalogue of over forty-seven thousand stars. By that time he had led an illustrious life, and this may be a good time to take a break from all these scientific details to talk about him personally because he was quite an enjoyable character. Here's a description of his appearance:

He was an extremely ugly man, and proud of it. His aubergine-shaped skull and shock of straggly hair trailing behind him like a comet's tail made him the favourite of portraitists and caricatur-ists. He claimed to stand five feet tall, but precise as he was at cal-culating the heights of stars he seems to have exaggerated his own altitude on earth. He loved women, especially brilliant women, and promoted them in word and deed.[5]

Lalande published a *Dictionary of Atheists*, in which he wrote: "It is up to the scholars to spread the light of science, so that one day they may curb those monstrous rulers who bloody the earth; that is to say, the warmongers. As religion has produced so many of them, we may hope to see an end to that as well." He amusingly remarked that his atheist views were his revenge against God who had made him so ugly.[6]

With initial approval, Lalande had included Napoleon in his dictionary. But, then, the emperor realized he needed the Church's support and tried to have the astronomer censured by the Institut de France. Lalande refused to stop presenting his atheistic views. Even under a dictator, France remained remarkably intellectually free.

Getting back to the science stuff, in 1802 the English physicist and chemist William Wollaston (1766–1828) observed dark lines in the optical spectrum of the sun. These *Fraunhofer lines* are named after the German physicist Joseph von Fraunhofer (1787–1826), who studied them systematically in 1814. Almost half a century later, German physicist Gustav Kirchhoff (1824–1887) and German chemist Robert Bunsen (1811–1899) associated the dark Fraunhofer lines with the bright lines observed in the emission spectra of various heated elements.

Thus began the science of spectroscopy, which would become a major tool enabling astronomers to identify the chemical elements in stars and the interstellar medium. The second element in the chemical periodic table, helium, got its name since it was identified by its absorption lines in light from the sun before being discovered on Earth.

The mechanism for the generation of line spectra would not be discovered until 1913 when Niels Bohr used the new quantum theory to calculate the spectrum of hydrogen (see chapter 5). In fact, these line spectra were a major anomaly that could not be explained by the wave theory of light and led to the development of quantum mechanics.

In the meantime, the spectroscopy of stars further established the universality of physics. Newton had made the first great leap in that direction with his universal law of gravity. Previously, it had been assumed that one set of laws applied on Earth and another set in the heavens. Newton proposed that the force that causes an apple to fall from a tree to the ground was the same force that caused the moon to fall around Earth. With the observation that the spectral lines from stars were the same as observed from hot gases in laboratories on Earth, the universality of physics was corroborated. Physics is the same throughout the universe.

Bessel functions are very familiar to students of physics, engineering, and mathematics. Although they were first introduced by physicist Daniel Bernoulli (1700–1782), they are named after astronomer Friedrich Wilhelm Bessel (1784–1846). Bessel was trained as an accountant and worked for a shipping company. His interest in navigation led Bessel into astronomy and in 1810 at age twenty-five he became the director of the Königsberg Observatory in Prussia.

Bessel was the first to use parallax to measure the distance to a star when, in 1838, he reported that 61 Cygni was 10.4 light-years away (1 light-year = 9.46×10^{12} kilometers). The current estimate is 11.4 light-years. Later that year, Friedrich Georg Wilhelm von Struve (1793–1864) and Thomas Henderson measured, respectively, the parallax of Vega (25 light-years away) and Alpha-Centauri (4.4 light-years away).

Figure 4.1 shows how parallax works. A star is viewed twice, six months apart. From the difference in the angles between the two lines of sight, θ (in radians) and the radius of Earth's orbit, r, the distance to the star (in small-angle approximation, which is good enough) is $d = 2r/\theta$.

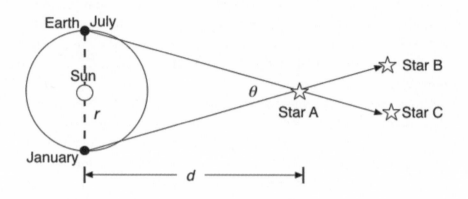

Figure 4.1. How parallax is used to determine the distance from Earth to a star or another astronomical object. The star A will appear at B in one time, and six months later at C. If the radius of Earth's orbit is r (distance to sun) and the measured parallax is θ, then the distance to the star is d = $2r/\theta$. Unlike depicted here, the angle θ is very small, so this small-angle approximation can be used. Image by the author.

In this way, humans began to learn about the immense distances of stars from Earth. Alpha-Centauri, a multiple-star system, is the closest. Well, except for the sun, which is 147 million kilometers or 1.55×10^{-5} light-years (8.17 light-minutes) away. At the time its distance was measured, the farthest known planet from the sun was Uranus. Although Uranus had been observed since ancient times, because of its dimness it was not established as a planet until 1781, by Herschel. Its distance from the sun at the farthest point in its orbit is 3.00 billion kilometers or 0.000317 light-years (2.78 light-hours).

This brings us to the next planet, Neptune. Galileo observed Neptune twice, in 1612 and 1613, but assumed it was a fixed star — although recent evidence suggests he may have noticed movement with respect to the stars. In the early 1800s, French astronomer Alex Bouvard (1766–1843) measured deviations from the then-standard tables on the orbit of Uranus. He proposed that an eighth planet beyond Uranus in the solar system was perturbing the orbit. British astronomer John Couch Adams (1819–1892), using various sources of data, made several estimates of where to search for the new planet.

Independently, the French mathematician Urbain Jean Joseph Le Verrier (1811–1877) was making similar calculations, and on August 31, 1846, he presented his final prediction to the French Academy. Two days later, Adams sent his final prediction to the Royal Greenwich Observatory. On September 18, Le Verrier sent his prediction to the Berlin Observatory. There, on September 23, the planet was found within one degree of Le Verrier's prediction and later identified with Neptune. Astronomers in Greenwich had dragged their feet, so Adams lost out. He graciously acknowledged Le Verrier's priority.

Neptune is 4.50 billion kilometers (0.000476 light-years or 4.17 light-hours) from the sun at its farthest point.

The next significant advance in nineteenth-century astronomy was made by English astronomer William Huggins (1824–1910) with his extensive measurements of the spectra of stars to determine their chemical composition. He demonstrated that the stars are composed of the same chemical elements as found on Earth. He discovered hydrocarbons in comets. And, most significantly, in 1868 Huggins was the first to measure the radial velocity (velocity component along the line of sight) of a star by assuming that the shifts observed in the spectral lines of stars were the result of the *Doppler effect*.

In 1842, Christian Andreas Doppler (1803–1853) had shown that the wavelength of a wave changes as the source of the wave moves toward or away from the observer. So, when a star is moving away from us, its visible light will be shifted toward the red end of the spectrum; when moving toward us, it will be shifted toward the blue. The amount of frequency shift enables the astronomer to calculate the radial velocity. Specifically, the redshift is defined as $z = 1 + \Delta\lambda/\lambda$, where $\Delta\lambda/\lambda$ is the fractional shift in the wavelength. The radial velocity then is $v = zc$, where c is the speed of light, for $v \ll c$. The exact formula valid for all speeds is more complicated and derived from the special theory of relativity.

As we will see, the spectral-line shifts from astronomical objects would have profound consequences in the twentieth century when it was discovered that most galaxies were moving away from us and their redshifts are a measure of their distances. This led to the discovery that the universe is far larger than the few light-years to stars that could be measured by parallax.

While nineteenth-century astronomers were learning about the size of the universe, physicists were pondering the problem of the age of the sun and Earth. In 1863, British physicist William Thomson, Lord Kelvin (1824–1907), estimated the age of Earth assuming it was initially molten and gradually thickened as it cooled, coming up with twenty million years. In 1856, German physicist Hermann von Helmholtz, who had formulated the principle of conservation of energy, considered the age of the sun and suggested its energy arose from gravitational collapse as the decrease in potential energy provided the energy of the emitted light. In 1862, Kelvin used Helmholtz's approach to conclude that the sun could be no more than twenty million years old. These calculations were very crude, and the fact that Kelvin obtained the same number from each method was no doubt due to some mutual fudging. However, the solar calculation was the more reliable.[7]

In any case, both estimates meant big trouble for the theory of evolution by natural selection announced jointly by Charles Darwin (1809–1882) and Alfred Russel Wallace (1823–1913) in 1858. Evolution required a huge timescale of at least hundreds of millions of years. Darwin himself fretted over this discrepancy and considered it the most serious threat to his theory.

On the other hand, geologists supported the hypothesis of evolution with their estimates of the age of Earth of around two billion years. The problem would not be solved until early in the twentieth century with the discovery of nuclear fusion, which will allow the sun to shine for another five billion years or more. The age of Earth is now well determined from radioactive dating and is 4.54 billion years with an error of 1 percent.

In the meantime, observational astronomy continued to advance. In 1888, American astronomer James Keeler (1857–1900) used the large 36-inch refracting (lens-based) telescope at the Lick Observatory on Mount Hamilton in California to observe a gap in Saturn's rings.

On a smaller side hill, Keeler installed a 36-inch reflecting telescope and, for the first time, Newtonian reflecting telescopes began to appear on mountaintops with the result of a significant improvement in capability, especially for spectroscopy, which was greatly improved with the elimination of the spherical aberration intrinsic to lens-based telescopes.

However, Keeler did not get to spend much time observing with this instrument. Disagreement with the director of Lick, a rigid West Point graduate, caused Keeler in 1891 to move to Alleghany Observatory. There, despite far inferior instruments and the industrial skies of Pittsburgh, he was able to make a major discovery that brought him international notice. Using spectroscopy, Keeler confirmed the theory of James Clerk Maxwell (1831–1879) that the rings of Saturn were composed of small objects that rotated around the planet at different rates.

Called back in 1898 to take over the directorship of Lick from its unpopular director, Keeler rebuilt another 36-inch reflector called the Crossley telescope that had been donated to Lick by a British politician, Edward Crossley, and was widely regarded as a "piece of junk." But Keeler fixed it up and when it was finally operative, he began taking beautiful photographs of spiral nebula. These would become the key to the next door to be opened on the cosmos. Unfortunately, Keeler was unable to walk through that door, dying in 1900 just short of his forty-third birthday.[8]

5.

HEAT, LIGHT, AND THE ATOM

THERMODYNAMICS

Two important developments in nineteenth-century physics that had immense significance, both practical and cosmological, were thermodynamics and electromagnetism. The Industrial Revolution had brought on the need to analyze heat engines, and the new science of thermodynamics was developed to describe the observations of heat phenomena. Based solely on observations of macroscopic mechanical systems involving the exchanges of heat and work, thermodynamics grew into a remarkably sophisticated mathematical science that applied directly to measurable quantities such as temperature, pressure, and density.

The two most important principles of thermodynamics are the first and second laws:

First law of thermodynamics

The change in internal energy of a system is equal to the heat into the system minus the work done by the system.

The first law follows from the principle of conservation of energy, which was developed in concert with thermodynamics. Heat was recognized as a form of energy, while work had been earlier defined as the useful application of a force. If you apply a force to a body to increase its speed, the work done on the body is equal to the increase in the body's kinetic energy (energy of motion). If the body experiences friction that slows it down, the loss of kinetic energy appears as the heat of friction.

Second law of thermodynamics

The *entropy* of a closed system must stay the same or increase with time.

The second law was originally cast in terms of heat engines and refrigerators, to account for the fact that neither can be perfectly efficient, even though perfect efficiency does not violate energy conservation. That is, an engine cannot convert all its input heat into work. Otherwise we could build a perpetual-motion machine that gets all its energy from the environment. Similarly, a refrigerator or air-conditioner cannot move heat from a lower to a higher temperature without work being done on it. Otherwise they would not have to be plugged into an electrical outlet. In 1865, Rudolf Clausius (1822–1888) recast these observed principles in terms of an abstract quantity called *entropy*, which is a measure of the disorder of the system.

Thermodynamics had a huge impact on the cosmological thinking of the nineteenth century, especially in regard to its connection with theology. At the time, many philosophers and theologians turned to the first and second laws of thermodynamics to argue for a finite, created universe. The first law was applied to show that the internal energy of the universe, composed of gravitational potential energy and kinetic (motional) energy, had to come from outside the universe.

The second law was applied to show that the universe cannot be infinitely old but had to have a beginning and, furthermore, would eventually die.[1] This was called *heat death*, which occurs when all motion stops and the universe is as cold as it can get, that is, absolute zero.

The thermodynamic case for a divine creation made by many authors followed along these lines: First, if the universe were infinitely old, we would have already reached heat death when everything is in complete disorder, that is, maximum entropy. Second, the entropy of the universe was lower in the past and at some point must have been minimum (zero), which marked the birth of the organized cosmos. This, it was argued, proved that the universe not only had a beginning but also was supernaturally created. This followed from the fact that the universe was in complete disorder or chaos at that time, so the order that exists had to be imparted from the outside.

Hermann von Helmholtz, who wrote the definitive treatise on conservation of energy in 1847, elucidated what he saw as the fate of the universe in a lecture in Königsberg in 1854:

> If the universe be delivered over to the undisturbed action of its physical processes, all force [by which he meant energy] will finally pass into the form of heat, and all heat come into a state of equilibrium. Then all possibility of a further change would be at an end, and the complete cessation of all natural processes must set in. . . . In short, the universe from that time onward would be condemned to a state of eternal rest.[2]

In 1868, Clausius formulated heat death in terms of entropy: "*The entropy of the universe tends toward a maximum*," at which point "the universe would be in a state of unchanging death.[3]

Not everyone was convinced. The eminent British physicist Lord Kelvin (William Thomson) acknowledged the idea of heat death, writing in 1862, "The result [of what he called the law of energy dissipation] would inevitably be a state of universal rest and death." However, he questioned the conclusion, saying, "Science points rather to an endless progress, through an endless space, of action involving the transformation of potential energy to palpable motion and thence to heat, than to a single finite mechanism, running down like a clock, and stopping forever."[4] In other words, the law of dissipation of energy did not hold for infinite space. But he had no convincing argument for space being infinite.

Other scientists, notably the Scottish engineer and physicist William Rankine (1820–1872), one of the founders of thermodynamics (the absolute temperature scale in Fahrenheit units is called *degrees Rankine*), sought ways that heat death might be avoided. He conjectured that "radiant heat" might have special properties that allowed it to be refocused instead of dissipated.[5]

As for the origin of the universe, it was believed that a *law of conservation of matter* existed, so that the matter of the universe also had to come from someplace. In an address in 1873, James Clerk Maxwell, the great unifier of electricity and magnetism about whom we will hear later, who also happened to be an evangelical Christian, expressed a common view:

> Science is incompetent to reason upon the creation itself out of nothing. We have reached the utmost limit of our thinking faculties when we have admitted that because matter cannot be eternal and self-existent, it must have been created.[6]

Basically, then, science as understood in the nineteenth century seemed to require that the universe began with a supernatural creation a finite time in the past and will for all purposes end a finite time in the future when everything comes to rest. And the reasons were good ones, resting on the best empirical and theoretical knowledge of the day. But, as we will see, none of the arguments hold up under the knowledge of today.

In the meantime, many physicists and philosophers of science, in particular Ernst Mach (1838–1916), adopted the doctrine of *positivism* in which anything unobservable is not physics but metaphysics and out of the purview of the experimental method. In an 1872 lecture, Mach asserted that no meaningful scientific statements can be made about the universe as a whole. Terms such as "energy of the world" and "entropy of the world" were meaningless because they could not be measured.[7]

In short, no consensus existed on the cosmological implications of thermodynamics. Most astronomers paid no attention to the matter. French philosopher and historian Pierre Duhem (1861–1916) made an interesting conjecture that turned out to be correct: Even if the second law requires that entropy increase with time, it need not have either a lower or an upper limit.[8]

ELECTROMAGNETISM

In the second significant development in nineteenth-century physics, electricity and magnetism joined gravity as basic forces of nature. Again, we find a combination of experiment and theory, in this case culminating in a set of equations written down in 1865 by Scottish physicist James Clerk Maxwell. Maxwell's equations combined a number of principles discovered by others:

- Faraday's law of induction, discovered empirically by Michael Faraday (1791–1867). This showed how a time-varying magnetic field produces an electric field.
- Ampère's law, discovered empirically by André-Marie Ampère (1775–1836). This showed how a magnetic field is produced by an electric current. An electric current is just a moving charge.
- Gauss's law, proposed theoretically by Johann Carl Friedrich Gauss (1777–1855). This shows how the electric field on a closed surface depends on the electric charge inside that surface.

Note that if an electric force is produced by a static charge and a magnetic force by a moving charge, then Galileo's principle of relativity requires that the two be equivalent. Which is which depends on the reference frame of the observer. This is a fundamentally important point rarely mentioned in physics classrooms or textbooks.

In physics, a *field* is a mathematical object that has a value at every point in space. That "value" can be a single number, such as the density or pressure in solids, liquids, and gases, in which case it is said to be a *scalar* field. Or, it can be a set of numbers. The Newtonian gravitational field, the electric field, and the magnetic field are *vector* fields that require three numbers to define each point in space—one number for the magnitude and two for the direction of the field. The gravitational field in Einstein's theory of general relativity is a *tensor* field requiring ten independent numbers to define.

Faraday and Ampère had earlier demonstrated that electricity and magnetism are the same phenomenon, thus unifying two forces that had previously been thought to be separate. Maxwell's equations codified that fact. Maxwell's theory provides a complete description of the classical electromagnetic field. Given any distribution of electric charges and currents in any medium, one can use Maxwell's equations to compute the electric and magnetic fields at any point in space or a material medium. With just one additional equation, provided by Hendrick Lorentz (1853–1928), one can then determine the electric and magnetic forces on a charged particle at any point in the field and, using Newtonian mechanics, predict the position and velocity of that particle at any time in the future (or the past, for that matter). Once, again, we find corroboration of the Newtonian world machine.

Dramatic as this was, the even more dramatic result of Maxwell's equations was their prediction that in the absence of electric charges and currents, an electromagnetic field can still be present in empty space. Furthermore, that field will propagate through space as a wave with a speed c exactly equal to the speed of light in a vacuum. This number was not in inserted in the model; it came out of the derivation. Thus, it was concluded that light is an electromagnetic wave, confirming the wave nature of light.

Another consequence of Maxwell's theory was that electromagnetic waves extend indefinitely below and above the spectrum of visible light, which extends in wavelength from the violet end at about

430 nanometers to the red end at about 700 nanometers. The region below violet is called *ultraviolet* and the region above red is called *infrared*. Below the ultraviolet we have *x-rays* and below that, *gamma rays*. Above the infrared we have *radio waves*. In 1887, the German physicist Heinrich Hertz (1857–1894) transmitted electromagnetic waves with a wavelength of 8 meters, a billion times longer than visible light, and determined that they moved at the speed of light.

Today, astronomy is conducted over the range of the electromagnetic spectrum from gamma rays with wavelengths as a short as 10^{-18} meter (I have participated in gamma-ray observations) to radio waves with wavelengths of several kilometers.

The wavelength of light is usually represented by the Greek symbol λ. It is the distance between crests of the wave. The frequency f of a wave is the time rate at which the crests of a wave pass a given point. For light, $f\lambda = c$, where c is the speed of light in a vacuum. This expression holds for waves in general, where c is the wave velocity.

ATOMS AND STATISTICAL MECHANICS

The nineteenth century saw not only the development of thermodynamics and electromagnetism but also the application of the atomic theory to the understanding of the behavior of bulk matter. Starting early in the century with John Dalton (1766–1844), chemists developed an atomic theory of matter that culminated in the periodic table of the chemical elements introduced by the Russian chemist Dmitri Mendeleev (1834–1907). However, chemists saw no empirical reason to associate their atoms with the particulate atoms proposed by the ancient Greeks that formed the basis of Newtonian mechanics, as described in chapter 2. The one feature the chemical atoms shared with those of the ancient atomists was that they were indivisible (*atomos* in Greek). They were called elements because chemists could not break them down into anything simpler.[9]

Meanwhile, physicists stuck with their predisposition toward particles. Austrian physicist Ludwig Boltzmann (1844–1906), along with Maxwell and American physicist Josiah Willard Gibbs (1839–1903), developed the theory of *statistical mechanics*, which was based on the particulate picture of matter. It was used to derive all the laws of thermodynamics by assuming that a macroscopic body is composed of a vast number of tiny particles moving around largely randomly, colliding with one another and the walls of any container according to the laws of Newtonian mechanics.

The laws of thermodynamics are thus said to be "emergent" principles — not fundamental principles of nature but derived from more fundamental principles. Indeed, the principles that govern all systems made of many particles such as fluid mechanics, condensed matter physics, chemistry, biology, neuroscience, and even the social sciences can be viewed as emergent. Even gravity is now being proposed as an emergent phenomenon rather than a fundamental force. (See chapter 15.)

No attempt is made in statistical mechanics to describe the motions of individual particles. This would be impossible. Instead, statistical methods were used to predict the average behavior of the ensemble of particles that constituted the system. Thus, the pressure on the wall of a container was associated with the average force per unit area produced by particles colliding with the wall, per unit time. Absolute (Kelvin) temperature was identified with the average kinetic energy of the particles in a system when the system is in equilibrium.

Statistical mechanics associated the chemical elements with particulate atoms. Chemical compounds composed from the elements were identified as molecules that formed when atoms combined.

Despite this success, the particulate theory of atoms was still challenged by a large number of influential chemists and philosophers, notably Ernst Mach. As mentioned earlier, Mach was a positivist who asserted that only objects that can be sensed can be subject to scientific study. He insisted he did not believe in atoms because he could not see them. Mach maintained that position until his death in 1916, by which time the indirect evidence for atoms was indisputable. Today, chemical atoms can be seen directly with the Scanning Tunneling Microscope.

Further evidence for the particulate nature of matter was provided by a series of laboratory observations culminating in an experiment by the British physicist J. J. Thomson (1856–1940)and colleagues in 1896 confirming that the rays emitted from the cathodes of vacuum tubes were charged particles far less massive than the previously known lightest object, the hydrogen ion. These were named *electrons* and were soon recognized to be the carrier of an electric current. Since they flowed in the opposite direction to the current flow that had been arbitrarily labeled "positive," electrons were specified to have negative electric charge. Today the electron is still considered to be one of the fundamental particles of matter.

VIOLATING THE SECOND LAW

Now, let's get back to the second law of thermodynamics. In 1872, Boltzmann derived what was called the H-*theorem*, which showed that a large ensemble of randomly moving particles will tend to move toward an equilibrium state where a certain quantity H, which is proportional to negative entropy, approaches minimum. That is, Boltzmann essentially proved that the second law of thermodynamics follows from the laws of statistical particle mechanics.

However, Boltzmann's good friend and colleague, Josef Loschmidt (1821–1895) saw a problem, which is called the *irreversibility paradox*. If a bunch of particles are moving around randomly, they can in principle accidentally form a state of lower entropy even when part of a closed system.

In 1890, Henri Poincaré published what is called the *recurrence theorem*, which says that dynamical systems will, after a sufficiently long time, return to their initial state. This directly contradicted

Boltzmann's theorem and, therefore, seemed to disprove the second law of thermodynamics.

In 1867, Maxwell had expressed similar misgivings about the second law with his famous "thought experiment" in which an imaginary creature others called "Maxwell's demon" redirects particles to produce a state of lower entropy.

But demons or angels aren't needed. As Boltzmann eventually realized, his *H*-theorem, and thus the second law, are not hard-and-fast principles but simply statistical statements. On average, a closed system of many randomly moving particles will move to state of maximum entropy, as Boltzmann proved, but statistical fluctuations can occasionally produce a state of lower entropy. In fact, if you have a system of just a few particles, that will happen often.

In everyday life, we are familiar with occurrences that are termed "irreversible." Puncture a tire, and the air flows out. We never witness a flat tire reinflate by outside air flowing back through the puncture. Broken glasses don't reassemble. The dead don't come back to life.

However, think about these from a particulate point of view. The molecules of air outside a flat tire are moving about randomly. Suppose a large number of them just happen, by chance, to be moving in the direction of the hole in the tire. Then the tire could reinflate!

The reason we don't see this happen is not because it is impossible but because it is highly unlikely that trillions upon trillions of air molecules will all be moving in the right direction to reinflate the tire.

But suppose we have a closed container of just three particles. Outside the container is an environment of many particles of the same type. Open the container, and the three particles escape. As long as we keep it open, the chance of three particles finding their way back into the container is very high.

In other words, the second law is not inviolate. It is just a statistical statement.

Boltzmann applied this realization to cosmology, speculating that if the universe were vast enough, entropic fluctuations could lead to pockets that deviate from equilibrium to produce other worlds such as ours with sufficiently low entropy to maintain and evolve order.[10] So a live universe could rise from the heat death of the second law. And if *one* could, so could any number. He did not, but might have called it a "multiverse."

THE ARROW OF TIME

Boltzmann also had another profound realization: the second law is not even a law! It's an arbitrary definition. The principle we are dealing with here is not that the average entropy of a closed system must increase with time, or, at best, remain constant. The principle is that the direction of time is *by definition* the direction in which the entropy of a closed system, namely the universe, increases. Arthur Eddington (1882–1944) would later dub this notion the *arrow of time*.

As we have seen, the fact that we do not observe certain time-reversed processes is that they are very unlikely, not that they are impossible. Our everyday experience of decay and death seems to confirm the second law, but that is because we, and the world around us, are composed of great numbers of particles in mostly random motion. But, when you deal with small numbers of particles, such as in chemical, nuclear, and elementary-particle reactions, events are observed to happen in either time direction.

THE END OF CLASSICAL PHYSICS

Physics as it stood at the end of the nineteenth century is generally referred to as "classical physics." By that time, physicists had developed an almost, but not quite, complete theory of the physical world. The matter that makes up that world is composed of elementary particles, called atoms, which are associated with the ninety or so elements of the chemical periodic table. These particles interact with one another by means of two fundamental forces: gravity and electromagnetism, which are mathematically described with unlimited precision by Newton's law of gravity and Maxwell's equations respectively. The motion of every particle in the universe is then completely determined by these laws applied to whatever the position and velocity of the particle is at a given time.

As evidenced from planetary mechanics and the spectroscopy of stars, these same atoms and the theoretical principles that describe their behavior are identical throughout the universe.

ANOMALIES

Still, there were remaining unsolved problems. Maxwell's equations predicted the existence of electromagnetic waves that travel through space at the speed of light. Visible light was identified as those electromagnetic waves with wavelengths within a specific narrow spectral band, strongly confirming Huygens's wave theory (see chapter 3). In addition, waves well outside that band were observed that also traveled at the speed of light. However, the wave theory of light offered no explanation for three observed features of light: (1) *line spectra*, (2) *blackbody radiation*, and (3) the *photoelectric effect*.

We have already discussed line spectra, the very narrow dark lines that are seen as light passes through matter and the bright lines that are seen in the emission of light from hot bodies. These could not be understood in terms of the wave theory of light.

Blackbody radiation refers to the electromagnetic emissions from everyday objects. A blackbody exhibits a smooth spectrum with a peak that depends on the body's temperature. In the case of the very hot sun, its spectrum peaks in the center of visible spectrum, at yellow. Advocates of the notion that the parameters of physics are fine-tuned so that humans would evolve would have us think that the spectrum of light from the sun was designed to peak in just the range where human eyes, patterned after God's, were most sensitive. More likely, our eyes evolved to be sensitive to the region around that peak, which is why we call it "visible." The spectra of cooler bodies, such as you and me, radiate in the infrared region, with wavelengths longer than that of red light. Rattlesnakes and other pit vipers evolved to see infrared light—the better to catch warm-blooded prey in the dark; so to them infrared is "visible." Unless they reflect light, these bodies appear black to us, hence the term "blackbody."

In 1905, Lord Rayleigh (John Strutt, 1842–1919) and James Jeans (1877–1946) used the classical wave theory of light to derive the spectrum of blackbody radiation. The calculation assumed that radiation results from oscillating charged particles inside the body. The shorter the wavelength, the greater the number of electromagnetic waves that can fit inside the body. Rayleigh and Jeans determined that the spectrum should fall off with the fourth power of wavelength.

However, the Rayleigh-Jeans model had a serious flaw. It predicted that the spectrum should increase indefinitely at shorter and shorter wavelengths. This is called the *ultraviolet catastrophe*. In fact, the measured spectra of all blackbodies fall off at either end (see figure 6.1)

The third observation that could not be explained by the wave theory involved ultraviolet light. Physicists, notably Hertz, discovered a wide range of phenomena in which ultraviolet light induces an electric current when the light is directed on various metals. The curious thing was, depending on the specific metal, a threshold wavelength of the light exists above which no current is produced. The wave theory of light offered no explanation for this.

And these were not the only problems with the wave theory of light. If light is an electromagnetic wave, what's doing the waving? The standard assumption was that electromagnetic waves are vibrations in an invisible, frictionless medium that smoothly pervades the universe and was identified with Aristotle's quintessence, the *aether*. But, as Maxwell himself noted, nothing in the theory of electromagnetism assumes the existence of the aether. Unlike the derivation of sound waves, which starts with the hypothesis of an elastic medium, Maxwell did not include any such medium in his prediction of electromagnetic waves. They just fell out of his equations of electromagnetism.

Starting in 1887, American physicists Albert Michelson (1852–1931) and Edward Morley (1863–1923) performed a series of experiments in which they attempted to detect the presence of the aether by measuring the difference in the speed of light expected between two perpendicular beams of light, which they anticipated would be carried by the motion of the Earth through the aether at different speeds by the principle of Galilean relativity. If Earth is moving at a speed v toward the aether, they should measure a light speed $v + c$. If away, they should measure $v - c$. They, and others who followed with increasingly more precise experiments, were unable to see the expected shift in the speed of light. Instead, they always obtained the same value, c.

And so, while by the dawn of the twentieth century, physics had scaled unimagined heights, problems remained that would result in further unimagined developments and newly scaled heights.

6.

THE SECOND PHYSICS REVOLUTION

EVERYTHING EXPLAINED? NOT QUITE

In the previous chapter, we saw that by the end of the nineteenth century, physics had just about explained everything there was to explain about the physical world. The world was just particles moving and colliding with one another according to the laws of mechanics, gravity, and electromagnetism.

The renowned physicist Lord Kelvin (William Thomson) is often quoted as having remarked in an address to the British Association for the Advancement of Science in 1900, "There is nothing new to be discovered in physics now. All that remains is more and more precise measurement." However, no evidence has been found to confirm this statement.[1] While Kelvin may not have said it, this expressed a common sentiment at the time. In any case, it wasn't factually true. As we have seen, in 1900 there were still several observations that could not be explained by the wave theory of light.

As was seen in the last chapter, nineteenth-century physics actually contradicted the notion of the ancient atomists of a universe composed of elementary particles moving around in an otherwise-empty void. The nineteenth-century cosmos was filled with a smooth, continuous, invisible medium called the *aether* in which particles moved .

Light was thought to be result from the vibration of the aether, just as sound waves arise from the vibrations of media such as air and water. The fact that air and water are not completely smooth and continuous but made of tiny atoms was not a concern since sound waves can be derived from Newtonian particle mechanics applied to discrete media. Indeed, Michael Faraday speculated that the aether might also be particulate, as did Isaac Newton before him. What's more, recall that James Clerk Maxwell did not assume the existence of the aether in deriving the existence of electromagnetic waves and that no one could find any evidence for the aether. Both theoretically and observationally, the aether did not appear to exist.

As for the rest of physics, the highly successful application of Newton's laws of motion and gravity to the motions of planets and apples demonstrated that those laws are universal. Similarly, the observation of the same spectral lines in stars as seen in laboratories on Earth demonstrated that they are based on equally universal laws.

In this case, since light is an electromagnetic wave, we can conclude Maxwell's equations are also universal. However, these equations do not provide a mechanism for the observed narrow-line spectra. And, on top of that, the wave theory of light could not explain the blackbody spectrum or the photoelectric effect.

As for the particulate nature of atoms, we have seen that many scientists remained unconvinced because the evidence was indirect.

In the following sections, I will briefly summarize the revolutionary developments in physics in the period from 1900 to the end of World War II in 1945, with an eye toward their particular relevance to cosmology. For more detailed explanations, you may refer to *God and the Atom*.

SPECIAL RELATIVITY

In 1905, Albert Einstein published his special theory of relativity and in the process revolutionized our concepts of space, time, and matter.

Albert Michelson and Edward Morley had failed to find evidence for the variation in the speed of light expected from Earth's motion through the hypothesized aether. While Einstein does not mention that result in his paper, he likely knew about it. However, instead of referring to any observations, Einstein made a purely theoretical argument—although it is important to remember that the theory was ultimately based on observed phenomena, namely, electricity and magnetism.

The electromagnetic waves derived from Maxwell's equations traveled exactly at the speed c in a vacuum, in apparent violation of Galileo's principle of relativity that, as we saw in chapter 2, asserts that all velocities are relative. So the relative velocity between a source of light and its observer should add or subtract from c to give an observed speed different from c. However, Maxwell's equations did not allow this and Michelson and Morley did not see it.

THE RELATIVITY OF TIME AND SPACE

Einstein was not ready to give up on the principle of relativity. So he asked what would be the consequences if (1) the principle of relativity is valid and (2) the speed of light in a vacuum is always c. From these two axioms, Einstein showed, among other results, that the time and distance intervals between two events are not invariant. That is, two observers in reference frames moving with respect to each other will measure different time and spatial intervals.

In other words, time and space are not absolute—as common sense dictates. A clock moving with respect to an observer will be seen by that observer to slow down (*time dilation*); an object moving with respect to an observer will be seen by that observer to contract in the direction of its motion (*Fitzgerald-Lorentz contraction*). That doesn't mean that they *really* do these things. Clocks don't slow down for observers sitting on the clocks, and objects don't contract for observers sitting on the objects. The just appear to do these weird things to outside observers in other reference frames.

Of all Einstein's revolutionary discoveries, his deconstruction of the commonsense view of time was probably the most profound. Nothing seems so ubiquitous—so absolute and universal—as time. Yet special relativity called into question some of our deepest intuitions of time. No moment in time can be labeled a universal "present." There is no past or future that applies to every point in space. Two events separated in space can never be judged to be objectively simultaneous, valid in all reference frames.

Time dilation is important only when the relative speeds of clocks are near the speed of light or when making highly precise measurements with atomic clocks. So these effects were not noticed in everyday life. However Einstein's theory has been confirmed by a century of experiments involving high-energy particles that move near the speed of light, as well as low-speed measurements with atomic clocks. Today everyone with a smartphone or Global Positioning System (GPS) in his car is relying on Einstein's theory, which, as we will see, also must take into account the general theory of relativity.

While we do not need to worry about the relativity of time in the social sphere, it is important not to make universal, philosophical, or metaphysical inferences from the limited range of everyday human experience.

Philosophers and theologians have introduced alternate "metaphysical times" more along the lines of common experience, but these have no connection with scientific observations and have no rational basis outside the realm of speculative theology. Scientific models uniformly assume that time is, by definition, what is measured on a clock and that time is relative.

TIME AND SPACE DEFINED

Until recently, the passage of time was registered by familiar regularities such as day and night and the phases of the moon, or more accurately by the apparent motions of certain stars. The second was defined by the ancient Babylonians to be 1/84,600 of a day. Our calendars are still based on *astronomical time* using the Gregorian calendar, introduced in 1582, in which the year is defined as 365.2425 days (see chapter 2). With the rise of science, the second has undergone several redefinitions to make it more useful in the laboratory.

The most recent change occurred in 1967 when the second was redefined by international agreement as the duration of 9,192,631,770 periods of the radiation corresponding to the transition between the two specific energy levels of the ground state of the cesium-133 atom. The minute remains sixty seconds, the hour remains sixty minutes, and the day remains twenty-four hours, following ancient traditions. The day is still taken to be 84,600 seconds, as in ancient Babylonia. Our calendars need to be corrected occasionally to keep them in harmony with the seasons because of the lack of complete synchronization between atomic time and the motions of astronomical bodies.

Note that we should not assume that the time measured on atomic clocks is some kind of "true time." It is no less arbitrary than astronomical time, time measured with a pendulum, or time defined, for example, by my personal heartbeats. However, by using atomic time rather than heartbeat time, the models that use time are much simpler and don't have to make constant corrections for my daily activity. More seriously, atomic time also avoids the corrections we have to make when we use astronomical time, which has its own small but significant irregularities.

Until Einstein came along, it was generally assumed that space and time were two independent properties of the universe. Until 1983, the standard unit of length, the meter, was defined as the length of a certain platinum rod stored under carefully controlled conditions in Paris. With the help of mathematician Hermann Minkowski (1864–1909), Einstein had formulated the special theory in terms of a four-dimensional space-time in which time is the added dimension. By 1983, relativity had become so well-established empirically that the meter was refined so that it, like time, was defined by what is read on a clock. Currently the meter is defined as the distance between two points in space when the time it takes light to go between the points in a vacuum is 1/299,792,458 second.

Note that this implies that the speed of light in a vacuum is c = 299,792,458 meters per second *by definition*. Many physicists and most other scientists still seem to think that c is a quantity that might be variable, changing with time or location. This is impossible because it has the value it has by definition.

THE RELATIVITY OF ENERGY AND MOMENTUM

Einstein found that energy and momentum are also relative. However, mass is invariant, that is, the same in all reference frames. In units where $c = 1$, we have a simple relationship between mass, energy, and momentum that define the inertial properties of a body and show their connection. The mass m of a body with energy E and momentum p is given by the equation: $m^2 = E^2 - p^2$. When a body is at rest, $p = 0$ and the rest energy of a body is just equal to its mass m. This is the result that is more famously written $E = mc^2$.[2]

Another famous result of special relativity is that a body with mass cannot be accelerated to or past the speed of light. That is, there exists a universal speed limit, c. Here I need to clear up another common misconception. Special relativity does not forbid particles from traveling faster than light, as long as they never go slower. Such particles are called *tachyons*. But they are still hypothetical; none have ever been observed.[3]

The special theory of relativity requires a different set of equations for most of the quantities associated with particle motion when the speeds of the particles are close to the speed of light. However, rest assured that these equations revert to the familiar Newtonian equations when the speeds are low compared to light.

With special relativity, Einstein eliminated the aether and restored Democritus's void to the cosmos, solving the empirical anomaly reported by Michelson and Morley and the theoretical problem with Maxwell's theory of electromagnetic waves. They are fully consistent with the principle of relativity. All speeds are still relative—except the speed of light. And that, as just shown, is an arbitrary number that just decides what units you want to use for distance and time. Throughout this book I mostly use $c = 1$ light-year per year.

GENERAL RELATIVITY

In November 1907, Einstein was sitting in his chair in the patent office in Bern when, he later described,

> All of a sudden a thought occurred to me. "If a person falls freely he will not feel his own weight." I was startled. This thought made a deep impression on me. It impelled me toward a theory of gravitation.[4]

Einstein's theory of gravity would not be published until 1916 as the general theory relativity. Special relativity only applied to reference frames moving at constant velocity. Einstein was able to include acceleration in a new theory of gravitation that predicted tiny effects that could not be understood with Newton's theory.

Let me put Einstein's insight in the following terms: An observer inside a closed capsule in free fall would not be able to distinguish that situation from one in which she is in the same capsule out in space far away from any planets or stars. Furthermore, if the capsule in space were accelerated, say by a rocket motor, with the same acceleration an object has when falling to the ground, like the apple that fell on Newton, then she could not tell the difference between that and just sitting still on the surface of the Earth. So acceleration and gravity are in a sense equivalent.

Now, the observer in the capsule could make delicate measurements of the paths of falling bodies, which would converge along lines pointing to the center of Earth when her capsule is sitting on the surface. If, on the other hand, the capsule were accelerating out in space, the lines would all be parallel. So the equivalence of the two pictures technically applies only for an infinitesimal region of space. This principle is what we call "local."

In the model of gravity developed by Einstein, the force of gravity is basically done away with. A body free of any forces follows a geodesic path through non-Euclidean space-time, analogous to an aircraft following a great circle from one point on Earth's surface to another in order to minimize the distance traveled. Earth falls around the sun in an ellipse because that's the geodesic path around a massive object.

Einstein wrote an equation that allowed him to calculate the shape of space-time and the geodesics from the distribution of matter in space:

$$\text{(curvature of space-time)} = \text{(density of matter)}$$

Einstein was troubled by the fact that purely attractive gravity would cause the universe to collapse. At the time, everyone thought the universe was a static firmament, as described in the Bible. So Einstein added another term, Λ, to his gravitational equation called the *cosmological constant* (CC).

$$\text{(curvature of space-time + cosmological constant)}$$
$$= \text{(density of matter)}$$

So the CC acts as another component of the curvature of space-time, which could be either positive or negative. When Λ is positive, the result is a gravitational *repulsion* that Einstein thought would stabilize the universe.

Note that the cosmological constant could just as well have been put on the right-hand side of the equation as part of the density of matter:

$$(\text{curvature of space-time}) = (\text{density of matter} - \text{cosmological constant})$$

But it's still the same equation, which works equally well either way. This is an example of why it is a mistake to take the elements of mathematical models and try to attribute metaphysical reality to them. Is the cosmological constant "really" part of the curvature of space-time or is it "really" part of matter? It doesn't matter. It's just a human contrivance and either choice gives the same empirical result.

General relativity predicted several phenomena that could not be explained with Newtonian gravity. One of these had already been observed for quite a while and was another empirical anomaly that could not be explained by nineteenth-century physics. In 1859, Urbain Le Verrier, who was mentioned in chapter 3 as the discoverer of Neptune, determined from already-recorded observations that the rate of precession of the perihelion of Mercury differed from that calculated from Newtonian theory by 38 seconds of arc per century, later revised to 43 seconds. In November 1915, Einstein applied his new general theory to the problem and obtained this exact number. He was so excited by this result that, he said, he had "heart palpitations."[5]

Einstein also predicted the light would be deflected by the sun. This was an old idea, going back to Newton. In a note at the end of *Opticks*, published in 1704, Newton suggested that the particles in his corpuscular theory of light should be affected by gravity just as ordinary matter. In 1801, German astronomer and physicist Johann Georg von Soldner (1776–1833) calculated from Newtonian physics that the deflection of a beam of corpuscles skimming along the surface of the sun would be 0.9 arc-seconds. However, a measurement of such a small angle was not technically feasible at the time and, as we have seen, in the early nineteenth century Newton's corpuscular theory of light was discarded in favor of the wave theory.

Einstein calculated an effect twice the Soldner value, in disagreement with Newtonian gravity. On May 29, 1919, two British expeditions took photographs of the region around the sun during a total solar eclipse and compared them with pictures taken from the same locations in July. The famous British astronomer Arthur Eddington led the expedition to Principe Island in Africa and claimed to confirm Einstein's calculation. The independent expedition to Sobral, Brazil reported a result closer to Soldner's value. However, the astronomy community sided with Eddington based on what they saw were defects in the Sobral telescopes, and based perhaps on a little more respect for Eddington's authority.

Eddington's announcement in 1919 made headlines around the world and, more than anything, transformed Einstein into the twentieth-century icon he became. He is the only scientist ever honored by a Manhattan ticker-tape parade, which occurred during his visit there in 1921.

Eddington's measurements were also challenged, but Einstein's calculation has been verified many times over since. One of the most popular science cruises today is to witness a total solar eclipse when it occurs, as they often do, in the middle of an ocean. Astronomers don't require much arm-twisting to join these cruises, all expenses paid, where they give a few lectures and make some observations while enjoying all the amenities.

Allow me to indulge in a little "what-if?" Suppose the gravitational deflection of light had been observable in 1804. Then the wave theory of light would have been falsified at that time since it produces no effect, while at least the Newtonian corpuscular theory is within a factor of two, quite respectable for such a tiny effect. Now the deflection of light by gravity joins line spectra, blackbody radiation, and the photoelectric effect in convincingly falsifying the wave theory of electromagnetic radiation.

Einstein also predicted that a clock in a gravitational field runs slower, as observed by someone outside the field. This is called *gravitational time dilation* and is derived directly from general relativity. This effect is also well confirmed. If the GPS in your car did not correct for gravitational time dilation, it would not always take you to where you want to go.

Gravitational time dilation also implies that light (or any electromagnetic "wave") will decrease in frequency when it moves away from a massive body. Viewed in terms conservation of energy, the kinetic energy of a photon is equal to hf, where f is the frequency of the corresponding electromagnetic wave, where h is Planck's constant, to be defined later. As a photon moves away from the body it gains potential energy and so it loses kinetic energy to compensate and so decreases in frequency.

Since Einstein's original predictions almost a century ago, general relativity has been submitted to many tests of ever-increasing sophistication. It remains consistent with all observations involving gravity today.[6]

BLACK HOLES

As far back as the eighteenth century, John Michell (1724–1793) and Pierre-Simon Laplace noted that a gravity field of a body could be so great that light could not escape. In 1916, Karl Schwarzschild showed from general relativity that if a body of mass M with a radius less than $R_s = 2GM/c^2$ will not allow light to escape. For an object with the mass of the sun, the Schwarzschild radius is about three kilometers. In 1967, physicist John Wheeler dubbed these objects "black holes." As we will see, plenty of evidence now exists for black holes and most, if not all, major galaxies, including our Milky Way, have supermassive black holes at their centers.

In 1974, Stephen Hawking showed that black holes actually radiate photons so that they are all ultimately unstable and eventually decay away.[7] However, the lifetimes for black holes of astronomical size is very long. A black hole of solar mass will last 10^{63} years. Microscopic black holes, on the other hand, have very short lifetimes, and although searches for the expected radiation have been conducted, none has yet been observed.

NOETHER'S THEOREM

On March 23, 1882, a girl named Emmy Noether was born in Erlangen, Bavaria. The daughter of a mathematician, she would turn out to be a mathematical genius and make one of the most important contributions to physics in the twentieth century. Its impact is only now beginning to be fully appreciated. Noether would be considered one of the foremost heroines of the twentieth century had more people understood mathematics and physics.[8]

In 1915, Noether published a theorem that completely transformed our philosophical understanding of the nature of physical law. Until I learned about it, I always thought, as most scientists still do, that the laws of physics are restrictions on the behavior of matter that are somehow built into the structure of the universe. Although she did not put it in those terms, Noether showed otherwise.

Noether proved that *for every continuous space-time symmetry there exists a conservation principle.*

Three conservation principles form the foundational laws of physics: conservation of energy, conservation of linear momentum, and conservation of angular momentum. Noether showed that conservation of energy follows from *time-translation symmetry*; conservation of linear momentum follows from *space-translation symmetry*; and conservation of angular momentum follows from *space-rotation symmetry*.

What this means in practice is that when a physicist makes a model that does not depend on any particular time, that is, one designed to work whether it is today, yesterday, thirteen billion years in the past, or thirteen billion years in the future, that model automatically contains conservation of energy. The physicist has no choice in the matter. If he tried to put violation of energy conservation into the model, it would be logically inconsistent.

If another physicist makes a model that does not depend on any particular place in space, that is, one designed to work whether it is in Oxford, in Timbuktu, on Pluto, or in the galaxy MACS0647-JD that is 13.3 billion light-years away, that model automatically contains conservation of linear momentum. The physicist, once again, has no choice in the matter. If she tried to put violation of linear-momentum conservation into the model, it would be logically inconsistent.

Similarly, any model that is designed to work with an arbitrary orientation of a coordinate system, to work whether "up" is defined in Iceland or Tasmania, must necessarily contain conservation of angular momentum.

Since these three principles form the basis of classical mechanics, it can be said that they are not "laws" that govern the behavior of matter. Rather, they are human artifacts that follow from symmetry principles that govern the behavior of physicists, forced on them if they want to describe the world objectively. There is no reason to think that the laws of physics are the construction of an extraphysical lawgiver.

As we will see in a later chapter, Noether's theorem can be generalized into a principle called *gauge invariance* from which most of the basic laws of physics can be derived. For example, charge conservation and Maxwell's equations result from the gauge symmetry of electromagnetism. I will defer the discussion of the philosophical implications of this notion until that time.

QUANTUM MECHANICS

The twentieth century began in 1900 with Max Planck providing a model that quantitatively described the blackbody radiation spectrum, illustrated in figure 6.1 for the specific case of the sun. (I know, the sun is yellow; but it is a blackbody by definition since it does not reflect light.) The model was based on the hypothesis that light is not continuous but occurs in bundles of energy Planck called *quanta*. These quanta carried energies that were proportional to the frequency f of the radiation. The proportionality h, now called *Planck's constant*, Planck obtained numerically by adjusting it to fit the spectral data. Recall that the frequency of light is related to its wavelength λ by $\lambda = c/f$, where c is the speed of light.

The ultraviolet catastrophe of classical wave theory discussed in chapter 5 is avoided by energy conservation. The lower-wavelength end of the spectrum corresponds to higher-energy quanta and, since a body contains only so much energy, the spectrum must fall off at short wavelengths. Also, the wavelength of the spectral peak decreases with temperature because, as we saw in the discussion of statistical mechanics in chapter 5, temperature is a measure of the average kinetic energy of a body. That is, the hotter a body, the lower will be the wavelength of the spectral peak — or the higher the frequency of that peak.

In the same amazing year of 1905 when he introduced relativity, Einstein carried Planck's idea further and proposed that light is composed of particles, later dubbed *photons*, whose energy is proportional to the frequency of the corresponding electromagnetic wave. That is, if f is the frequency of the wave, the energy of each photon in the wave is $E = hf$, where h is Planck's constant. Using this assumption, Einstein as able to explain the photoelectric effect. An electric current is produced when photons kick electrons out of a metal. In order to do this, they need a minimum energy, which is why there is a threshold — a minimum frequency at which the current is produced. In 1914, American physicist Robert Millikan) confirmed Einstein's proposal in the laboratory.

Einstein showed that light is not the vibration of the aether or any other medium but a stream of particles, just as Newton claimed in his corpuscular theory of light. But, if light is corpuscular, then how can it produce the wavelike effects seen in interference and diffraction experiments?

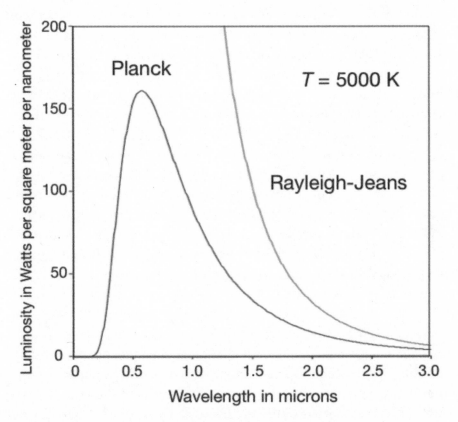

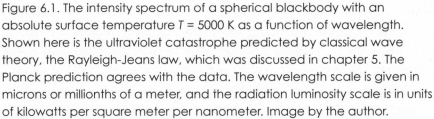

Figure 6.1. The intensity spectrum of a spherical blackbody with an absolute surface temperature $T = 5000$ K as a function of wavelength. Shown here is the ultraviolet catastrophe predicted by classical wave theory, the Rayleigh-Jeans law, which was discussed in chapter 5. The Planck prediction agrees with the data. The wavelength scale is given in microns or millionths of a meter, and the radiation luminosity scale is in units of kilowatts per square meter per nanometer. Image by the author.

French physicist and aristocrat Louis de Broglie provided the answer in 1924 in his doctoral thesis: *all* particles have wavelike properties. De Broglie noted that a photon of momentum p has an associated wavelength $\lambda = h/p$. He hypothesized that this relation holds for all particles, in particular, electrons. This is called the *de Broglie wavelength*.

De Broglie's conjecture was verified in 1927 when American physicists Clinton Davisson and Lester Germer observed the diffraction of electrons beamed onto the surface of a nickel crystal.

And so not only photons but also electrons and ultimately all particles have wavelike properties. This is known as the *wave-particle duality*. However, here we have another physics result that is widely misunderstood, even among physicists. You often hear, "An object is a particle or a wave depending on what you decide to measure." This is wrong. No one has ever measured a wavelike property associated with a single particle. Interference and diffraction effects are only observed for beams of particles and only particles are detected, even when you are trying to measure a wavelength. The statistical behavior of these ensembles of particles is described mathematically using equations that sometimes, but not always, resemble the equations for waves.

If you do an interference or diffraction experiment in which you detect individual photons, you will not see the effect until you accumulate a large number of detections. For example, you could do a double-slit interference experiment with a beam of photons of one per day. Watch it for a year, and you will see the interference pattern develop. Note that the photons can hardly be said to be "interfering with each other," which is often the way the effect is described.

If you object to me calling one photon a day a "beam," where do you draw the line where a beam suddenly appears? One per hour? One per second? One per nanosecond?

Let me make this is explicit as I can: It is incorrect to say, "this photon has a frequency f" or "this electron has a wavelength λ." The correct statements are: "this photon is a member of an ensemble of photons that can be described statistically as a wave of frequency f" and "this electron is a member of an ensemble of electrons that can be described statistically as a wave of wavelength λ."

In 1926, Austrian physicist Erwin Schrödinger produced a mathematical theory called *wave mechanics* in which he associated particles with a complex number called the *wave function*.[9] The same year, German physicist Max Born gave the now generally accepted interpretation in which the square of the magnitude of the wave function gives the probability per unit volume for finding a particle at a particular position at a given time. Nothing in quantum mechanics predicts the behavior of a lone particle — consistent with the interpretation of the wave-particle duality given previously.

A year earlier, in 1925, German physicist Werner Heisenberg had produced the first version of what has come to be called *quantum mechanics* that makes no use of waves and instead applies the mathematics of matrix algebra. After some initial controversy as to which formulation was better, Schrödinger showed that they were mathematically equivalent. Both the Heisenberg and Schrödinger formulations applied only to nonrelativistic particles, that is, those moving at speeds much less than the speed of light. That is, they could be used to describe slow-moving electrons but not photons.

In 1927, British physicist Paul Dirac, whose genius came close to matching Einstein's, formulated a quantum theory of photons. The next year, he produced a relativistic theory of electrons that predicted the existence of antimatter. In 1932, American physicist Carl Anderson reported the detection of particles in cosmic rays that looked like electrons but curved opposite to the direction of electrons in a magnetic field and so had to have positive electric charge. Anderson associated the particles with Dirac's antimatter and dubbed the antielectrons *positrons*.

In 1930, Dirac published the definitive work on quantum mechanics titled *The Principles of Quantum Mechanics*.[10] In the book, which has gone through many editions and printings since, he does away completely with the wave function, replacing both wave mechanics and matrix mechanics with more-powerful linear vector algebra. While most chemists and those physicists who deal just with low-energy phenomena can get by with less-sophisticated Schrödinger wave mechanics, Dirac's quantum mechanics is necessary for understanding the behavior of elementary particles and high-energy phenomena in general.

While special relativity has been successfully reconciled with quantum mechanics, general relativity has not been. In particular, important for our cosmological story, general relativity does not apply to the earliest moments of our universe when quantum effects predominate. As we will see, this has not stopped religious apologists from using general relativistic arguments to claim evidence for a divine creation of the universe.

THE PLANCK SCALE

At this point I would like to introduce a concept that will be increasingly important as we get deeper into cosmology. As I have emphasized, every quantity in physics that directly relates to observations is operationally defined by how it is measured with a well-prescribed measuring apparatus. We have seen that both time and distance intervals are defined by what is measured on a clock, where the distance between two points is the time measured for light to pass between these points in a vacuum.

It can be shown that the smallest possible time interval that can be measured is the *Planck time*, 5.391×10^{-44} second, and the shortest possible distance interval that can be measured is the *Planck length*, 1.616×10^{-35} meter.[11]

Another important quantity is the *Planck mass*, 2.177×10^{-8} kilogram. The Schwarzschild radius for a sphere with a mass equal to the Planck mass is twice the Planck length, which implies that it is a black hole (see earlier discussion on black holes). The *Planck energy* is then defined as the rest energy of a body with the Planck mass, which is 1.221×10^{28} electron volts. An electron volt is the energy gained by an electron in dropping through an electric potential of one volt. We will use this unit of energy frequently, abbreviated eV.

ATOMS AND NUCLEI

In *De rerum natura*, mentioned in chapter 1, Lucretius described the zigzag motion of dust motes in a sunbeam and argued that it is caused by atoms colliding with the motes in the beam. In 1857, the Scottish botanist Robert Brown (1773–1858) observed pollen grains zigzagging in water and the effect came to be known as *Brownian motion*. In a third paper in 1905, Einstein derived equations that showed how the existence and scale of atoms could be demonstrated from the jaggedness of the paths of Brownian grains. In 1909, French physicist Jean Baptiste Perrin applied Einstein's theory as well as other techniques to measure *Avogadro's number*, an important constant in chemistry that, for our purposes, we can simply view as the number of atoms in a gram of hydrogen gas. The current value is 6.022×10^{23}, the reciprocal giving the mass of the hydrogen atom, 1.66×10^{-24} gram. Although this measurement was still indirect, at this point only the most obdurate holdouts, such as Ernst Mach, could still deny that matter is composed of huge numbers of tiny particles.

In 1896, French physicist Henri Becquerel (1852–1908) discovered that the element uranium emitted previously unobserved penetrating radiation. Further laboratory experiments by Becquerel, Ernest Rutherford, Pierre Curie, and Marie Curie identified three types of radiation from a number of different elements: α, β, and γ.

In 1909, Hans Geiger and Ernest Marsden bombarded gold foil by α-radiation from radon gas, which showed surprisingly large deflections of the paths of the rays as they scattered off the foil. In 1911, Rutherford inferred from these observations that the atom, tiny as it already is, has an even tinier nucleus, much smaller than the atom itself, containing most of the mass of the atom. In the model, electrons circle about the nucleus in planetary-like orbits.

In 1913, Danish physicist Niels Bohr proposed that electrons in atoms could exist only in certain orbits. Each orbit corresponded to a specific *energy level*, with the ground state the lowest energy. When an electron in an atom drops from a higher to a lower level, a photon of a very precise energy equal to the energy difference between the levels is emitted, which appears as a very narrow line in the emission spectrum. Bohr was able to calculate the observed emission spectrum for hydrogen. The absorption spectrum followed since only photons with energies exactly equal to the differences between levels will be absorbed. In this way, the final anomaly from nineteenth-century physics that could not be accounted for in the wave theory was solved; the narrow line spectra of atoms was explained.

Bohr's theory was very crude but remarkably agreed with the data. The quantum mechanics of both Heisenberg and Schrödinger obtained the same hydrogen energy level formula as Bohr's and offered the opportunity to be applied to other atoms. Dirac's relativistic quantum theory did even better by calculating the small splittings of spectral lines called *fine structure* that were observed as spectroscopic instruments improved.

Dirac's theory also proved that the electron has half-integer *spin*, where spin is the name for a particle's intrinsic angular momentum. This notion had been earlier conjectured by Austrian physicist Wolfgang Pauli.[12] In 1925, Pauli had proposed what is now called the *Pauli exclusion principle*: two identical particles of half-integer spin cannot simultaneously occupy the same quantum state. With this principle, it became possible to explain the periodic table of the chemical elements.

While we still call the chemical elements "atoms," they are no longer irreducible when we move from low-energy chemical reactions to high-energy nuclear reactions. The chemical atoms are not pointlike particles but composite structures of more-elementary objects, nuclei and electrons. Furthermore, they can be transmuted from one type to another in nuclear reactions, fulfilling the alchemist's dream.

In the 1930s, nuclei themselves were shown to be composed of more-fundamental objects, *protons* and *neutrons*, where the neutron is just slightly heavier than the proton and electrically neutral—

although it is a tiny magnet, as are the proton and electron. The proton is positively charged. Hydrogen is the simplest element, made of a proton and an electron. Add a neutron to its nucleus, and you have heavy hydrogen or *deuterium*. Add another neutron, and you have *tritium*. Add another proton to tritium, and you have *helium*.

Each element in the periodic table is identified by an *atomic number* Z that is equal to the number of protons in its nucleus. It is also the number of electrons in a neutral atom. Atoms with more or fewer electrons are electrically charged *ions*.

Changing the number of neutrons in the nucleus does not change the atom's position in the periodic table but produces a new *isotope* whose chemical properties are generally not much different than the original isotope but whose nuclear properties can be dramatically different. The standard notation for an isotope is X^A, where X is the chemical symbol that uniquely specifies the atomic number Z. The quantity A is usually called the *atomic weight* but is more accurately described as the *nucleon number*, the number of protons and neutrons in the nucleus, where nucleon is the generic term for either a proton or a neutron.

The three types of nuclear radiation were identified: α-rays are helium nuclei, β-rays are electrons or positrons, and γ-rays are high-energy photons.

In 1932, English physicist James Chadwick confirmed the existence of the neutron. And so, as of that moment in time, the universe appeared to be reducible to just four elementary particles: the electron, the proton, and the neutron that form atoms, and the photon that composes light.

However, we have already seen that, in the same year, Anderson verified Dirac's prediction of the antielectron or positron. The implication was that a whole separate realm of matter called antimatter existed. For example, antihydrogen is comprised of an antiproton and a positron. However, antiprotons, antineutrons, and antihydrogen would not be confirmed until the 1950s.

In 1936, Anderson and his collaborator Seth Neddermeyer discovered another particle in cosmic rays that looked like an electron except it was heavier. We now call this particle the *muon*. It is basically a heavier electron. It would be the first of a whole new zoo of particles that would be discovered in the 1950s and 1960s.

Protons are packed close together in the nucleus and, being positively charged, strongly repel one another. So what holds the nucleus together? Because the neutron has zero charge and only a very weak magnetic field, and because gravity is so much weaker than the electric force for particles of such low mass, something else has to be holding the nucleus together. This is called the *strong nuclear force* since it must be strong enough to overcome the electrical repulsion between the positively charged protons in the nucleus. While the electromagnetic force acts over great distances (light reaches us from galaxies billions of light-years away), the strong nuclear force acts only between particles less than a few femtometers (10^{-15} meter) apart.

Furthermore, it was discovered that the force responsible for the radioactive decay of nuclei that give β-rays (electrons) was a separate, even shorter-range force called the *weak nuclear force*. And it turns out that the weak force drives the primary reaction that supplies the energy of the sun and other stars.

In 1930, Pauli proposed that the missing energy that was being observed in β-decay was being carried away by a very low-mass neutral particle, which Italian physicist Enrico Fermi dubbed the *neutrino*. The existence of the neutrino would not be confirmed until 1956.

I will not go into the well-known story of the development of nuclear energy prior to World War II and its use in incredibly powerful bombs and as a troublesome source of energy that ultimately may be the only viable solution to the world's energy needs.

As of 1945, then, we had the proton and neutron making up nuclear matter, the electron joining these to make atomic matter, and the photon providing light. In addition, we had the positron, the muon, and the suggestion of a neutrino. These particles interacted with one another by means of four fundamental forces: (1) gravity, (2) electromagnetism, (3) the strong nuclear force, and (4) the weak nuclear force.

We also had relativistic quantum theories for photons and electrons, but these all involved only electromagnetic interactions. Dirac and others had developed a relativistic quantum field theory that could be solved only in a first-order perturbation approximation and produced infinities when pushed further. Fermi had made a start toward a theory of the weak interaction, and a Japanese physicist named Hideki Yukawa had produced a rudimentary theory of the strong interaction that did not work very well. So, a lot of problems remained.

Once the war was over, a new generation of physicists assumed command who would carry the work of the great early-century pioneers on to the next level, joining with them to leave a legacy in 1999 that, unlike 1899, successfully described all that was known at the time about the structure of matter.

Before we get to that, however, let us move back to cosmology. During the first half of the twentieth century, the ever-improving quality of telescopic observations had combined with the new physics of that period to push our view of the universe far beyond our own galaxy of stars, the Milky Way, and to provide a new picture of a universe expanding from the initial explosion called the big bang, which we will now cover.

7.

ISLAND UNIVERSES

THE CEPHEID SCALE

By the end of the nineteenth century, astronomers had begun to recognize that the universe extended well beyond the solar system. The planets were far from Earth, but the stars were much farther. Still, they had no idea how far they really were. With the instruments of the day, parallax could be used to measure distances out to no more than a few tens of light-years.

Despite the Copernican revolution, astronomers still imagined Earth to be near the center of the universe. This was not just typical human self-centeredness. When astronomers counted stars, they found the number to fall off fairly uniformly in all directions from Earth; so it seemed we were near the center. They were unaware of the existence of interstellar gas that dims the light from farther objects equally in all directions and leaves the appearance of isotropy.

In 1908, Henrietta Leavitt was one of a small group of young, female "computers" working for Charles Pickering at Harvard College Observatory in Cambridge, Massachusetts. Pickering realized that professional astronomers were not needed to perform all the detailed work that was involved in poring over the many photographic plates exposed at the observatories of Harvard's several telescopes. In fact, young women who could be hired for a fraction of the wage did a very good job of analyzing what was on the plates and recording the exact positions, spectral classes, and brightness magnitudes of stars and other astronomical objects. Naturally, the women did no actual observing, given their "delicate natures."

Pickering assigned Leavitt to look at the photos of variable stars—stars whose brightness magnitudes change over time. Harvard had an observatory in Peru, and Leavitt looked at photos taken there of a sample of stars within the Small Magellanic Cloud (SMC), which along with those in the Large Magellanic Cloud (LMC) are only visible in the Southern Hemisphere.

Comparing plates of sixteen giant stars called Cepheid variables, Leavitt made a discovery that would prove to be of huge significance. She noticed that the period of variation, that is, the time between peak luminosities, was longer when the star was brighter. Being at roughly the same distances from Earth, she reasoned that the observed luminosities of the Cepheids in the SMC should be directly related to their intrinsic luminosities. Leavitt had uncovered a connection between peak luminosity and period that would ultimately provide a means to measure far greater distances than is possible with parallax.

Leavitt's efforts were delayed by chronic illness. However, by 1912 she had gathered nine more Cepheids in the SMC and published a three-page article, No. 173, in the *Harvard College Observatory Circular*. The article contained a graph, on a logarithmic scale, of Cepheid brightness versus period and showed a clear relationship between the two. The graph was termed a *period-luminosity relation*.[1]

OFF CENTER

In 1908, the most powerful telescope of the time, a 60-inch reflector on Mount Wilson in California, built by George Ellery Hale, began looking up into the crystal-clear night skies above Los Angeles. In 1912, Harlow Shapley, a former crime reporter from rural Missouri who had gone on to earn a doctorate in astronomy from Princeton, joined the observatory staff. Shapley took an interest in "globular clusters," spherical systems containing hundreds or even thousands of stars.

He (or his wife, Martha, who dabbled in astronomy) discovered that the clusters included many faint giant blue stars. Comparing their observed luminosities with the stars of that type near the sun, he estimated that they were at least fifty thousand light-years away — two orders of magnitude farther than the greatest distances of stars then known from parallax.

Unfortunately, globular clusters contain few or no Cepheids. Furthermore, Shapley found that some Cepheids had much shorter periods than the ones in the SMC used in Leavitt's sample. For this reason, at first Shapley hesitated to use Leavitt's period-luminosity relation on the Cepheid variables with shorter periods. However, he was not your typical cautious scientist and plowed ahead anyway. He used Leavitt's period-luminosity relation to measure the distances to those Cepheids he could see in his globular-cluster photos.

Here's how it works: You measure the observed period of a Cepheid variable. From the period-luminosity relation you obtain the intrinsic luminosity of the star. As the star radiates outward on the surface of an imaginary sphere of ever-increasing radius r, the energy per unit area will fall off with the sphere's area as $1/r^2$ because of energy conservation. Thus, by comparing the measured luminosity with the apparent luminosity, you can determine the distance to the star.

For the clusters with no Cepheids, or those too faint to measure, Shapley estimated their distances using the brightest stars as distance markers. When the stars were not resolvable, he used the size of the cluster to judge distance.

By these methods, Shapley determined that the Milky Way is shaped like an ellipsoid and some three hundred thousand light-years across. He concluded that it must constitute the entire universe. Imaginative as he was, he could not imagine a larger universe.

Shapley noticed that his distribution of globular clusters was not centered on Earth. Rather the distribution appeared to be centered on the region of the constellation Sagittarius. (He was not the first to suggest this.) He estimated that the sun is sixty-five thousand light-years from the center.

Shapley's distances were in fact overestimates. Today we know the Milky Way is 100,000 to 120,000 light-years in diameter, with the sun 27,000 light-years from the center.

HIGH-SPEED ASTRONOMY

Percival Lowell was a descendent of the patrician Massachusetts family that first landed in Boston in 1639. He built an observatory in Arizona to pursue his obsession with Mars and the possibility that it contained artificial waterways or *canals* built by a past civilization on the red planet. In 1877, Italian astronomer Giovanni Schiaparelli (1835–1910) had reported seeing dark streaks across the surface of Mars, which he called *canali*, meaning "channels" (not "canals"). Unlike most astronomers, Lowell took them seriously and wrote three books on the subject that popularized the notion of life on Mars. The observatory was built on "Mars Hill," which, at 7,250 feet, is three thousand feet higher than Mount Hamilton and fifteen hundred feet higher than Mount Wilson.

Lowell purchased a spectroscope that was an improvement over the one at Lick. However, Lowell had only a 24-inch refractor, which made it less suitable for spectral work.

Lowell hired the young Vesto Slipher right out of Indiana University in 1909. Slipher remained at the observatory until retiring in 1954, serving as director for thirty-eight years.

With great patience and skill, Slipher significantly improved the tricky instrument and pretty much did whatever Lowell asked, which was mainly planetary astronomy. However, in 1909, Lowell directed Slipher to take a spectrogram of what he called a "white nebula," by which he meant a spiral nebula. Slipher did not get around to it until 1912, when he began a series of exposures of Andromeda, the largest spiral nebula in the sky, on four different plates. Then, in January 1913 he obtained his result: The spectrum of Andromeda was *blueshifted*, that is, shifted to shorter wavelengths.

Assuming the mechanism was a Doppler shift, Slipher calculated that Andromeda is moving toward us with a speed or *radial velocity* of 300 kilometers per second.[2] He wasn't far off. Currently astronomers measure 301.

At the time, this was the highest velocity ever measured in nature. The radial velocity of Andromeda is ten times the speed of Earth around the sun, thirty kilometers per second, which is about the same as the typical speed of a star in the Milky Way.

Slipher's result was extraordinary, and other astronomers, especially those at the nearby competitor, Lick, were properly skeptical. Undaunted, Slipher continued his measurements, which were that much tougher for smaller nebulae. He found the spectrum of M87 (Messier catalog) was *redshifted*, indicating it is moving away from Earth with a radial velocity of one thousand kilometers per second — three times faster than M31, Andromeda, which is moving toward us and will someday merge with the Milky Way. By the summer of 1914, Slipher had measured the velocities of fourteen spiral nebulae and found most to be receding. In August, he gave a report to the meeting of the American Astronomical Society at Northwestern University. A tall, handsome young man just elected to the society named Edwin Hubble (1889–1953) was in the audience.

Slipher reported the spiral nebula move at an average speed of twenty-five times the average speed of stars in our galaxy. He received a standing ovation and congratulations from his colleagues at Lick. However, the evidence that the spirals were separate island universes was still not conclusive. There had to be a way to measure their distances.[3]

As mentioned in chapter 4, today astronomers define a quantity z called the redshift that equals the fractional amount by which an observed wavelength of spectral lines exceeds the wavelengths as measured in the laboratory. If it is negative, we have a blueshift. From the Doppler effect, $z = v/c$ where v is the recessional velocity. This only applies for $v \ll c$; a more-complicated formula must be used at relativistic speeds.

A SPIRALING DEBATE

As we saw in chapter 3, near the end of the nineteenth century, James Keeler had taken beautiful photographs of hundreds of spiral nebulae with the Crossley telescope on Mount Hamilton. Nine years after Keeler's death in 1900, Heber Curtis (1872–1942) resumed Keeler's pursuit with the Crossley. By 1913, he had assembled two hundred photos of spirals.

On July 19, 1917, George Richey (1864–1945), observing with the 60-inch reflector on Mount Wilson, three hundred miles from Mount Hamilton, took a long-exposure photo of the spiral nebula NGC 6946 (NGC designates *The New General Catalogue*, compiled in 1888). Comparing with three earlier pictures, Richey saw a pinpoint of light hear the edge. He concluded it was a *nova*, one of the points of light that occasionally appear in the sky and quickly fade away.

Curtis noticed the phenomenon in three different nebulae. Searching through plates, he found separate novae had appeared in the spiral NGC 4321 in 1901 and 1914. It seemed very strange to him to have so many novae appear in spiral nebulae.

Furthermore, Curtis noticed that some nebulae occasionally, though rarely, exhibited relatively bright bursts. For example, such bursts were seen in Andromeda in 1885 and in the constellation Centaurus in 1895. Curtis placed these bursts in another class, which are now called *supernovae*.

Intrigued, other astronomers began searching their plates. Continuing his own search, Curtis observed that most of the novae in spirals resembled novae elsewhere but were much dimmer. To account for their faintness, he concluded that they had to be *millions* of light-years away.

Curtis became the champion of the notion that spiral nebulae were "island universes," galaxies of stars located far beyond our own Milky Way. But most astronomers remained skeptical. In the meantime, World War I intervened and most astronomers, including Curtis (but not Shapley), joined the war effort.[4]

After the war, the spiral controversy continued, with Harlow Shapley holding strong to his view that they were within our galaxy. There he appeared to have strong support from the Dutch-born astronomer Adriaan van Maanen , who worked on Mount Wilson. Van Maanen, who had a good reputation as a meticulous researcher, claimed to have measured rotational periods for spiral nebula that implied that if they were separate galaxies as large as the Milky Way, then their spiral arms moved faster than the speed of light. His results were purportedly corroborated by other observers on Mount Wilson, Lowell Observatory, and in Russia and the Netherlands.[5]

On April 26, 1920, Shapley and Curtis engaged in what is called the "Great Debate" at an evening public meeting during a three-day National Academy of Sciences convention in Washington, DC. The day before, the *Washington Post* announced that "Dr. Harlow Shapley, of the Mount Wilson solar observatory, will discuss evidence which seems to indicate the scale of the [Milky Way] to be many times greater than is held. . . . Dr. Heber D. Curtis, of the Lick Observatory, will defend the *old* [my emphasis] theory that there are possibly numerous universes similar to our own, each of which may contain three billion stars."[6]

If we are to judge from a newspaper article, which is always a risky business, the general tenor at the time may have been that the "old" idea that there were other galaxies besides our own was being superseded by Shapley's more recent discoveries.

Actually, it wasn't much of a debate. Rather there were two back-to-back lectures that really did not address each other, with any rebuttals left to the discussion afterward.

Shapley spent most of the debate describing his estimate of the size of our galaxy, which was already ten times what had been earlier thought. He aimed at a public audience and gave few technical details. Curtis gave a more technical talk, focusing on spiral nebulae, emphasizing the Andromeda nova, which was too bright to be within our galaxy, and the high speeds of the spirals.

Curtis also questioned the size estimate for the Milky Way,

saying it was ten times smaller than Shapley estimated. Here he was three times too small while Shapley was three times too big, so we can split the difference.

Curtis had to admit that if the spiral nebulae were separate galaxies and one reasonably assumed they were on the same order of magnitude of size as the Milky Way, they would have to be at least three hundred thousand light-years away.

No good record of the actual debate survives. But the two astronomers, competent scientists that they each were, agreed to write articles on their positions for the *Bulletin of the National Research Council*, which appeared a year later apparently substantially modified from what they presented orally after they exchanged a series of drafts.[7]

The public began to take notice. On May 31, 1921, the *New York Times* reported on its front page that Shapley had proved that "the little speck of light around which a tiny shadow called the earth revolves, is 60,000 light-years from the centre of the universe." Shapley had replaced Pickering as director of the Harvard Observatory that year. He is quoted in the article as saying, "Personally, I am glad to see man sink into such physical nothingness, and it is wholesome for human beings to realize of what small importance they are in comparison with the universe."[8]

In 1925, the Royal Swedish Academy contacted Harvard Observatory about the possibility of awarding the Nobel Prize for Physics to Henrietta Leavitt. They were unaware that she had died four years earlier, on December 12, 1921, of stomach cancer at the age of fifty-three. The prize is not awarded posthumously.

THE REALM OF HUBBLE

On September 11, 1919, thanks to the tireless efforts of George Ellery Hale, a 100-inch reflecting telescope came into operation on Mount Wilson. A week later, Edwin Hubble joined the staff after serving as a captain in the army during the Great War, where he saw no combat. In 1923, Hubble was using both the 60-inch and 100-inch to survey nebulae, and he turned his attention to NGC 6822, a nebula in Sagittarius that resembles the LMC.

Hubble found five variable stars in NGC 6822 and asked Shapley in Harvard if he could look at the object in his store of plates. Shapley had used the observed magnitudes of the brightest stars to estimate that NGC 6822 is about a million light-years away. He admitted that it was "probably quite beyond the limits of the galactic system," but it wasn't a spiral so Shapley continued to insist that spirals were "neither galactic in size nor stellar in composition."[9]

Hubble found eleven Cepheids in NGC 6822 and used them to obtain a distance of seven hundred thousand light-years. However, the great discovery that would bring him world fame occurred in early 1924 when he found a Cepheid variable in Andromeda and determined it to be nine hundred thousand light-years away. Hubble wrote Shapley, who told a colleague when he received the letter: "Here is the letter that destroyed my universe."[10]

Actually, in 1922 the Estonian astronomer Ernst Öpik had published a more-accurate measurement of the distance to Andromeda by means of a novel technique in which he used the rotational velocity of a galaxy, which depends on the mass of the galaxy, and assumed its luminosity is proportional to its mass. His estimate was 1.5 million light-years, compared to Hubble's estimate of nine hundred thousand light-years. This was closer to the modern value of 2.5 million light-years.

Nevertheless, Shapley played devil's advocate for a while, which was a perfectly proper position to take as the prime expert supporting the opposing theory. When new discoveries are made in science, they are not so obvious at the time to those in the trenches, and Hubble was also properly being very cautious and conservative.

In the midst of it all, Hubble married the daughter of a wealthy Los Angeles banker, and the two embarked on a three-month honeymoon that included a tour of Europe.

After returning, Hubble turned to other nebula, in particular the beautiful face-on spiral M33 in the Triangulum constellation. There he found twenty-two Cepheids implying again a distance of at least a million light-years. Measuring the periods of all these variable stars in those days before computers or even hand calculators was tedious and laborious. And Hubble was bothered by the measurements on spirals made by his senior colleague at Mount Wilson, van Maanen. If they were correct, the spirals could not be extragalactic, so he was reluctant to announce his results publicly or express his own doubts about van Maanen's conclusions.

However, Hubble's results quickly made it around the world by what was already a remarkably efficient astronomical grapevine (it's instantaneous today). Even the *New York Times* got wind of it, and on November 23, 1924, it headlined, "Dr. Hubbell [sic] Confirms View That There Are 'Island Universes' Similar To Our Own."[11]

Despite Hubble's hesitation, the astronomical community took his results seriously because they were based on what was by then well-established, the Cepheid-distance scale. Van Maanen was assumed to be wrong, and eventually flaws in his methods were uncovered. Hubble shared a $1,000 prize from the American Association for the Advancement of Science with parasitologist Lemuel Cleveland for his discovery of protozoa in the digestive tracks of termites.[12] Hubble published his results in 1925 in the *Publications of the American Astronomical Society*.[13]

Shapley regretted his move to Harvard. He believed he would have made the same discoveries as Hubble had he remained at Mount Wilson. But, in the end he was gracious, saying Hubble earned his fame and was "an excellent observer, better than I."[14]

Hubble, though, regarded van Maanen's findings to be a stain on his great accomplishment, nursing a personal grudge as they continued to work on the same mountain. Van Maanen only reluctantly admitted he may have made some errors and promised to follow up. He never did.[15]

8.

A DYNAMIC COSMOS

RELATIVISTIC COSMOLOGY

If Edwin Hubble had gone no further than his convincing demonstration that the universe extended far beyond the Milky Way and that many galaxies of stars exist besides our own, his name would have been etched in history. Instead he had it etched a second time when he provided evidence that most galaxies are moving away from Earth at speeds that increase linearly (at least as indicated by the best data of the time) with their distances.

As far back as 1912, Vesto Slipher had used the shifts of spectral lines to infer that spiral nebulae are speeding away from us, or, in the case of Andromeda, toward us, at enormous speeds. While this was not direct proof of their extragalactic nature, it was at least an early indication that the universe extended beyond the Milky Way.

It was at this point that observation and theory began to gel as a common unit, although few theorists knew anything about data and few observers knew anything about theory. As we saw in chapter 6, Einstein added a term to his gravitational equation, called the cosmological constant, to provide a repulsive gravity. He felt the term was needed to stabilize the universe; otherwise, all the stars would collapse onto one another.

In 1917, Einstein was able to come up with a solution to his equation in which the universe was a spatially finite (closed), static, four-dimensional hypersphere.[1] This is sometimes called the "cylindrical universe" since when you suppress one of the spatial axes, the universe at any given time is a circle and when you add the time dimension, it is a cylinder in space-time (see figure 8.1).

It should be noted that while Einstein's solution was technically static, it was unstable, like a rock sitting on top of a mountain peak. Any slight change in one of the parameters of the model, such as the cosmological constant or matter density, would result in the universe expanding forever or collapsing in a big crunch.

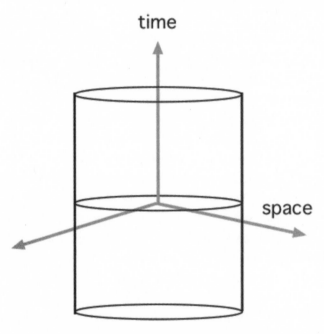

Figure 8.1. Einstein's static universe. A hypersphere in four dimensions, shown with one spatial dimension suppressed so it appears as a cylinder. It has no beginning or end. Image by the author.

Also in 1917, Dutch astronomer Willem de Sitter (1872–1934) showed that another static cosmological solution of Einstein's equation existed in which the universe is empty of matter, containing only the energy implicit in the cosmological constant.[2] It is illustrated in figure 8.2. In Einstein's solution, the gravitation attraction of the mass of the universe is exactly balanced by the repulsion of the cosmological constant. In the de Sitter solution, there is no matter or radiation—just a constant, positive curvature of space given by the cosmological constant and a gravitational repulsion that causes the universe to expand exponentially.

Of course, you will say that an expanding universe is not "static." De Sitter's universe is called static since it always expands at the same exponential rate and always will. In this model, the energy density is constant as the universe expands so the total internal energy increases with time. This does not violate energy conservation since the internal pressure corresponding to the cosmological constant is negative. Viewed as a thermodynamic system, it does work on itself.

In the diagram shown in figure 8.2, the de Sitter universe has no beginning or end. The line below the vertex represents the ever-decreasing diameter of the cone as you go in the negative-time direction. However, as we will see later it has since been proven that inflationary expansion must have a beginning, although that beginning can have been preceded by a prior contraction.

Einstein was not happy with the de Sitter solution. Besides, the universe is not empty.[3] De Sitter suggested that his solution might be a good approximation if the matter density is low. As we will see, he was not far off. Our universe is only about 26 percent matter and negligible radiation, as measured by energy or mass density. (In chapter 10 we will clarify the distinction cosmologists make between matter and radiation.)

FRIEDMANN'S UNIVERSE

In 1922, Russian physicist and mathematician Alexander Friedmann showed that space and time do not necessarily comprise a static manifold but can also constitute a dynamic one. Rather than presenting his original formulation, I will use the current conventional way to describe Friedmann's concept.

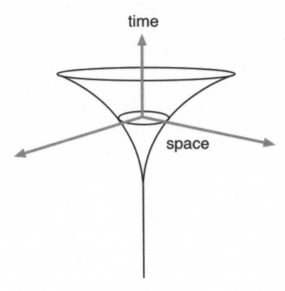

Figure 8.2. The de Sitter universe with one space dimension suppressed. The universe is an exponentially expanding sphere empty of matter. It has a positive cosmological constant that is equivalent to a constant energy density. Image by the author.

In 1929, American physicist Howard Robertson wrote a definitive paper, "On the Foundations of Relativistic Cosmology," that introduced what is now called the *Robertson-Walker metric*, also derived by Arthur Walker in 1935, that specifies all the possible line elements in four-dimensional space-time for a homogeneous, isotropic universe. He showed that the Einstein and de Sitter solutions are the only ones that are static and Friedmann's equations provided for all general, dynamic solutions.[4]

Working from Einstein's gravitational equation, Friedmann derived two new equations describing how the universe can evolve with time.[5] Assuming a homogeneous and isotropic universe, Friedmann's equations allow one to calculate the time-dependence of a quantity $a(t)$, called the *scale factor* in the Robertson-Walker metric, which describes how space expands or contracts.

Friedmann's concept is often described in popular terms as an expanding balloon. Paint two spots on the surface of a partially inflated balloon. When you inflate the balloon further, the spots will move apart; if you deflate it, they will move together. In Friedmann's picture, the two-dimensional surface of a three-dimensional balloon is analogous to three-dimensional space in four-dimensional Minkowski space-time.

Friedmann found three possible general scenarios for cosmic evolution depending on a parameter k, called the *curvature parameter*, which determines the overall geometry of three-dimensional space. If $k = 0$, space is *flat*, that is, its geometry is Euclidean. If $k = +1$, the universe is closed and space is non-Euclidean with positive curvature, like the surface of a three-dimensional sphere. If $k = -1$, the universe is an open three-dimensional hyperboloid and space has negative curvature, like a saddle. We can think of each of these geometries in terms of the sum of interior angles of a triangle: 180° for $k = 0$, greater than 180° for $k = +1$, less than 180° for $k = -1$.

The specific solutions to Friedmann's equations depend on the nature of matter in the universe and the values of k and the cosmological constant, Λ.

Einstein did not welcome Friedmann's model. He thought he had found a mathematical error in Friedmann's paper, but then admitted it was mathematically correct but had "no physical meaning." Unfortunately, Friedmann was not able to pursue his work further, dying in 1925 at the young age of thirty-seven. A recent article claims that his contributions to cosmology are not well understood and are often misrepresented.[6] Perhaps Friedmann died before he had time to recognize them himself, since he did not connect his calculations to astronomical observations.

LEMAÎTRE'S UNIVERSE

About the only scientist of the time who seemed to grasp the developing connection between mathematical cosmology and the remarkable observations that were coming in at the same time was a Belgian Jesuit priest and physicist, Georges Lemaître. In 1927, Lemaître published an article in French titled, "Un Univers homogène de masse constante et de rayon croissant rendant compte de la vitesse radiale des nébuleuses extra-galactiques" ("A Homogeneous Universe of Constant Mass and Growing Radius Accounting for the Radial Velocity of Extragalactic Nebulae").[7] There he showed that Einstein's equation leads to an expanding universe that accounts for galactic redshifts. Lemaître did not cite Friedmann nor mention other cosmological solutions since he was interested in describing what was being observed at the time. Lemaître's formulation is now known as the *Friedmann-Lemaître solution*.

However, having been written in French and published in a minor journal that few read, Lemaître's paper lay unrecognized for several years. Lemaître himself did not promote it, although he had sent a copy to Eddington, who did not respond.

Just six months after his paper came out, Lemaître met with Einstein in a park in Brussels. Einstein was attending one of the historic Solvay Conferences that were held regularly in Brussels. The fifth conference in 1927 became legendary, having been attended by every physicist who was anybody (Einstein, Bohr, Planck, Schrödinger, Heisenberg, Born, Pauli, Dirac, Lorentz, Perrin, de Broglie, Rutherford, Jeans, Poincaré, Brillouin, and so on . . .) and featuring the beginning of the momentous debate on quantum mechanics between Einstein and Bohr that continued for years.[8] Seventeen of the twenty-nine attendees were or would become Nobel Prize winners.[9] (No astronomers were included).

In the park, Lemaître, who was also not invited, briefly explained his model to Einstein, who replied, "Your calculations are correct, but your physical insight is abominable."[10]

HUBBLE'S LAW

In the meantime, Hubble and his capable, meticulous assistant, Milton Humason (1891–1972) were working away. Neither knew much about general relativity. What they knew was how to observe the skies. Humason had an eighth-grade education at the time (he eventually earned a doctor of science degree) and learned how to make astronomical observations after driving mule trains of supplies up the mountain. At Hubble's request, Humason spent long, cold, boring nights at the 100-inch telescope, taking long-exposure spectrographs of faint galaxies.[11] In that day before computers, the observer had to sit high up in a small cage near the mirror's focal point, open to the cold outside air, peering through a viewer at the object of interest, constantly adjusting the mirror manually as Earth rotated so that the object stayed in the crosshairs.

In 1929, Hubble published a seminal paper in the *Proceedings of the National Academy of Sciences* titled "A Relation between Distance and Radial Velocity among Extra-Galactic Nebulae."[12] There he proposed what became known as *Hubble's law:* the radial velocity of a galaxy is proportional to its distance. Humason was not left out. He authored the paper just preceding Hubble's in which he reported that the elliptical galaxy NGC 7619 was traveling away from Earth at 3,779 kilometers per second, twice as fast as any previous measurement and over a hundred times faster than Earth's orbital speed around the sun.[13]

Hubble's paper presents the graph shown in figure 8.3, which plots radial velocities of galaxies against their distances. Although the scatter of points is large, a clear trend can be seen for the more-distant nebula to be moving away from us as higher speeds, as expected if the universe is expanding. This was not a surprise; others had already noticed this tendency. However, Hubble now had the conclusive evidence. As for the fit to a straight line, that's hardly demonstrated, but, with the given data, it's the best you can do.

The caption in figure 8.3 is quoted exactly from the paper. The velocity axis on the graph is mislabeled and should be kilometers per second. The distance axis is in parsecs, where one parsec equals 3.26 light-years.[14]

While most of the velocities plotted are positive, a few negative points indicate some closer galaxies, such as Andromeda, moving toward us. The distances to most of the galaxies in Hubble's sample were not determined by Cepheids, which were too faint, but by the brightest stars or the luminosity of the galaxy as a whole.

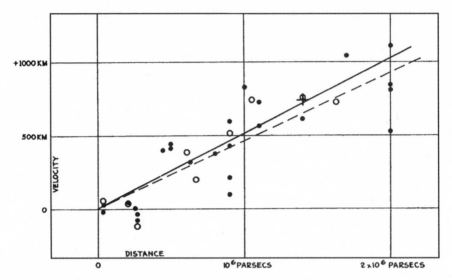

Figure 8.3. "Velocity-Distance Relation among Extra-Galactic Nebulae. Radial velocities, corrected for solar motion, are plotted against distances estimated from involved stars and mean luminosities of nebulae in a cluster. The black discs and full line represent the solution for solar motion using the nebulae individually; the circles and broken line represent the solution combining the nebulae into groups; the cross represents the mean velocity corresponding to the mean distance of 22 nebulae whose distances could not be estimated individually." (Direct quotation from caption in Hubble's paper.) Image from Edwin Hubble, "A Relation between Distance and Radial Velocity among Extra-Galactic Nebulae," *Proceedings of the National Academy of Sciences* 15, no. 3 (1929): 168–73.

In those days, scientific papers were not as scrupulous about references as they are today. Indeed, some can be called sloppy and unscholarly by modern standards. So, Hubble was not unusual in failing to cite the sources of his data. This has left the impression, found in many popular astronomy books, that all the data were taken with the 100-inch telescope on Mount Wilson by Hubble and Humason. Actually, only four points were from Mount Wilson, recorded by Humason. The bulk of the data in Hubble's paper was from Slipher, taken with smaller telescopes at Lowell Observatory.[15] Nevertheless, by 1931 Hubble and Humason had added forty galaxies of their own.

The slope of that line on the velocity v vs. distance r plot, $K = v/r$, is called *Hubble's constant*, now designated H. Hubble's law then is $v = Hr$. Hubble quotes two values based on different analyses of the data: $K = 500$ kilometers per second per million parsecs for individual nebulae and 530 for nebulae combined into groups.

With these values of K, Humason's NGC 7619 would be twenty million light-years away. As we will see, Hubble's estimate of H was a factor of seven too high, putting Humason's galaxy actually 140 million light-years distant.

Note that, technically, H is the rate of expansion of the universe, which need not, and indeed we now know does not, have the same value throughout the life of the universe. So I will follow convention and define the value H has today, designated H_0, as Hubble's constant; and the more general rate of expansion H we will refer to as the *Hubble parameter*.

LEMAÎTRE IS NOTICED

In his 1927 paper, two years earlier than Hubble's publication, Lemaître had made an estimate for K (or H) of 625 kilometers per second per million parsecs based on what seems to have been the same data sample used by Hubble.[16] In the paper Lemaître says explicitly, "The receding galaxies are a cosmical effect of the expansion of the universe."

Lemaître's article began to receive attention 1931 when it was translated into English with the help of Eddington, who, along with Shapley, had been one of Lemaître's mentors and had finally taken notice.[17] Lemaître, however, left out of the English translation his calculation of Hubble's constant.[18] In any case, even in the French version Lemaître did not include the crucial plot of velocity versus distance that is really needed to see the effect.

Nevertheless, cosmologists saw the value of Lemaître's work. Eddington noted that Einstein's static universe depended on the cosmological constant having exactly the correct value and any small perturbation would cause the universe to expand or contract. Eddington wrote to de Sitter that Lemaître had provided a "brilliant solution" to the problem, and de Sitter agreed.

Einstein finally came around, and by 1933 so had the astronomical community. Eventually, Einstein abandoned the cosmological constant, calling it his "biggest blunder." Little did he suspect that it (or something like it) would turn out to be carrying three-quarters of the energy of the universe. Neither de Sitter nor Lemaître discarded the constant from their models, although it would be years before most cosmologists would see any need for the term.

Hubble and Humason continued to measure the redshifts of galaxies, the limit of their spectrograph being reached at forty thousand kilometers per second—fast enough to go to the moon in ten seconds. Hubble did not theorize much and, while he is widely regarded as the discoverer of the expanding universe, he never fully accepted it, keeping his careful, empirical mind open to alternatives, while Shapley became a true believer.[19]

Today there is no dispute that Lemaître was the first to associate the redshifts of galaxies with the expansion of the universe. However, Lemaître was not an observer, and theories in science are useless without data to verify them. Hubble's role, with Humason's assistance, was to provide the clinching observations.

In 1935, Hubble gave the Silliman Lectures at Yale University that are reproduced in his classic *The Realm of the Nebulae*.[20] The Silliman Lectures were instituted "to illustrate the presence and providence of God as manifested in the natural and moral world."

LEMAÎTRE'S PRIMEVAL ATOM

Popular books on astronomy usually use the discovery that the universe is expanding to conclude that the universe therefore had an origin sometime in the finite past. But it doesn't follow. As historian Helge Kragh points out, "When the idea of a finite-age universe was first proposed — and that occurred only two years after Hubble's discovery — most astronomers and physicists resisted it."[21]

Eddington was particularly horrified with the idea of an origin to the universe, saying, "philosophically, the notion of a beginning of the present order of nature is repugnant to me."[22] Here's how he imagined it:

> I picture ... an even distribution of protons and electrons, extremely diffuse and filling all (spherical) space, remaining nearly balanced for an exceedingly long time until its inherent instability prevails.... There is no hurry for anything to happen. But at last small irregular tendencies accumulate, and evolution gets underway.... As the matter drew closer together in the condensations, the various evolutionary processes followed — evolution of stars, evolution of the more complex elements, evolution of planets and life.[23]

Lemaître begged to differ. During a conference in London in 1931 on the relation between physics and spirituality, Lemaître proposed that the universe expanded from an initial glob of nuclear matter in an explosion that the British astronomer Fred Hoyle would, in 1948, derisively term the "big bang." Lemaître announced his idea in a one-page article containing no equations, published in *Nature* that year.[24] He wrote:

> If we go back in the course of time, we find all the energy of the universe packed in a few or even in a unique quantum.... If this suggestion be correct, the beginning of the world happened a little before the beginning of space and time. I think that such a beginning of the world is far from repugnant.... We could conceive the

beginning of the universe in the form of a unique atom, the atomic
weight of which is the total mass of the universe. This highly unstable atom would divide in smaller and smaller atoms by a kind
of super-radioactive process.

Note that, unlike some modern theists we will visit later, Lemaître
does not argue from general relativistic cosmology because he knew
full well that it does not provide a unique scenario for the origin of the
universe. Instead, he relies on quantum mechanics, which to this day
has not been reconciled with general relativity, although it has with
special relativity. Lemaître imagines the universe starts as something
like an atomic nucleus containing all the universe's matter, which
then spontaneously decays to produce the universe, as we know it.

In 1941, Lemaître published a book in French titled *L'hypothèse
de l'atome primitif* (*The Primeval Atom*), which was translated into
English in 1950.[25] Although containing a few equations and a mathematical appendix, it is not a technical manuscript but a wide-
ranging discussion of cosmogony and cosmology aimed for the
French-speaking public. Lemaître spends a long chapter arguing
that his hypothesis in which the universe began as a single, giant
nucleus is consistent with all the knowledge of the day.

It is worth noting that Lemaître anticipated a claim that in recent
years has become a staple argument among theologians: Not only
did the universe itself come into being with the big bang, but so did
space and time.

Much has been made of the fact that Lemaître was a Jesuit priest
and his notion that the universe began a finite time ago with a giant
explosion was informed by his religious beliefs. However, he always
insisted that the primeval atom was purely a scientific hypothesis
not yet confirmed by the data. Indeed, Lemaître's scenario sounds
more like the Chinese creation myth of a "cosmic egg" than what is
described in Genesis.

The original typescript of his *Nature* paper ended with the following sentence that Lemaître crossed out before submitting:

> I think that every one who believes in a supreme being supporting every being and every acting, believes also that God is essentially hidden and may be glad to see how present physics provides a veil hiding his creation.[26]

In other words, the big bang should *not* be taken as evidence for a creator God since that God is hidden. Many like Lemaître who choose to believe in God despite the fact that his existence is far from obvious have little recourse but to assume that he must have reasons to hide from us. However, this "hiddenness argument" has been shown to fail.[27]

Lemaître's proposal was largely speculation. He provided no quantitative model. The most that could be said at the time was that Hubble's law, $v = Hr$, is exactly what you would expect for the particles in an explosion. If they do not collide with one another after the explosion, the faster moving will be farther away at any given time and r should be linearly proportional to v. Certainly the best one can do with the data points in figure 8.3 is fit them to a straight line. That's the way things stood for much of the twentieth century until just two years before the new millennium when the data became far more precise. But we will get to that later.

In the linear model, where H is a constant, then its reciprocal is the age of the universe. Hubble's value of five hundred kilometers per second per million parsecs converts to $T = 1/H =$ two billion years. At the time, this was less than the age of Earth determined from radiative decay, which was estimated to be at least three billion years. You would expect the universe to be older than Earth.

And there were other problems. Judging from their distances as determined from Hubble's law, the galaxies outside our own are all smaller than the Milky Way. In particular, Andromeda was smaller and its stars fainter than expected. It seemed unlikely that we were again so special. For years, these were major defects in Hubble's law.

It was not until 1948 when the 200-inch Hale reflector went in operation at the Palomar Observatory in California that the quantitative problem with Hubble's law was solved. Using Hale, in 1952 Walter Baade showed that there were two distinct kinds of Cepheids.[28] Hubble had assumed one kind to measure the distance to Andromeda, while Shapley had used another kind for his measurements of the distances to globular clusters in the Milky Way. So, Andromeda was not smaller and its stars fainter than those in our galaxy. It was just farther away than Hubble thought. His value of H was seven times too large and so his distance estimates were seven times too small.

The current best value of Hubble's constant is about seventy kilometers per second per million parsecs, giving an age for the universe of $T = 1/H$ = fourteen billion years. This estimate is based on the assumption that H is a constant. As we will see later, this is now known not to be true and, as of this writing in 2014, the best estimate of the age of the universe is 13.8 billion years, with a small and statistically insignificant (despite media hype) disagreement on the precise number among observers using different techniques.

"THEREFORE, GOD EXISTS!"

Against his expressed wishes, Lemaître's big bang was and still is being used by theologians and believing scientists to claim scientific evidence for a divine creation. On November 22, 1951, Pope Pius XII spoke before the Pontifical Academy of Sciences where he announced that modern science and the Church are converging on the same fundamental truths.[29]

Introducing a theme that would be expanded upon by religious apologists in succeeding decades, the pope said: "According to the measure of its progress, and contrary to affirmations advanced in the past, true science discovers God in an ever-increasing degree — as though God were waiting behind every door opened by science." He identified two essential characteristics of the cosmos "which science has, in a marvelous degree, fathomed, verified and deepened beyond all expectations (1) the mutability of things, including their origin and their end; and (2) the teleological order which stands out in every corner of the cosmos."

Pius XII gave examples of mutability, that is change, in both the microcosm and macrocosm, such as phase changes and chemical changes in matter, emission of radiation by atoms, and radioactivity of nuclei.

Moving to the cosmos, the pope says: "Everything seems to indicate that the material universe had in finite times a mighty beginning, provided as it was with an indescribably vast abundance of energy reserves, in virtue of which, at first rapidly and then with increasing slowness, it evolved into its present state."

The pope concludes:

[Modern science] has followed the course and the direction of cosmic developments, and, just as it was able to get a glimpse of the term toward which these developments were inexorably leading, so also it has pointed to their beginning in time some five billion years ago. [Note: This was a reasonable estimate at the time]. Thus, with that concreteness which is characteristic of physical proofs, it has confirmed the contingency of the universe and also the well-founded deduction as to the epoch when the cosmos came forth in the hand of the Creator.

And, in the lines that have been quoted frequently in the decades since:

Hence creation took place in time. Therefore, there is a Creator. Therefore, God exists.

Interestingly, in no place in his speech does the pope refer to specific Catholic doctrines or to the Bible. While he mentions "cosmic developments" that point to the "beginning of time some five billion years ago," he does not mention that Genesis implies a much more recent creation — on the order of ten thousand years. He also does not mention the doctrine of the immutability of species unmistakably present in Genesis.

Of course, that's the advantage of being pope. It is he who determines the official doctrines of the Catholic Church, not the Bible. Poor Protestants. They do not have available to them the same continuous line of authority going back to Christ and must rely on an obviously flawed book of myths thousands of years old.

Even so, in 1951 there still was no convincing evidence for the big bang, so it wasn't proof of anything. As Kragh points out, "At the time the pope gave his presentation of cosmology, the field was not characterized by harmonious agreement, but on the contrary, by a fierce controversy." The pope's message was actually quite misleading, leaving the impression among laypersons that "the biblical Genesis had literally been proved by big-bang cosmology." Lemaître knew this was not the case and managed to get the pope to slightly moderate his views in later speeches.[30]

TIRED LIGHT

If Hubble wasn't totally sold on the expanding universe, you can bet others weren't as well. They were, after all, just doing their jobs as scientists: question everything until the data supporting a theory are completely convincing. So, cosmologists sought other explanations for the undeniable empirical fact of redshifts in the spectra of galaxies. If the redshifts weren't the Doppler effect of receding velocities, then what were they?

A plausible explanation was proposed by one of the more interesting characters in the history of astronomy, Caltech astronomer Fritz Zwicky. Zwicky was born in 1898 in Varna, Bulgaria, where his father was Swiss ambassador. In 1925 he joined the Caltech faculty and worked most of his life at the Palomar and Mount Wilson Observatories. Zwicky had many ideas that were ahead of his time such as galaxies occur in clusters.

In 1929, Zwicky suggested that the redshift was caused by *tired light*: photons coming from great distances simply lose energy by interacting with matter, including perhaps other photons, along the way. As the photons lose energy, their corresponding wavelengths will increase toward the red.

Although this idea stuck around in various forms for decades, it was eventually falsified by a surface-brightness test proposed by American theoretical physicist Richard Tolman. In a static universe, the intensity of light from a star or a galaxy will fall off as $1/r^2$, where r is the distance to the galaxy, while the apparent area also drops off in the same way; so the observed brightness per unit area will be constant. In an expanding universe, the brightness falls off more rapidly with distance for two reasons. First, because the object emitting the photons is moving away, each photon has to travel a greater distance than the one before, resulting in a lower rate and decreased intensity. Second, the object appears larger than it really is because the light we observe was emitted when it was closer. The best measurements of surface brightness supported an expanding universe and ruled out the tired-light hypothesis.

VARIABLE CONSTANTS

Another suggestion made for the observed redshifts was that the speed of light is decreasing with time. Eddington pointed out that this contradicted special relativity, which assumes c is constant. As we saw in chapter 6, the assumption that the speed of light in a vacuum is a constant is now built into physics. But, as always, we must rest our case on data rather than on theory. Since special relativity has been consistent with many thousands of experiments conducted for over a century, we can sit comfortably on its foundation until the data tell us to do otherwise.

In 1938, Paul Dirac proposed a model in which Newton's gravitational constant G varies with time.[31] This contradicted the assumptions of general relativity and led to a universe only seven hundred million years old. When he realized this, Dirac abandoned the idea. Today we recognize that G and c are arbitrary constants whose numerical values just set the units with which we are working.[32]

MILNE'S COSMOLOGY

In the 1930s, Edward Arthur Milne, brother of A. A. Milne, creator of Winnie-the-Pooh, proposed a cosmology of his own that did not depend on general relativity.[33] He denied that space is curved or expands. In fact, space was not something physical but simply a system of reference.[34]

In my 2006 book *The Comprehensible Cosmos*, I described the procedure Milne outlined by which we can make observations solely with a clock.[35] I thought Milne presented a precise and objective implementation of the way we actually find out what is out there in the world beyond our individual selves and describe it in terms of a space-time model. All we do as observers is send and receive signals. Following Milne, I showed how with just a clock, a pulsed light source, and a detector we can develop a picture of the world. We simply send out signals and receive replies analogous to radar echoes. Without any meter sticks or other devices to measure distances, just measuring times, we can determine a quantity we call "distance" from which we can formulate a model of the outside world. In this scheme, the speed of light is constant by convention and so special relativity falls right out.

Milne's model universe was made of randomly moving particles like molecules in a gas. He further made the reasonable assumption that the universe should look the same to all observers. This *cosmological principle* can be traced back to Nicholas de Cusa in the fifteenth century (see chapter 2).[36]

From this model, Milne was able to derive Hubble's law $v = Hr$ and, thus, the expansion of the universe without general relativity. However, he did not derive much else. He predicted that Newton's gravitational constant G would increase with time, but so slowly as to be undetectable. Actually, he did not regard this time variation to be subject to experimental test since it involved another kind of time he invented, *kinematic time*, which is not measured with clocks.

Although highly unconventional, Milne's cosmology attracted

considerable attention in the 1930s. Hubble, in particular, was favorable of it and mentions it near the end of *The Realm of the Nebulae* where he spends a few pages on theory.[37]

However, Milne turned off a lot of people with his mixing of cosmology and a rather (also) unconventional theology. In his 1935 book, he says that ultimate questions of cosmology need reference to God. He frequently mentioned God in his speeches and other writings. Unlike Lemaître, who was careful to leave religion out of his science, Milne believed that theological and metaphysical arguments had a role to play in science, especially in regard to the creation of the universe.[38] He wrote: "Investigators who leave out God, the *raison d'être* of the universe, find themselves lamentably handicapped in dealing with cosmological questions."[39]

Even worse than Milne's theology, from the standpoint of the empirical scientific community, was his extreme rationalism. Milne placed logic and reason over observation and experiment. He wrote:

> The philosopher may take comfort from the fact that, in spite of the much vaunted sway and dominance of pure observation and experiment in ordinary physics, world-physics propounds questions of an objective, non-metaphysical character which cannot be answered by observation but must be answered, if at all, by pure reason; natural philosophy is something bigger than the totality of conceivable observations.[40]

Milne's theory suffered the fate of most theories that put too much emphasis on reason, logic, and mathematics (and, as in his case, revelation) and not enough on data: it failed to provide a falsifiable empirical test. In science, a theory that is not falsifiable is history (or at least should be), and so Milne's cosmology was ultimately ignored as general relativistic cosmology and the big bang, aided by developments in submicroscopic physics came to the fore. But, as we will now see, that did not happen overnight.

MISSING MASS

The 1930s also provided the first hints of what would become another unexpected and enormously significant feature of the universe. In 1932, Dutch astronomer Jan Oort discovered that the masses of luminous objects such as stars and dust in the galaxy were insufficient to account for the observed orbital motions of stars. The following year, Zwicky noticed the same effect in the orbital speed of galaxies inside galactic clusters. From his measurements and applying Newton's laws, he calculated that the mass of the Coma cluster is four hundred times its luminous mass. He named this invisible gravitating stuff *dunckle materie* — dark matter.[41]

In 1939, American astronomer Horace Babcock measured the *rotation curve* of the Andromeda galaxy, which is a plot of the velocity of stars against their distance to the center of the galaxy. Based on Newton's law of gravity, that velocity should fall off with distance, as it does for planets in the solar system. However, as later measurements in the 1960s and 1970s verified, the rotation curve stays flat well past the main luminous part of the galaxy, indicating that the stars are moving through a roughly uniform distribution of invisible matter.

We now know that dark matter constitutes 26 percent of the mass of the universe, while all the luminous matter we see with optical telescopes comprises only 0.5 percent. Furthermore, as we will see next, the great bulk of dark matter is composed of something besides familiar atomic matter, still not identified.

RADIO ASTRONOMY

The 1930s also saw the opening up of a new window on the universe. In 1933, Bell Telephone Laboratories' engineer Karl Jansky detected radio waves from beyond Earth.[42] This would lead to a new field of radio astronomy that explores the universe in a spectral region far from the visible region and plays an important role in cosmology.

9.

NUCLEAR COSMOLOGY

FILLING IN MORE DETAILS

L et us review the status of cosmology in the mid-twentieth century. By the early 1930s, the great discovery that we live in a vast, expanding universe composed of galaxies of stars hurtling away from one another at great speeds was well confirmed and astronomers busied themselves with filling in more details. The most powerful telescope in the world remained the 100-inch reflector on Mount Wilson, which had gone into operation in 1908. It retained that position for forty years, until 1948 when it was finally supplanted by the 200-inch reflector at the Palomar Observatory. Of course, these were not the only telescopes and many others had special designs that made them very useful for particular types of observations.

With these instruments, astronomers began surveying the skies and cataloging galaxies, a process that would continue for years with many surprising and dramatic results. One of the most productive catalogers was an idiosyncratic astrophysicist, Fritz Zwicky, already noted for proposing the unsuccessful tired-light mechanism for the redshift of galaxies and his observation of the missing mass of galaxies. He also conjectured that high-energy cosmic rays come from outside the solar system and result from the explosions of supermassive stars. He dubbed these *supernovae*. Zwicky used the Palomar 18-inch Schmidt telescope, invented in 1930 by German optician Bernard Schmidt, to search for supernovae. The Schmidt could accurately survey large patches of sky. After it went into operation in 1936, Zwicky discovered a dozen supernovae.[1]

In 1948, the larger 48-inch Schmidt telescope was used for the National Geographic Palomar Sky Survey. It spectacularly verified Zwicky's proposal, mentioned in chapter 8, that galaxies gather together in clusters. In 1958, UCLA astronomy professor George Abell produced a catalog of 2,712 clusters in the northern sky and by 1989 had catalogued 4,073 rich clusters over the entire sky. By the 1970s astronomers began to see the clusters themselves organizing into "spongy" structures of their own, with holes, filaments, and walls.

A HOT, DENSE PAST

In the meantime, the focus of cosmological physics shifted from general relativity to nuclear physics. It began to be realized that if the universe is expanding, it must have grown from a very small, hot, and dense region in the past when nuclear reactions played a major role. Georges Lemaître may have been the first to recognize this. However, his proposal that a primordial supernucleus decayed radioactively into the universe we have today was pure speculation and had no empirical or theoretical foundation.

As a result, few at the time took the notion seriously. What Lemaître did provide was a cosmological solution to general relativity for an expanding universe. This is now called the *Friedmann-Lemaître solution* since it was already implicit in Friedmann's equations. The model itself became known as the *Lemaître-Eddington model*, since Eddington developed it further.

In any case, Lemaître pictured a finite universe with a definite beginning. While this may have been suggested to him by his religious belief in a creator, as mentioned earlier he never based his arguments on theology and, indeed, objected to such an interpretation.

On the other hand, we saw in chapter 7 that Eddington found the idea of a beginning to the universe "repugnant." His universe was expanding but eternal and most physicists of the time were inclined to agree. As for observational astronomers, they stuck to their telescopes.

Lemaître continued to develop his model and was cognizant of the need to make it testable. He realized that if the universe had once been hot, dense, and radioactive, it must have left a remnant of radiation that might be observable today. However, he did not think this radiation was electromagnetic, that is, photons, but instead would be composed of charged particles. Once again, most physicists were not convinced, although Einstein expressed mild interest. But the evidence simply was not there.[2]

Furthermore, a major problem for the finite-universe hypotheses lay in the timescale implied by the data. Recall, Hubble's law implies that the age of the universe is equal to the inverse of the Hubble constant. That value was coming out under two billion years, less than estimates of the age of Earth from geology and nuclear physics. In fact, it may come as a surprise to hear that Hubble himself questioned the expanding universe, which would ultimately bring him great fame. He wrote, "No effects of expansion—no recession factor—can be detected. The available data still favor the model of a static universe rather than a rapidly expanding universe."[3]

However, the assumption that the age of the universe $T = 1/H$ is based on zero cosmological constant. Lemaître's model included a cosmological constant and allowed for a much-older universe. Unfortunately, Einstein had disowned the cosmological constant and would have nothing further to do with it.[4]

Up until this point, the late 1930s, the only physics involved in theoretical cosmology was general relativity. Lemaître's primeval atom was mainly speculation, with some tentative attempts to develop a quantitative model. But then, a great breakthrough occurred in 1938–1939 when the German physicists Hans Bethe (working in America) and Carl Friedrich von Weizäcker independently proposed nuclear-fusion mechanisms responsible for the energy of stars. Bethe's process was particularly simple. Four protons unite to form helium by a series of two-body collisions involving only fundamental particles: protons, neutrons, electrons, photons, and (as we know today) neutrinos. Weizäcker's mechanism was more complex, involving carbon, oxygen, and nitrogen isotopes.[5]

Weizäcker also suggested that his theory might account for the formation of the elements.[6] However, his model failed to give an accurate account of cosmic abundances of the elements.[7] Still, nuclear physicists were sufficiently intrigued to begin taking part in cosmological research.

YLEM

A major step toward establishing the validity of the big-bang model was made by a Russian-Ukrainian physicist who emigrated to the United States, George Gamow.[8] In 1924, Gamow was in the classroom in Leningrad when Alexander Friedmann gave a series of lectures, "Mathematical Foundations of the Theory of Relativity." Gamow wanted to study under Friedmann, who sadly died young just a year later.

After earning a doctorate in quantum theory from Göttingen with a thesis on the atomic nucleus, Gamow worked in Copenhagen with Niels Bohr, in Cambridge with Ernest Rutherford, and in 1931, at age twenty-eight, became a member of the Academy of Sciences of the USSR. His many accomplishments in nuclear physics include demonstrating quantitatively that nuclear alpha decay (emission of helium nuclei, also called *alpha particles*) can be understood as resulting from *quantum tunneling*. This process also plays an important role in fusion reactions in stars. We will see later how this uniquely quantum mechanical process has been recognized in cosmology as the mechanism by which our universe may have come into existence.

In 1934, Gamow moved to the United States and worked with Edward Teller at George Washington University in Washington, DC. During World War II, Teller went off to work on the Manhattan Project. However, Gamow was not cleared to work on the nuclear bomb because he had been made a commissioned officer in the Red Army in order that he could teach at the Soviet military academy. He remained in Washington and consulted for the US Navy. After the war, Gamow was cleared for nuclear research at Los Alamos.[9]

Besides continuing to work on nuclear physics, with a developing interest in astrophysics, Gamow became famous as the author of bestselling, popular books including *One, Two, Three . . . Infinity*, *The Birth and Death of the Sun, Mr. Tompkins in Wonderland* (a series of six books), and many others. I devoured them as a teenager and they undoubtedly contributed to my becoming a physicist. This was just one more indication of Gamow's immense genius; he didn't even speak English until graduate school.

In 1948, Gamow, Ralph Alpher, and Hans Bethe published a short letter in the *Physical Review*, "The Origin of the Chemical Elements," reinvigorating the notion that the nuclei of the atoms that make up the periodic table of the chemical elements were produced in the early universe.[10] Bethe's name was added so the paper could be referred to as "Alpher, Bethe, and Gamow." While not an original author, Bethe nevertheless made valuable contributions. The paper does not cite Lemaître or Weizäcker.

Alpher, Bethe, and Gamow proposed that, in the beginning there existed a compact core of primordial stuff, which they would later dub "ylem," composed of neutrons at high density and temperature. Some neutrons decayed into protons by beta decay, emitting an electron and (we now know) an antielectron neutrino, leaving a bath of these particles.

The elements then formed as the protons and neutrons recombined by a process known as *radiative capture* in which photons are emitted, adding to the mix. In this manner, a proton and a neutron fuse to make a deuteron (a hydrogen nucleus with two neutrons). Add another neutron and you get a triton (a hydrogen nucleus with three neutrons). A triton and a proton, or two deuterons, can fuse to make helium, with a substantial release in energy.

As an aside, attempts at controlled nuclear fusion utilize these reactions, which require lower temperatures than those reactions taking place in the cores of stars. Even so, that temperature is still incredibly high, on the order of 100 million degrees, and despite over a half century of effort, this source of energy is still far from being realized.

Gamow and collaborators envisaged building up the complete periodic table by a series of nuclear reactions in the early universe. However, despite their intense efforts, the process was found not to continue. Adding a neutron or a proton to helium does not produce a stable nucleus with five nucleons. Fusing two helium nuclei together also does not produce a stable state of eight nucleons.

As we will see, Fred Hoyle and his collaborators were able to show later how the heavier nuclei form inside stars by what is called *stellar nucleosynthesis*, a process first suggested by Arthur Eddington and also studied by Bethe. When Gamow's *primordial nucleosynthesis* failed to account for the full periodic table of the elements, the big bang once again fell out of favor.

However, the Alpher-Bethe-Gamow model, also contributed to by Ralph Herman, had another implication that was realized in a remarkable paper by Alpher and Herman published in 1949. The reaction rates involved are sufficiently fast compared to the expansion rate of the universe, given by the Hubble parameter H, that a kind of quasi-thermal equilibrium with slowly decreasing temperature occurs in the plasma of interacting particles. Alpher and Herman estimated that the temperature "when the neutron capture process became important" would be about 600 million degrees.[11] Including in the mix at the time were photons. Applying the theory of cosmological expansion, they estimated that today this temperature would have cooled to "on the order of 5 K," that is five degrees Kelvin.[12]

Although they were not quite so explicit, Alpher and Herman had made the world-shaking prediction that the universe should be bathed in thermal radiation, that is, radiation following the spectrum of a blackbody at 5 K, which is in the microwave region. This only applies to the photons, which remained in thermal equilibrium as the universe expanded while matter did not. A 5 K blackbody spectrum peaks at a wavelength of about one meter. By comparison, the optical spectrum of the sun (5000 K) peaks at 550 billionths of a meter.

This prediction failed to excite any interest among physicists and astronomers, probably because it was tied to the big-bang nucleosynthesis mechanism that was unable to reach beyond the first few elements of the periodic table.

Furthermore, there still was the age paradox. So, yet again, the big bang was forgotten. Despite the enthusiasm of Pope Pius XII, after 1953, only one paper on the big bang was published over the period of a decade.[13] In its place was a model of an eternal, unchanging universe that attracted far more attention than it deserved, perhaps because of the eminence of its proposers.

THE STEADY-STATE CHALLENGE

As mentioned, "big bang" was a derisive term coined by Fred Hoyle in an interview on the BBC in 1948. That year, Hoyle,[14] along with Hermann Bondi and Thomas Gold,[15] developed an alternative to the big bang called the *steady-state universe*. James Jeans had first suggested a steady-state cosmology in 1928.[16] With the aid of Margaret and Geoffrey Burbidge, and Jayant Narlikar, Hoyle continued to promote the steady-state model well after the empirical evidence for the big bang had become overwhelming.[17]

These authors felt strongly that the universe had to obey what they called the *perfect cosmological principle*, which they interpreted to mean that the universe must look the same everywhere and everywhen, that is, at every place and at every time. Not only is there no special place in space that we can regard as center of the universe, there is no special moment in time when it all began.

Now, in an expanding universe with a fixed total mass, the mass density must decrease with time. Since, in the steady-state authors' minds, the average mass density of the universe must remain constant or else the universe will "look different," matter must be continuously and uniformly created. The rate was very small, however, only 10^{-43} grams per cubic centimeter per second.

This would seem to violate energy conservation — as the authors realized. However, various versions of the model that were proposed over the years provided for energy conservation with the introduction of a field of negative pressure. It was noted that this special field was equivalent to the cosmological constant. A positive cosmological constant produces a constant negative pressure in an expanding gas, resulting in an increase in internal energy fully consistent with conservation of energy. The energy to create the new mass from the work is done on the system by its own negative pressure. However, this interpretation seems to have been rejected by the steady-staters.[18]

In any case, numerous observations would soon prove conclusively that the universe looked quite different at different times. In the 1950s, radio astronomer Martin Ryle and his group at Cambridge showed, after some controversy (Hoyle's office was down the hall), that the density of radio sources at large distances, and thus early times, was greater than it is now. When, as we will discuss later in this chapter, quasars and other active galaxies were discovered, it was found that they also were far more abundant in the deep past.

Ryle would share the 1974 Nobel Prize for Physics with his Cambridge colleague Antony Hewish, whose student Jocelyn Bell discovered the first pulsar (more about this in a moment). This was the first Nobel Prize ever awarded for astronomical accomplishments. It would not be the last.

The fact that the universe has changed its appearance over billions of years constitutes a failure of the perfect cosmological principle as it was defined by Hoyle. However, instead of asserting that the universe must appear the same at all times and places, we can introduce a cosmological principle that serves the same purpose as that proposed by Hoyle, which was to extend the Copernican principle to include time as well as space. We simply require that the models scientists construct to describe the universe apply at all times and places. The most distant objects such as quasars, which are also being viewed in the deep past, exhibit the same spectral lines and other basic physics properties as we measure in the laboratory here and now on Earth, confirming this version of the cosmological principle.

STELLAR NUCLEOSYNTHESIS

Hoyle's views received a tremendous boost when he and his team of collaborators were able to develop a successful theory of element formation in stars called stellar nucleosynthesis, published in 1957.[19] This undermined the big-bang model because of the failure of big-bang nucleosynthesis.

In 1952, physicist Edwin Salpeter had found a way to jump the instability gap of five and eight nucleons in building up the elements heavier than helium. In his so-called *triple alpha process*, two alpha particles, that is, helium-4 nuclei (He^4) first unite to from beryllium-8 (Be^8), a nucleus containing four protons and four neutrons.[20] The beryllium nucleus is unstable, however, which we have seen is the major bottleneck that prevents big-bang nucleosynthesis from going further. Salpeter showed that at sufficiently high temperature and density, a Be^8 nucleus can capture another He^4 nucleus to form stable carbon-12 (C^{12}) before decaying. Here are the reactions:

$$He^4 + He^4 \rightarrow Be^8$$
$$He^4 + Be^8 \rightarrow C^{12}$$

Of course, carbon is the most important element in the formation of life as we know it. Other nuclei important to life, such as oxygen and calcium, can also be formed from He^4 fusing with other nuclei:

$$He^4 + C^{12} \rightarrow O^{16}$$
$$He^4 + O^{16} \rightarrow Ca^{20}$$

These processes can, in principle, occur in the big bang. But the temperature has to drop below one billion degrees because above that temperature nuclei break apart by photodisintegration as fast as they synthesize. At that temperature, the density of the early universe has dropped to 10^{-4} grams per cubic centimeter, far too low for the Salpeter process to occur.

In 1954, Hoyle showed that when a star burns up all its hydrogen and collapses under gravity, its core will reach a temperature on the order of one hundred million degrees and density of about ten thousand grams per cubic centimeter, allowing the triple alpha process to take place.[21]

THE STEADY STATE AND GOD

One of the strong objections Hoyle, Bondi, Gold, and other proponents of the steady-state universe had for the big-bang model was what Hoyle termed "esthetic." Any explanation for an abrupt origin would have to rely on "causes unknown to science."[22] By this he meant metaphysical. Hoyle was outspokenly atheist and as late as 1982, he was assailing those scientists who supported the big bang, by then the majority, unfairly attributing to them religious motives:

I have always thought it curious that, while most scientists claim to eschew religion, it actually dominates their thoughts more than it does the clergy. The passionate frenzy with which the big-bang cosmology is clutched to the corporate scientific bosom evidently arises from a deep-rooted attachment to the first page of Genesis, religious fundamentalism at its strongest.[23]

In his 1994 autobiography he declared, "Big-bang cosmology is a form of religious fundamentalism."[24]

However, as we will now see, the big-bang model emerged triumphant while the steady-state model failed, both for the best of reasons. The big-bang model agreed with the data and the steady-state model did not. To the everlasting credit of Hoyle and his collaborators, stellar nucleosynthesis also triumphed. But it never had anything to do with the steady-state universe in the first place and its success should never have been held against the big-bang model.

Perhaps Hoyle's atheism may have prejudiced him against the big bang and motivated him to seek another explanation for nucleosynthesis than element production in the early universe. In any case, today stellar nucleosynthesis is an integral part of cosmology, with big-bang nucleosynthesis providing for just the right amount of the lighter elements to supplement what is not produced in stars. In science and Denver Bronco football, things do not always work out the way you expect.

ACTIVE GALAXIES

Perhaps the most important discovery of cosmological importance prior to the discovery of the cosmic microwave background in 1964, which will be discussed in the next chapter, was the observation of *quasi-stellar objects*, now popularly known as *quasars*. By 1960, radio astronomy had begun to flourish and a hundred or so curious radio objects were recorded that seemed to have a small angular size. One, 3C 48, was tied to an optical source by John Bolton. In 1963,

Maarten Schmidt using the 200-inch Hale telescope at the Palomar Observatory optically identified the radio source 3C 273.[25] Actually, 3C 273 is visible with relatively small, amateur telescopes—looking just like a star (hence "quasi-stellar").

Measuring the optical spectrum, Schmidt found that the spectral lines of hydrogen were shifted by 47,400 kilometers per second or 15.8 percent of the speed of light. If the speed measured from the redshift of the source is used to determine its distance by Hubble's law, 3C 273 was two billion light-years away when it emitted the observed light. Note that it is much farther away now since the universe has been expanding during all that time. It obviously was not a single star.

Schmidt estimated that the object, if at the distance implied by Hubble's law, had to be one hundred times brighter than any of the galaxies that had so far been associated with radio sources, and that the light was coming from a nucleus less than three light-years across. Having somewhat more than an amateur instrument at Palomar, Schmidt also observed an optical jet about 150 light-years long that he associated with a radio feature. He inferred that the object was of galactic dimensions.

In the paper immediately following the one published in *Nature* by Schmidt, Jesse Greenstein and Thomas Matthews reported a redshift for 3C 48 corresponding to a recessional speed of 110,200 kilometers per second, 37 percent of the speed of light and implying a distance of almost five billion light-years.[26]

While alternate explanations for quasars that placed them much nearer Earth were debated for a while, it was soon established that they are indeed distant objects with enormous, unprecedented energy emission.[27]

Eventually quasars were identified as members of a class of astronomical objects called *active galaxies*. These are galaxies that are far more luminous than normal galaxies with strong emission lines usually associated with a central core as well as radio and x-ray emissions. They often have radio jets thousands of light-years long pointing out from the center. They also often exhibit a highly variable luminosity that can change by a factor of two in a few days, which is very unusual for an object of galactic dimensions. This implies that the sources of the enormous energies of active galaxies are confined to regions of just a few light-days in size, tiny by comparison with the size of a galaxy such as the Milky Way, which is one hundred thousand light-years in diameter. It now appears that these sources are supermassive black holes.

Active galaxies fall into three subclasses:

Seyfert galaxies

These are named after astronomer Carl Seyfert, who first noticed them in 1943. Seyferts are galaxies with very bright nuclei containing broad emission lines of hydrogen, helium, nitrogen, and oxygen. The broadening is interpreted as the Doppler shifts from gases moving at speeds from five hundred to four thousand kilometers per second.

Radio galaxies

These are very bright radio sources emitting huge double-lobed structures of radio emission usually shooting out in opposite directions from an optical core. When the jets point along a line toward Earth so we can't see them, they are called *blazars*. *BL-Lac objects* form a subclass of blazars. As we will see, blazars send to Earth very high-energy gamma-rays and, perhaps, neutrinos.

Quasars

It is now realized that quasars are active galaxies that are so distant that they appear as point sources.

All of the galaxies that form our local group are "normal." The nearest active galaxy is Centaurus A, which is ten million light-years away. Most are quite distant. In fact, the preponderance of active galaxies existed around two billion years after the big bang. Their population was reduced to less than 10 percent by six billion years.[28]

If ever there was conclusive evidence against a steady-state universe, this is it. The universe that we see when we study active galaxies looks quite different than the universe we see when we study nearby galaxies. Apparently, when galaxies first formed they went through a period known as the *bright phase* when they were far more luminous than are the galaxies of today. These earlier galaxies died off while later galaxies formed that were much more benign. The explanation is simple. The early galaxies contained many more giant stars than do current galaxies. These emit more light but are shorter-lived.

PULSARS

Although not of great cosmological significance from the perspective of this book, another unexpected discovery in the 1960s bears mention. In 1967, astronomy graduate student Jocelyn Bell, mentored by Antony Hewish in Cambridge, discovered the first *pulsar*, an astronomical object emitting pulses of radio emission 1.33 seconds apart.[29] Soon many more were observed, some with pulse periods as low as milliseconds, and they were eventually associated with *neutron stars*.

A neutron star is the remnant of a supernova that has blown off most of its mass, leaving behind a highly dense sphere that is mostly composed of neutrons formed by gravitational collapse. Typically it is about 1–3 solar masses with a radius of about twelve kilometers. Its density is comparable to that of an atomic nucleus. If it has a strong magnetic field and is rotating rapidly, it will emit the short period electromagnetic pulses as are observed.

10.

RELICS OF THE BIG BANG

STATIC FROM THE SKY

As we saw in the previous chapter, in 1949 Ralph Alpher and Robert Herman predicted that if the early universe was as hot and dense as implied by the big-bang model, then the universe today should be bathed in thermal radiation that has been cooled to about five degrees Kelvin because of the universe's expansion in the billions of years since. Most physicists and astronomers at the time ignored the prediction. They did not take the big bang seriously, perhaps because of its failure to account for the production of chemical elements beyond helium except for, at most, trace amounts of lithium and beryllium.

Or, as with Fred Hoyle and his collaborators, they may have objected to the implication being drawn by religious apologists, including the pope, that the big bang provided scientific evidence for divine creation.

This lack of enthusiasm for the big bang was further fortified by the remarkable success of Hoyle, William Fowler, Margaret Burbidge, and Geoffrey Burbidge in explaining the formation of the heavier elements in the cores of dying stars. These stars collapse gravitationally after they have exhausted all the hydrogen they use for fuel. The combination of high temperature and high density needed for nucleosynthesis exists in these collapsing cores, but not in the early universe.

Nevertheless, the idea of a cosmic background radiation kept cropping up in the minds of researchers apparently unaware of Alpher and Herman's work. In the early 1960s, Princeton physicist Robert Dicke was investigating the possibility that the universe might oscillate in a series of big bangs and big crunches, which is one of the allowed solutions of Friedmann's cosmological equations. He reasoned that, during the current phase of oscillation, our universe was once much hotter than it is now. In that case, it might generate thermal radiation that has been cooled off since by the universe's expansion.

In 1964, Dicke's former student, James Peebles, estimated that the temperature of this relic radiation should be on the order of ten degrees Kelvin, in the microwave region.

Also at Princeton at the time were Peter Roll and David Wilkinson, who had built a radiometer capable of measuring microwaves of three centimeters' wavelength. So a collaboration was formed to seek evidence for a cosmic microwave background.

However, in an oft-told tale, notably in the bestseller *The First Three Minutes* by Nobel laureate physicist Steven Weinberg,[1] the Princeton physicists were beaten to the punch serendipitously by two researchers working at the Bell Laboratories forty miles away in Holmdel Township, New Jersey. In 1963, radio astronomers Arno Penzias and Robert Wilson had started to work with a highly sensitive, 7.35-centimeter microwave transmitter-receiver with a horn antenna and state-of-the-art cryogenic technology, including a maser amplifier. The instrument was originally designed for communication with the Bell System satellite Telstar. Bell's European partners in the project had developed a similar system, and the Bell instrument was a backup. As it turned out, the backup wasn't needed so Penzias and Wilson could use it for radio astronomy.

The radio astronomers decided to try to measure the radiation intensity from our galaxy outside the galactic plane, that is, away from the Milky Way. In the spring of 1964, they detected an excess of static that could not be blamed on their antenna and electronic circuits. They did everything they could to find an instrumental source of the noise, including cleaning up "white dielectric material" left by pigeons on the antenna, and concluded the static was from an external source. Since the signal was coming in uniformly from all directions, this ruled out Earth's atmosphere or the Milky Way as sources.

Radio engineers express noise at a given wavelength in terms of the temperature of a blackbody that would give the same amount of noise. The Penzias and Wilson results were equivalent to 3.5 degrees Kelvin (3.5 K) antenna temperature.

By means of the grapevine, the two New Jersey groups found out about each other and began to communicate. In 1965, Penzias and Wilson presented their observations in a paper in *Astrophysical Journal* with the unassuming title, "A Measurement of Excess Antenna Temperature at 4080 Mc/s."[2] As to the interpretation of the data, the authors referred to the preceding article in the same issue by Dicke, Peebles, Roll, and Wilkinson titled "Cosmic Black-Body Radiation."[3] Penzias and Wilson would receive the 1978 Nobel Prize for Physics.

It did not take the scientific community long to recognize the significance of what would eventually amount to the most important cosmological discovery since the redshift of galaxies. However, more measurements were needed to confirm that what was being detected was in fact thermal radiation. Only one wavelength had been observed. In early 1966, Roll and Wilkinson reported observations at 3.2 centimeters, confirming that microwave radiation is in fact out there. But more was needed.

THE SPECTRUM

The Planck spectrum for blackbody radiation at 3 K is shown in figure 10.1. Note that the shape is different from figure 6.1 since that plot was made against frequency whereas here the plot is against wavelength. Also, here both axes are logarithmic.

Several points are to be noted: The measurements by the two New Jersey teams fall in what is called the Rayleigh-Jeans portion of the spectrum, which is given by classical wave theory described in chapter 5. To be certain they were seeing blackbody radiation, it was necessary to map the full spectrum. However, while the atmosphere is relatively transparent to microwaves, it becomes increasingly opaque as we move down into the infrared region of the spectrum — just where at 3 K the deviation from classical wave theory takes place and the spectrum falls off because of quantum effects.

As far back as 1941, two optical astronomers, Walter Adams and Andrew McKellar, observed a splitting into three lines of an interstellar absorption line identified with the molecule cyanogen (CN). McKellar estimated that two of the split lines were the result of some unknown perturbation that corresponded to a temperature of 2.3 K. After the discovery of the microwave background, it was realized that this splitting was caused by blackbody photons with a wavelength of 0.263 centimeters, right at the peak of the Planck spectrum in figure 10.1.[4]

The big bang was back in business. The remnant of the explosion that produced the universe presumably had been detected. Of course, there were still a few doubters. Fred Hoyle and his loyal band of steady-staters did their honest best to play devil's advocate in seeking other explanations for the observed radiation. But these were all ad hoc and lacked empirical support. Gradually, the steady-state model lost its remaining credibility, although Hoyle and Geoffrey Burbidge remained unconvinced that the big bang had been confirmed. They demanded even more evidence.

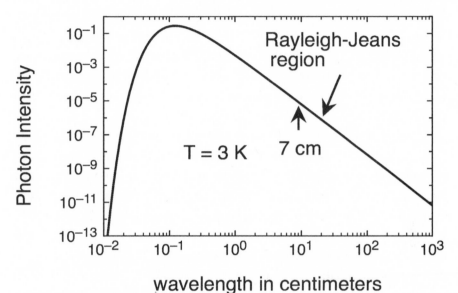

Figure 10.1. The predicted Planck blackbody spectrum at 3K. The vertical scale is given in photons per cubic centimeter volume per centimeter wavelength interval. The linear portion is the classical Rayleigh-Jeans spectrum that falls off as the fourth power of wavelength. The Penzias-Wilson measurement at 7.35 centimeters is indicated. The Roll and Wilkinson measurement at 3.2 centimeters is just below that. Neither covered the quantum region. Image by the author.

BIG-BANG NUCLEOSYNTHESIS REBORN

Back in 1953, Alpher, Herman, and James Follin had published a lengthy paper detailing what the best physics knowledge at the time had to say about the conditions of the early universe.[5] Weinberg called it "the first thoroughly modern analysis of the early history of the universe."[6] However, the authors did not consider nucleosynthesis nor mention microwave radiation.

With the discovery of the cosmic microwave background (CMB) in 1965, the big bang moved to the front burner and new, independent calculations of big-bang nucleosynthesis were performed by Yakov Zel'dovich in Russia, Hoyle and Roger Taylor in England, and Peebles in the United States. (Another little insight into how science works: Even though Hoyle bitterly opposed the big bang, he was not beyond making serious, unbiased calculations assuming the model).

While by then it was clear that the elements above helium are largely produced in stars, these represent only about 1 percent of the total mass of atoms in the universe. The rest is about 75 percent hydrogen and 25 percent helium, and Hoyle admitted that the conditions in stars were not such as to produce a 25 percent abundance of helium. Still, he was not ready to accept the big bang and continued to look for other explanations. But a new generation of cosmologists would prove beyond a reasonable doubt that the big bang happened.

DAVID SCHRAMM: THE GENTLE GIANT OF COSMOLOGY

In *The First Three Minutes*, Weinberg makes a diversion, spending a chapter trying to understand why the discovery of the CMB took so long and happened by accident, when the knowledge and technology were there earlier.[7] One reason he mentions is the lack of communication between theorists working on the big-bang model and radio astronomers. We saw a similar situation earlier with Edwin Hubble and other astronomical observers having little knowledge of relativistic cosmology, while relativistic cosmologists such as Alexander Friedmann paid no attention to observations. Georges Lemaître was an exception. Although he made no observations himself, at least he related his theories to the data.

In the exciting physics era in which I participated that flourished in the 1960s and 1970s, I saw quite the opposite. By then, interdiscipline communications had greatly improved. Although we still did not have the Internet, we had phones, faxes, photocopiers, and jet liners that enabled us to meet and talk to one another. Theorists and experimenters went to each other's seminars, ate lunch together, and shared the same coffee pot in the lab. In the two sabbaticals I spent at the Nuclear Physics Laboratory at Oxford University, everybody stopped work once in the morning for coffee and once in the afternoon for tea (best and cheapest tea in town), sitting together at low tables in a big common room, talking the latest physics. Often a distinguished visitor would join us.

By 1979, the lines between areas of expertise in physics and astronomy had also begun to crumble. One of the primary creators of a new way of doing science was a larger-than-life, six-foot-four, 240-pound, redheaded mountain-climbing, Olympic-trials Greco-Roman wrestling finalist named David Norman Schramm, to whom this book is dedicated. Schramm received his doctorate in 1971 from Caltech under Willy Fowler, the future Nobel Prize winner who, as we have seen, was one of Fred Hoyle's collaborators on stellar nucleosynthesis.

Joining the faculty of the University of Chicago in 1974, Schramm was chair of the Department of Astronomy and Astrophysics from 1978 to 1984 and became Vice President for Research in 1995. In Chicago, Schramm was the mentor and leader of a team composed of new brand of theoretical physicist, one who has mastered nuclear physics, elementary particle physics, astrophysics, and relativistic cosmology. Members of this team would put this combined knowledge to work in providing a remarkably detailed picture of the early universe and see it successfully tested against the increasingly precise astronomical data that was coming in, especially from space telescopes. A new field was born called *particle astrophysics*, and David Schramm was its father.

With the help of the Leon Lederman, another future Nobel Prize winner and director of Fermi National Laboratory (Fermilab) near Chicago, in 1982 Schramm organized a particle-astrophysics center at the lab that would, working with the University of Chicago, pioneer much of this remarkable work. I got to know Schramm and his team when I became involved in very high-energy gamma-ray and neutrino astrophysics after working on neutrino experiments at the Fermilab accelerator. He was always kind, patient, and good-humored, and both his public and technical lectures were joys to experience. I considered him a friend; but then he had hundreds.

Schramm had bought a house in Aspen, Colorado, in 1980 and became involved with the Aspen Center for Physics, which had regular summer workshops that I attended on occasion. He usually flew his own plane, incorporated as Big Bang Airlines, between there and Chicago. (He was a licensed commercial pilot.) On December 19, 1997, he was on his way to Aspen for the holidays when his plane stalled and crashed just after refueling in Denver. Dave was just fifty-two years old. Lederman said he was in heaven having a "terrific argument with God about what happened before the big bang."

I borrowed the title of this section from an essay in the *New York Times* by Dennis Overbye.[8] A loving but still highly informative biographical memoir was written by one of Dave's outstanding postdocs, Michael Turner. Now possessor of an endowed chair at the University of Chicago, Turner became a major leader in particle astrophysics and was the 2013 president of the American Physical Society.[9] Turner worked closely with Edward ("Rocky") Kolb at Fermilab and Chicago, and in 1990 they coauthored the classic monograph on particle astrophysics called *The Early Universe*.[10]

Schramm and collaborators demonstrated that rare processes seen only in elementary particle physics play a crucial role in both the current and early universe. For example, in 1975 they showed that the just-discovered *weak neutral current* interaction is involved in the collapse of massive stars that produce supernova explosions.[11]

Schramm always had an interest in neutrinos, which was also my main field of study in those days, so I followed his work closely. The observation by two underground experiments in 1987 of neutrinos from a supernova (SN1987A) in the Large Magellanic Cloud confirmed the important role that these elusive particles play in supernovae.

In a significant paper published in 1977, Schramm, Gary Steigman, and James Gunn showed that cosmology places a limit on the number of types of leptons.[12] *Lepton* is the generic name given to the negatively charged electron, e, and its two heavier companions, the muon, μ, the tauon, τ, and their associated neutrinos, v_e, v_μ, and v_τ. Each lepton is accompanied by an antiparticle. At this time, these three "generations" of leptons, along with related generations of quarks, were established and physicists had no reason to think there might not be more.

The 1977 paper by Schramm and collaborators showed that additional neutrinos would speed up the production of He^4 and placed the limit at five neutrino types based on then-measured helium abundance. By 1989, the abundance measurements had improved sufficiently to place the limit at three generations. This agreed with measurements of missing energy in colliding-beam accelerator experiments. More neutrinos means more missing energy, that is, energy not accounted for in the detected particles. The amount missing was consistent with three generations and no more.

And so the standard model of elementary particles and forces, which will be discussed in the next chapter, settled on a scheme in which both quarks and leptons are found in just three generations. This was just as Schramm had anticipated. Who would ever have dreamed, except such a dreamer, that cosmology would tell us fundamental facts about the nature of matter on the smallest scale? And that process is continuing to the present day and is likely to carry on for years.

TEMPERATURE IN THE EXPANDING UNIVERSE

Before we delve into the details of big-bang nuclear physics, it will help to get an idea of the energies involved at various stages in the universe's history as they apply not just to nuclear physics but also to the physics at each stage.

Although the universe is expanding very rapidly, the particles present at early times were interacting with one another even more rapidly so that they still thermalized sufficiently to be in a state of *quasi equilibrium*. This means that the particles can be described as having an absolute temperature T that nevertheless decreases with time as the universe expands.

Most authors writing on this subject will give you the values of the temperatures at various stages in degrees Kelvin, presumably because they figure that these units are more familiar to the reader. But the actual temperature numbers in the early universe are so high as to not be very meaningful by our experience.

More informative is the average kinetic energies of the particles in the universe at any given time, which to an accuracy sufficient for our purposes is given by $K = k_B T$ where k_B is Boltzmann's constant. That is, the temperature of a body is simply the average kinetic energy of the particles in the body. Since k_B is an arbitrary constant that just converts degrees Kevin to energy units, we can set $k_B = 1$ and measure temperature in energy units.

When we are dealing with atomic, nuclear, and subnuclear processes, the most useful unit of energy and temperature is the electron-volt (eV), which is the kinetic energy an electron gains when it falls through an electric potential of one volt. Atomic processes are characterized by energies of a few eV or keV, where 1 keV = 1,000 eV. Nuclear processes are characterized by energies in the MeV region, where 1 MeV = 1 million eV. Subnuclear processes occur in the GeV and TeV regions, where 1 GeV = 1 billion eV (10^9) and 1 TeV = 1 trillion eV (10^{12}).

Note that colliding-beam accelerators enable us to study the physics of the very early universe. For example, when the Large Hadron Collider (LHC) is upgraded to a total energy of 14 TeV in 2015, it will enable physicists to measure the properties of matter 10^{-15} second after the big bang when the temperature was that high.

Figure 10.2 plots the average kinetic energy or temperature of the universe from 10^{-43} second, the Planck *time*, until the present. We will have much to say later about the Planck time and what might have gone on before it. For now, let's begin things at that point.

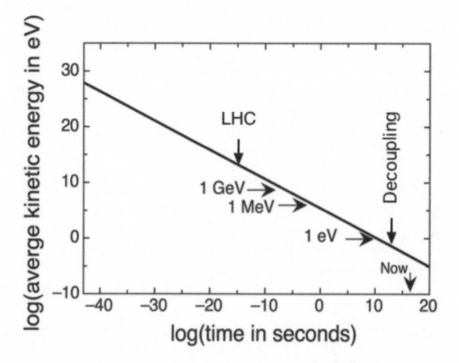

Figure 10.2. Average kinetic energy of the particles of the universe as a function of time after the big bang, on a log-log plot. This can also be considered a plot of the absolute temperature against time, with the temperature in electron volts (eV). Image by the author.

Until 380,000 years after the big bang, all the particles in the universe were in quasi equilibrium at the same decreasing temperature as the universe expanded and cooled. At that time, called *decoupling*, atomic matter fell out of equilibrium while photons and neutrinos remained in quasi equilibrium. Don't let the log scale fool you. The time between decoupling and now is still 13.8 billion years for all practical purposes.

As the universe expands and cools, various particles drop out of equilibrium. Let me illustrate this with the example of antiprotons. They collide with protons and the two disintegrate into photons and other lighter particles. Consider annihilation into photons. The reaction is

$$p + \bar{p} \rightarrow \gamma + \gamma$$

where $\bar{p}$ is an antiproton and γ is a photon. The photons will carry off the rest energy of the proton and antiproton, plus whatever their initial kinetic energies may be. The reverse reaction can also occur and regenerate antiprotons

$$\gamma + \gamma \rightarrow p + \bar{p}$$

However, since the photons have zero rest energy, their total kinetic energy must at least equal the total rest energy of the proton and antiproton, which is 1876 MeV. So, as long as the temperature of the universe is more than this, antiprotons and protons will remain in equilibrium with about the same number of each. However, once the temperature of the universe drops below 1876 MeV, which happens when it is about 10^{-7} second old, photons no longer have sufficient energy to produce proton-antiproton pairs and the two will gradually decrease in number from the mix.

Now, as we will see in chapter 11, a slight asymmetry exists between matter and antimatter so when the antiprotons have all annihilated, some protons are left over. These amount to one in a billion of the photons, electrons, positrons, and neutrinos that remain. If it were not for this asymmetry, all the protons would have annihilated, leaving nothing to make the atoms of stars, planets, and you and me.

Just as antiprotons disappeared from the early universe when the temperature dropped below the value needed to regenerate them, so did the positrons after the universe had cooled further. Let's consider the analogous process, electron pair annihilation into photons:

$$e^+ + e^- \rightarrow \gamma + \gamma$$

In order to regenerate positrons, the reverse reaction must occur:

$$\gamma + \gamma \rightarrow e^+ + e^-$$

Here the photons must have a total energy at least equal to the total rest energy of the positron and electron, which is 1.022 MeV. When the temperature of the universe dropped below this value, which happens when it is about 0.15 second old, photons no longer had sufficient energy to produce the pair and the positrons annihilated. As was the case for protons, one in a billion electrons were left over because of particle-antiparticle asymmetry. Eventually, but not for another 380,000 years, those electrons combined with the protons to form hydrogen atoms. Before that happened, however, the light nuclei formed.

THE LIGHT NUCLEI

He^4 was not the only light nucleus formed in the big bang. In fact, substantial amounts of H^2 (the deuteron), H^3 (the triton), and He^3 were produced, as well as smaller amounts of Li^7, Be^7, and Li^6. In the 1970s, Schramm and his growing stable of nuclear and particle astrophysicists, joined by others worldwide, began an intense program of calculating the primordial abundances of the light elements and comparing them with the data. They obtained remarkable agreement. This program continues to the present day and has been especially advanced by the accompanying great strides in observations.[13]

To form nuclei we need neutrons. The neutron is 0.782 MeV more massive than the proton and is produced by the following weak interactions:

$$e^- + p \leftrightarrow v_e + n$$
$$\bar{v}_e + p \leftrightarrow e^+ + n$$

where v_e and $\bar{v}_e$ are the electron neutrino and electron antineutrino respectively. We will discuss weak interactions and neutrinos, and other fundamental particles in the next chapter. Note that I have used double arrows to indicate that these reactions are reversible.

Because the total mass (rest energy) on the right side of the reactions is greater than that on the left by 0.271 MeV and 1.293 MeV respectively, neutron production for each reaction ceased when the average kinetic energy of the universe dropped below these values. The second reaction, with the higher energy difference, dropped out first at about 0.1 second, while the first reaction continued to produce neutrons until about 2 seconds. After that, the number of neutrons began to diminish to about one-sixth the number of protons as they converted to protons by way of beta decay:

$$n \rightarrow p + e^- + \nu_e$$

The mean lifetime of the neutron is about 880 seconds, with the exact value still in dispute. Big-bang nucleosynthesis is very sensitive to this number.

With the temperature now below 1 MeV, nuclei were able to form without being immediately split apart by the many high-energy photons swarming about. By then, all the positrons had annihilated, as described above, and the neutrinos (and antineutrinos) no longer had anything to do and so turned into a relic thermal cloud—like that of photon background that would form much later. Today this cloud forms the cosmic neutrino background (CvB) with a temperature of 1.95 K, and with little hope of being directly detected in the foreseeable future.

Now let's see how the lighter nuclei form. A proton and neutron can collide to produce a deuteron and photon:

$$p + n \rightarrow H^2 + \gamma$$

At first, the lightly bound deuterons were split by the reverse of the above reaction. But when things cooled sufficiently, two deuterons could stick around long enough to produce a neutron and He^3,

$$H^2 + H^2 \rightarrow He^3 + n$$

or a triton and a proton,

$$H^2 + H^2 \rightarrow H^3 + p$$

He^4 was then produced by

$$H^2 + H^3 \rightarrow He^4 + n$$

or by

$$H^2 + He^3 \rightarrow He^4 + p$$

while Li^7 was produced by

$$H^3 + He^4 \rightarrow Li^7 + \gamma$$

and Be^7 was produced by

$$He^3 + He^4 \rightarrow Be^7 + \gamma$$

And so on. This is not a complete list of reactions, but it should give the idea.

Note in all these reactions, both the atomic number (given by the elemental symbol) and nucleon number are conserved. The first is just charge conservation. The second is a special case of the more general conservation of baryon number that we will discuss later.

The time evolution of the mass fraction of the various light elements relative to protons is shown in figure 10.3. This figure is from Edward Wright's online cosmological tutorial[14] and is based on the paper by Burles, Nollett, and Turner.[15] As we see, the peak in production is at about two hundred seconds and most abundances have leveled off by about one thousand seconds. Li^6 makes just a brief appearance and neutrons fade rapidly as they are used to form nuclei or decay. Only He^4 is produced in any substantial quantity.

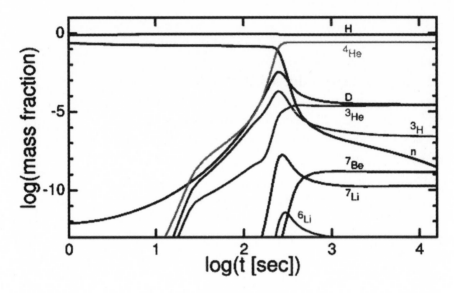

Figure 10.3. The mass fraction of nucleons and nuclei relative to protons in the early universe as function of time. Image courtesy of Edward L. Wright.

Further nucleus building was cut off by the absence of stable nuclei with five or eight nucleons. As we have seen, the heavier nuclei would later be formed in the higher density and temperature of collapsing stars.

The conventional model of big-bang nucleosynthesis used by most nuclear cosmologists depends on only a single parameter, η, the ratio of baryons to photons, which is on the order of 10^{-9}. *Baryon* is a generic term in particle physics for a certain class of particles that includes protons and neutrons (see chapter 11). At this stage of the early universe, these and the nuclei made from them are the only baryons that were present.

The He^4 abundance, about 25 percent of protons by mass, depends only weakly on the conditions of the early universe. This is why even the earliest estimates, made when these conditions were poorly known, were in the right ballpark. On the other hand, the remaining light nuclei, especially deuterons (H^2) depend very sensitively on the baryon mass density, ρ_B, which at this time was simply the nucleon density.

The baryon density is usually expressed as the ratio $\Omega_B = \rho_B / \rho_c$, where ρ_c is the *critical density*, the average density of the universe when it is exactly balanced between positive kinetic energy and negative gravitational energy. The current best value is $\rho_c = 9.467 \times 10^{-30}$ grams per cubic centimeter. In the Friedmann model described in chapter 8, this is the situation where the curvature parameter $k = 0$ and the universe is Euclidean, although, as we will see, $k = \pm 1$ are not ruled out.

In figure 10.4, the theoretical and measured abundances of the elements are given by their numbers as a fraction of the number of protons. The bands represent the measured quantities with the widths of the bands indicating measurement uncertainties.[16] Rather than look at old theories and data, this graph shows the latest as of this writing, where data from the Wilkinson Microwave Anisotropy Probe (WMAP) satellite have enabled a significant improvement over previous results.[17] While even more precise results from the Planck satellite are now coming in, the WMAP data suffices for our purposes.

The abundances are plotted as a function of $\Omega_B h^2$, where h is a dimensionless factor that allows for possible changes in the empirical value of the Hubble constant H_0 (h here is not to be confused with Planck's constant). That is, cosmologists write $H_0 = 100h$ kilometers per second per megaparsec. The current estimate is $h = 0.71$.

Estimating the primordial abundances of elements is no easy task. One has to rely on measurements made in the current universe and then figure out how much is primordial.

He^4 is also produced in stars in the basic fusion processes going on at their cores, but the stars don't disperse any unless they explode into supernovae, and this only happens for the heaviest stars. He^4 is observed in hot, ionized gas in other galaxies and in so-called metal-poor stars, where astronomers use the term "metal" to refer generically to any element above helium, so this is likely mostly primordial. Some disagreement still exists on the exact fraction of He^4 relative to protons and calculations are continuing to improve.[18] Indeed, as with the constraint placed by cosmology on the number of neutrino species mentioned earlier, the helium abundance is sensitive to the exact lifetime of the neutron so, once again we see the important role subatomic physics plays in cosmology and vice versa.[19]

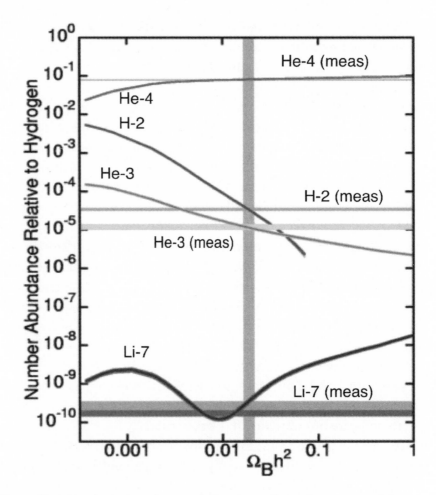

Figure 10.4. Number abundance relative to protons of various nuclei as a function of the baryon density. The bands show the latest measured quantities. The widths of the bands indicate the measurement errors. Image courtesy of Edward L. Wright.

Deuterium, H^2, is a very loosely bound nucleus composed of a proton and a neutron; it is easily destroyed by many astrophysical processes. The best recent estimate of its primordial abundance comes from observing the absorption lines in very distant intergalactic clouds where the sources are quasars.

Li^7 is both produced and destroyed in stars. Its primordial abundance has been estimated from its abundance in the atmospheres of the oldest stars in the halo of our galaxy, which are believed to not have depleted much of their lithium.

Stars burn primeval H^2 into He^3 but measurements indicate that the sum of the two is relatively constant so the He^3 abundance is estimated by subtracting out the H^2 abundance estimated elsewhere.

As we can see from figure 10.4, the agreement of the big-bang nucleosynthesis model with the data is truly remarkable. The primordial abundances of four nuclei are precisely calculated from a single parameter, the baryon density. All except He^4 are highly sensitive to that parameter, although the correct abundance is calculated. All five quantities are in complete agreement with the data.

Thanks to David Schramm and his students and collaborators, the big-bang model is strongly validated by these results. None of the alternatives that are still being tossed around at this time even come close to matching this result. Indeed, they have no idea how to do it. Let's face it. The big bang happened.

ON TO ATOMS

Thirty minutes after the universe first appeared, things quieted down. At that time the temperature was three hundred million degrees and the average kinetic energy about 25 keV and falling. Nuclear processes stopped because the temperature was now too low. Electrons, which were once as common as photons, almost completely annihilated with positrons except for one in a billion left over as a result of a slight asymmetry between matter and antimatter. This asymmetry, without which we would not have the universe as we know it is yet to be completely understood. It will be discussed further in chapter 11.

At this point, most of the universe (excluding dark matter) was now photons (69 percent) and neutrinos (31 percent), with a billion times fewer protons, electrons, and He^4 nuclei, and even much smaller numbers of deuterons, He^3, Li^7, and Be^7. Neutrons have disappeared, either by beta decay or by being absorbed into nuclei.

It is no coincidence that the number of protons (plus other nuclei) and electrons match, although some religious apologists have claimed this is yet another fine-tuning coincidence arranged by God.[20] In fact, the equality follows from conservation of electric charge, on the reasonable assumption, verified by observations, that the total charge of the universe is zero.[21]

The period of the universe we have described so far is called the era of *radiation dominance* because photons continue to outnumber everything else, except neutrinos. Unlike neutrinos, these photons were still interacting importantly with the rest of matter.

A note of clarification: Cosmologists make a distinction between what they call "radiation" and what they call "matter." A particle is called *radiation* when its kinetic energy is much greater than its rest energy, so that it travels at or very near the speed of light. In that case, its motion must be described by the kinematics of Einstein's theory of special relativity. Since the highest mass particles in the early universe are nuclei with rest energies less than a few GeV, everything was radiation until about 10^{-10} second, according to figure 10.2.

Since it has zero rest energy, a photon is always radiation. A neutrino is radiation until its kinetic energy drops below its rest energy, which is no more than on the order of 0.1 eV, five million times smaller than the electron's mass (see chapter 13).

When a particle's rest energy is much greater than its kinetic energy, it is called *matter*, although even photons and other stuff termed radiation are fully material so the distinction is quite misleading. These particles haves speeds much less than c.

Once the temperature of the universe has dropped below about 100 MeV, which happens when the universe is about 10^{-5} second old, the kinetic energies of particles such as protons and nuclei become much lower than their rest energies and can be adequately described by nonrelativistic Newtonian kinematics.

For thousands of years, radiation in the form of photons would continue to dominate the universe. As mentioned, this era is referred to as radiation dominance. However, the energy density of radiation (all kinetic energy) falls off as $1/a^4$, where a is the scale factor of the universe, while the energy density of matter (all rest energy) falls off as only $1/a^3$. So, by about seventy thousand years after the big bang, the energy density of matter caught up to and exceeded the energy density of radiation and the universe entered the era of *matter dominance*.

Up until this time, the universe was opaque because photons had all these electrically charged electrons and nuclei with which to interact. An observer inside the universe (obviously not human) would not be able to see by means of light at any wavelength since photons did not go very far, as in a thick fog. Then, when the temperature had cooled to a few thousand degrees and the kinetic energy dropped to a few tenths of an electron volt, electrons and nuclei started sticking together into atoms.

This process is called *recombination*, which is silly because nuclei and electrons were never combined into atoms in the first place. But, that's what it is called in chemistry, where atoms usually come first. In any case, the oppositely charged particles all combined into neutral atoms (remember there were equal numbers) and the photons no longer had anything but themselves with which to interact. This important moment in history, which occurred 380,000 years after the big bang, we have seen is called *decoupling*. The universe became transparent and the photons formed a thermalized cloud that, in another 13.8 billion years would cool to 3 K and form the cosmic microwave background.

11.

PARTICLES AND THE COSMOS

THE VISIBLE UNIVERSE

Astronomy in the 1970s was headlined by the space program, with the first manned moon landing by Apollo 11 on July 20, 1969, followed by five more, ending with Apollo 17 on December 11, 1972. The unmanned vehicles Voyager 1 and 2, launched in 1977, visited Jupiter and Saturn before continuing into the outer solar system and are now entering interstellar space. In 1974, Mariner 10 passed Venus and visited Mercury. In 1976, the Viking spacecraft landed on Mars.

The International Ultraviolet Explorer (IUE) was launched in 1978 to observe astronomical objects in the UV region of the spectrum, which is impossible from Earth because of atmospheric absorption. Lasting almost eighteen years, IUE made over 104,000 observations of every kind of object, from planets to quasars.

Three satellites in the NASA High Energy Astronomy Observatory program explored three additional new windows on the cosmos: x-rays, gamma rays, and cosmic rays. HEAO 1, launched in 1977, surveyed the x-ray sky and discovered 1,500 sources. HEAO 2, launched the following year and renamed the Einstein Observatory, was an x-ray imaging telescope that discovered several thousand more sources and pinpointed their locations. The principal investigator of the Einstein project, Riccardo Giacconi, had previously led a team that in 1962 discovered a strong x-ray source that was designated Scorpius X-1 and was later identified as a neutron star. Its x-ray emission is ten thousand times as powerful as its optical emission. Giacconi would receive the 2002 Nobel Prize for Physics. HEAO 3, launched in 1979, measured the spectra and isotropy of x-ray and gamma-ray sources and determined the isotopic composition of cosmic rays.

On the ground, new large reflecting telescopes materialized on mountaintops in Arizona, Chile, Australia, Hawaii, and Russia. The photographic plate was gradually replaced by the *charge-coupled device* (CCD) as the main detector, greatly improving photon-collection sensitivity and efficiency while enabling automatic digital readouts. The new high-speed digital computers processed data rapidly and in large quantities, and provided automatic control of the mirrors. No longer would astronomers have to spend long, cold hours in the cages of their telescopes, manually keeping them pointed at their targets.

By decade's end, the collection area of telescopes was increased above what was possible with a single mirror by connecting many mirrors together with computers synchronizing the light gathered. The first such device, called the Multi-Mirror Telescope (MMT), was operating on Mount Hopkins in Arizona at the time I was collaborating on another project on the mountain measuring very high-energy gamma rays.

During my years at the University of Hawaii I witnessed the steady installation of international telescopes on Mauna Kea, the 13,796-foot mountain on the Big Island of Hawaii, creating the finest ground-based astronomical site in the world. Because of its high altitude with exceptionally dry air above, Mauna Kea not only is a superior location for observations in the visible band but also enables the infrared sky to be explored as well.

I have no need to catalog, nor can I give justice to, the spectacular observations made with these wonderful instruments. The photographs that fill astronomy books and NASA's web sites demonstrate that nature can compete with any human art form or religion in creating awe and beauty. For my purposes at this point, it suffices to say that the contrast between luminous matter in the universe and the CMB could not be starker. The optically visible universe is complex, varied, and irregular. To one part in one hundred thousand, the CMB is simple, smooth, and regular. To that level of precision, it requires only one number to describe—its temperature, 2.725 K. Yet it would turn out that observations of the tiny CMB deviations from smoothness would tell us how all these complex wonders came about.

THE STRUCTURE PROBLEM

Long before the discovery of the CMB, astronomers puzzled over the origin of the structure in the universe. The prominent British physicist and astronomer James Jeans had worked out the mechanism by which a smooth cloud of gas will contract under gravity to produce a dense clump. He derived an expression for the minimum mass at which gravitational collapse overcomes the outward pressure of the gas. Called the *Jeans mass*, it depends on the speed of sound in the gas and the density of the gas.

The Jeans mechanism worked fairly well for the formation of stars, but not for galaxies. In 1946, Russian physicist Evgenii Lifshitz extended Jeans's calculation to the expanding universe and proved that gravitational instability alone is unable to account for the formation of galaxies from the surrounding medium.[1] Basically, expansion combines with radiation pressure to overwhelm gravity. The failure to understand galaxy formation remained a bugaboo in astronomy until the 1980s.

When the decade of the 1970s began, a number of authors proposed that fluctuations in the density of primordial matter in the early universe acted to form the galaxies. Since the pressure of a medium is related to its density by an *equation of state*, density fluctuations produce pressure fluctuations that are just what we call sound. You often hear it said that "big bang" is a misnomer, since explosions in space make no noise. Well, in fact, the big bang did make an audible bang.

As first pointed out by Pythagoras, the sounds from musical instruments can be decomposed into harmonics where each harmonic is a pure sound of a single frequency or pitch. The same is true for any sound, although the harmonics are usually not so pure as with musical instruments. The distribution of sound intensity among the various frequencies is given by its *power spectrum*.

A mathematical method called *Fourier transforms*, developed by the great French mathematician Jean Baptiste Fourier (1768–1830), is widely used by physicists and engineers in many applications besides sound. The Fourier transform enables one to take any function of spatial position or time and convert it to a function of wavelength or frequency. If the function is periodic in space or time, the spectrum contains peaks at specific wavelengths or frequencies.

In the 1970s, Edward Harrison[2] and Yakov Zel'dovich[3] independently predicted that the spectrum of sounds produced by density fluctuations in the universe should exhibit a property known as *scale invariance*. In general, scale invariance is a principle that is applied in many fields from physics to economics. It refers to any feature of a system that does not change whenever the variables of the system are changed by the same factor. For example, Newton's laws of mechanics do not change when spatial units are converted from feet to meters. Scale invariance is another symmetry principle.

But scale invariance does not always apply. Given similar structure and biology, the height an animal can jump is almost independent of how big the animal is. That is, it does not scale. This is known as *Borelli's law*, proposed by Giovanni Alfonso Borelli (1608–1679). In his 1917 classic *On Growth and Form*, D'Arcy Wentworth Thompson wrote, "The grasshopper seems as well planned for jumping as the flea . . . but the flea's jump is about 200 times its own height, the grasshopper's at most 20–30 times."[4]

Although they put it in more technical terms in their papers, Harrison and Zel'dovich basically pointed out that the density fluctuations inside the horizon of the universe must be independent of the scale of the universe as the universe expands. If the density fluctuation were greater at either an earlier time or at a later time, parts of the universe would collapse into black holes.

The Harrison-Zel'dovich power spectrum is expressed in terms of the *wave number* (also called the *spatial frequency*) $k = 2\pi/\lambda$, where λ is the wavelength. (Do not confuse this k with the cosmic curvature parameter k). The power spectrum is assumed to be proportional to k^n, where n is the *spectral index*. Scale invariance requires $n = 1$.

Now, how could we expect to "listen" to those primordial sounds? In 1966, after the discovery of the cosmic microwave background, Ranier Sachs and Arthur Wolfe showed that density variations in universe should cause the temperature of the CMB to fluctuate across the sky as photons traveling through higher gravitational potentials are redshifted and those moving through lower potentials are blueshifted.[5]

Sachs and Wolfe weren't thinking of primordial fluctuations, but in one of the most remarkable achievements in scientific history, it would turn out that the CMB would be able to read those primordial fluctuations all the way back to when the universe was only 10^{-35} second old and observe exactly how the galaxies and other lumps of matter came together billions of years later as a consequence of these fluctuations. It was estimated that to account for galaxy formation the fractional variation from temperature of the currently observed radiation must be at least $\Delta T/T = 10^{-5}$.[6]

GRAVITATIONAL LENSING

One of the dramatic predictions of general relativity was the deflecting of light by the gravitational field of the sun. In 1936, Einstein pointed out that light bent by astronomical bodies could produce multiple images. In 1937, Fritz Zwicky suggested that a cluster of galaxies could act as a *gravitational lens*. However, it was not until 1979 that astronomers at the Kitt Peak Observatory in Arizona observed the effect. They photographed what appeared to be two quasars unusually close to one another with the same redshift and spectrum, which indicated they were actually the same object. Many examples of lensing have since been identified.

In 2013, a telescope at the South Pole called, appropriately enough, the South Pole Telescope, observed a statistically significant curl pattern called *B*-Mode in the polarization of the CMB that results from the lensing of the intervening structure in the universe.[7] This observation was confirmed in 2013 and 2014 by an experiment in Chile called POLARBEAR.[8] We will talk more about gravitational lensing in chapter 14 and also discuss the latest results from the South Pole on gravitational waves, which also involved the detections of *B*-mode polarization in the CMB.

THE INVISIBLE UNIVERSE

We have seen how astronomers in the 1930s discovered that there was a much larger amount of matter in the universe than could be accounted for by the luminous matter—stars and hot gas—in galaxies. Observations just did not agree with Newton's laws of mechanics and gravity, but few were ready to say these were in any sense invalidated. Fritz Zwicky had dubbed this invisible gravitating stuff *dunckle materie*, or dark matter.

Not much was done on the question until the 1970s when radio astronomers in Groningen, Netherlands, studied the 21-centimeter hyperfine spectral line emitted by neutral hydrogen molecules from various galaxies. Their measurements showed flat *rotation curves* for a large sample of galaxies.[9] In a rotation curve, one graphs the rotational speeds of stars, which Doppler-shifts the observed spectral line, against their distances to the center of their galaxies. By Newton's laws, these should fall off for stars farther from the center, just as the speeds of the planets in the solar system fall off with distance from the sun, where most of the mass of the solar system is located. Instead the speeds remained largely constant.

The explanation for this observation was that the galaxies have halos of unseen dark matter that extend well beyond the denser regions nearer the centers that give off light. This invisible matter is hardly negligible. In fact, we now know it constitutes 90 percent of the masses of the galaxies studied. As we will see in a later chapter, the gravitational lensing described in the previous section has been used to provide direct evidence for the existence of dark matter.

In the meantime, American astronomer Vera Rubin and her collaborators were making a systematic study of the rotations of spiral galaxies in the optical spectrum and observing the same effect.

Astronomers were well aware that many astronomical bodies do not directly emit much, if any, detectable light—for example, planets, brown dwarfs, black holes, neutron stars. However, it seemed unlikely that these were sufficient to account for the total mass inferred from Newtonian dynamics.

Furthermore, there was independent evidence that most of the dark matter could not be composed of familiar atomic matter but must be something not yet identified. This evidence came from the same source that we saw in chapter 10 provided strong confirmation of the big bang—big-bang nucleosynthesis.

In figure 10.4 we compared the theoretical and measured abundances of light nuclei as a function of Ω_B, the ratio of the baryon density to the critical density. While the precise numbers are still changing, the latest measurements indicate that Ω_B is less than 5 percent while 26 percent of the mass of the universe is dark matter that cannot be composed of familiar atoms.

THE RISE OF PARTICLE PHYSICS

The discovery of the CMB in 1964 coincided with the rise of the new field of elementary particle physics in which I participated, first as a graduate student at UCLA (University of California at Los Angeles) and, after receiving my doctorate in 1963, as physics professor for thirty-seven years at the University of Hawaii with visiting professorships at the universities of Heidelberg, Oxford, Rome, and Florence. Eventually particle physics would play an important role in cosmology, so let me redirect our attention for a moment from the very big to the very small.

Using particle accelerators of ever-increasing energy, and particle detectors of ever-increasing sensitivity, a whole new subatomic world was uncovered that culminated in the 1970s with the development of the *standard model of elementary particles and forces*. A place was found in this model for all these particles, which successfully described how they interact with one another.

On April 10, 2014, as this book was in production, the CERN laboratory in Geneva (European Organization for Nuclear Research) confirmed with high significance the existence of an "exotic" negatively charged particle called Z(4430), reported earlier by another collaboration. Stories in the media suggested a violation of the standard model. However, this is not the case. The Z(4430) is evidently composed of four quarks, the first of its kind. However, this no more refutes the standard model than the existence of the helium nucleus with four nucleons refutes the nuclear model.

Refer to table 11.1, which shows the elementary particles and their masses according to the standard model. The mass of each particle is given in energy units, MeV (million electron volts) or GeV (billion electron volts), which is just the rest energy of the particle, equivalent to its mass by $E = mc^2$ since c is just an arbitrary constant.

Consider the group of particles labeled *fermions*. These all have intrinsic angular momentum, or *spin*, of ½.[10] They are found in three "generations," the columns headed by u, c, and t. Each generation has two *quarks* and two *leptons*. The first generation on the left is composed of a u quark with charge $+2e/3$, where e is the unit electric charge, and a d quark with charge $-e/3$. Below the quarks are the first generation leptons: the electron neutrino v_e, which has zero charge, and the electron e, which has charge $-e$. Each of the fermions is accompanied by an antiparticle of opposite charge, not shown in the table. (Antineutrinos, like neutrinos, are uncharged.)

I will discuss the masses of neutrinos in chapter 13. For now, suffice it to say that at least one neutrino has a mass on the order of 0.1 eV. This is to be compared with the next lightest nonzero mass particle, the electron, whose mass is 511,000 eV.

Fermions (antiparticles not shown)			Bosons
Quarks	u 2.3 MeV	c 1.27 GeV	t 173 GeV	γ 0
	d 4.8 MeV	s 95 MeV	b 4.18 GeV	g 0
Leptons	v_e *see text*	v_μ *see text*	v_τ *see text*	Z 90.8 GeV
	e 0.511 MeV	μ 106 MeV	τ 1.78 GeV	W 80.4 GeV

Table 11.1. The particles of the standard model. The masses are given in energy units. Antiparticles and the Higgs boson are not shown.

The second and third generations have a similar structure of quarks and electrons, except they are all heavier and unstable, decaying rapidly into lighter particles. For example, the muon, μ^-, with a mean lifetime of 2.2 microseconds, is essentially just a heavier electron with a mass of 106 MeV. Its main decay process is

$$\mu^- \rightarrow e^- + \bar{v}_e + v_\mu$$

where $\overline{\nu}_e$ is an antielectron neutrino. The antimuon, μ^+, has a similar decay:

$$\mu^+ \rightarrow e^+ + \nu_e + \overline{\nu}_\mu$$

Note that the t quark is 184 times as massive as the proton (mass 938 MeV).

The standard model contains three forces: electromagnetism, the weak nuclear force, and the strong nuclear force. Gravity, which has a negligible effect at the subatomic level and is already well described at the large scale by general relativity, is not included. Only when we get down to the Planck scale, 10^{-35} meter, does general relativity break down. We will talk about that later.

The particles in the column on the right of table 11.1 are the so-called mediators of the forces. They are *bosons*, which are defined as particles of integer spin. In this case, they all have spin 1. The bosons of the standard model are sometimes referred to as "force particles" because, in the quantum field theories of forces that are at the heart of the standard model, these particles are the *quanta* associated with the force fields. For example, the photon γ (for gamma ray) is the quantum of the electromagnetic field.

The force particles are usually pictured in the standard model as being exchanged between interacting quarks or leptons, carrying momentum and energy from one to the other. Figure 11.1 shows two electrons interacting by exchanging a photon. This is the canonical example of a *Feynman diagram*, introduced by physicist Richard Feynman in 1948.[11] Feynman diagrams are basically calculational tools and should not be taken too literally.[12]

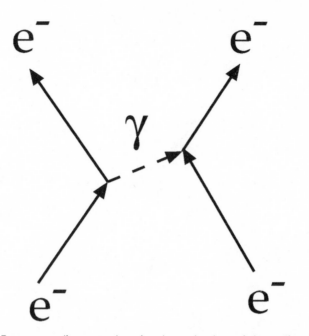

Figure 11.1. Feynman diagram showing two electrons interacting by exchanging a photon. Image by the author.

So, in the standard model, the photon is the force carrier of the electromagnetic force. All the elementary particles except neutrinos interact electromagnetically. The quantum field theory called *quantum electrodynamics* (QED) that successfully describes the electromagnetic force was developed in the late 1940s by Sin-Itiro Tomonaga, Julian Schwinger, Richard Feynman, and Freeman Dyson.[13] The first three would share the 1965 Nobel Prize for Physics, which is limited to three honorees.

In figure 11.2, an electron and positron collide and annihilate into a Z-boson, which then re-creates the pair. These are just two examples of the many diagrams that describe particle interactions.

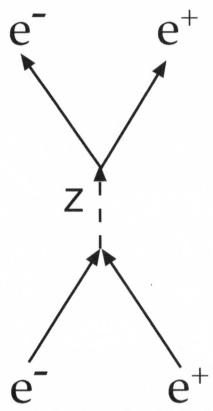

Figure 11.2. An electron and positron pair annihilating into a Z-boson that then re-creates the pair. Image by the author.

The W boson appears in two electrically charged states, $+e$ and $-e$. Together with the Z boson, which has zero charge, these "weak bosons" mediate the weak nuclear force that acts on all elementary particles except the photon and the *gluon*. The gluon will be discussed in a moment.

The most familiar weak interaction is the beta decay of nuclei in which an electron and antineutrino are emitted. In the standard model, the fundamental process involves the quarks inside nucleons (protons and neutrons), which are themselves inside nuclei,

$$d \rightarrow u + e^- + \bar{v}_e$$

To see the role that the W boson plays, refer to figure 11.3.

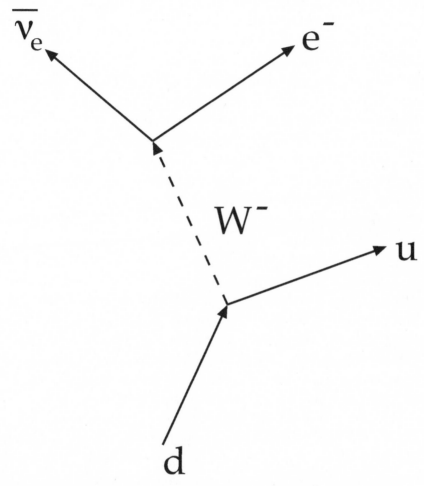

Figure 11.3. The beta decay of the d quark. In this picture, the d quark decays into a W^- boson and a u quark. The W^- propagates a short distance in space (about 10^{-18} meter) and then decays into an electron and an antielectron neutrino. Image by the author.

The particle labeled g in table 11.1 is the *gluon*. It is found in eight different states of what is called *color charge*, which is analogous to electric charge but comes in eight varieties that are metaphorically called "colors." The gluon mediates the *strong nuclear force* responsible for holding nuclei together. Only quarks interact strongly. In the standard model, the strong force is described by a quantum field theory, dubbed by Feynman *quantum chromodynamics*.

Most of the hundreds of new particles that were discovered in the 1960s were strongly interacting particles. These are generically referred to as *hadrons*. Two types of hadrons were identified: *baryons*, which have half-integer spin and *mesons*, which have integer spin. The proton and neutron are baryons. The lightest meson is the *pion* or pi-meson, which comes on three charge states: π^+, π^0, and π^-. I did my doctoral thesis work on the K-meson or *kaon*, which comes in four varieties: K^+, K^-, K^0, $\bar{K}^0$. They are composed of quark-antiquark pairs where one of the quarks is an s or its antiparticle. I need not elaborate further.

All the hadrons except nucleons are highly unstable, some with lifetimes so short that they can barely traverse the diameter of a nucleus before decaying. The neutron is unstable to beta decay, with a mean lifetime of about fifteen minutes. Although containing neutrons, most nuclei are stable since energy conservation prevents them from disintegrating. You won't find many free neutrons (or other hadrons except protons) floating about in outer space now, apart from a few being produced momentarily by high-energy cosmic-ray collisions.

The proton is obviously highly stable or else we wouldn't have so much hydrogen in the universe after 13.8 billion years. However, as we will see shortly, the possibility of proton decay even with a very long lifetime has enormous cosmological implications.

The standard model of elementary particles and forces arose, in part, from the attempt to fit these many new particles into a simple scheme. It succeeded spectacularly. Here's the scheme: The baryons are all composed of three quarks. Antibaryons are composed of three antiquarks. Mesons are composed of a quark and an antiquark. No hadron has ever been observed that cannot be formed from the quarks in table 11.1 and their antiquarks.

The nuclei of the atoms that constitute familiar matter are composed of just u and d quarks. The proton is uud and the neutron is udd. Every physical object that we deal with in our everyday lives, and every object that scientists in all disciplines except particle physics and particle astrophysics deal with, is made of just three elementary particles: the u quarks and d quarks that compose atomic nuclei and the electrons that swarm in clouds around them to make atoms.

The discovery in 2012 of what is almost certainly the long-expected Higgs boson was the final icing on the cake for the standard model. The Higgs boson is a spin-zero particle, H, that is responsible for generating the masses of leptons and weak bosons. The quarks get some of their masses this way, but most by another mechanism involving the strong force that I need not get into. The photon and gluon are massless.

Let us now look at the theoretical ideas behind the standard model. We will see they go well beyond that specific example to our whole concept of the meaning of physical law.

SYMMETRY AND INVARIANCE

The central concepts of modern physics, from relativity and quantum mechanics to the standard model, are symmetry principles and the manner in which those symmetries are broken. These have greatly aided us in our understanding of both the current and the early universe.

Symmetry is related to another concept—*invariance*. A perfect sphere is invariant to a rotation about any axis. That is, it looks the same from every angle. So we say it possesses *spherical symmetry*.

If you take a spherical ball of squishy matter (like Earth) and start to spin it rapidly, it will begin to bulge at the equator and its spherical symmetry will be broken. However, the ball will still possess rotational symmetry about the spin axis.

Here I am less interested in the symmetries of geometrical objects than those embedded within the mathematical principles called the "laws of physics." These are principles that appear in the models physicists devise to describe observations.

When an observation is invariant to some operation, such as a change in the angle at which the observation is made, then the model properly describing that operation must possess the appropriate symmetry. In particular, the model cannot assume an x, y, z-coordinate system in which the axes point in specific directions.

In the 1950s, it was shown that the weak nuclear force violates mirror symmetry, technically referred to as *parity*. That is, weak processes are not invariant to the interchange of left and right, just like your hands (or face, for that matter). Mathematically, an operator P, called the *parity* operator, changes the state of a system to its mirror image.

Particle physicists define another operator C that exchanges a particle with its antiparticle and an operator T that reverses time. In the 1960s, it was discovered that the combined symmetry CP is very slightly violated in the decay of neutral kaons. The combined symmetry CPT is believed to be fundamental. In that case, the violation of CP implies the violation of T. Direct T-violation has been empirically confirmed while CPT has never been violated in any observed physical process.

Note that T-violation is not to be interpreted as the source of the arrow of time since the effect is small, about 0.1 percent, and does not forbid time reversal. It just makes one direction in time slightly more probable than the other.

CPT invariance says specifically that if you take any reaction and change all its particles to antiparticles, run the reaction in the opposite time direction, and view it in a mirror, you will be unable to distinguish it from the original reaction. So far, this seems to be the case.

In short, not only do the laws of physics obey certain symmetries but a few (not all) can also violate some of these symmetries — usually spontaneously, that is, by accident.

By analogy, consider the teen party game "spin the bottle." A boy sets a bottle spinning on a floor in the center of a circle of girls. The spinning bottle has rotational symmetry about a vertical axis. But when friction brings it to a halt, the symmetry is spontaneously broken and the bottle randomly points to the particular girl the boy gets to kiss.

SYMMETRIES AND THE "LAWS OF PHYSICS"

As described in chapter 6, in 1915 Emmy Noether proved that the three great conservation principles of physics—linear momentum, angular momentum, and energy—are automatically obeyed in any theory that possesses, respectively, space-translation symmetry, space-rotation symmetry, and time-translation symmetry. The conservation principles are not, as usually taught in physics classes and textbooks, restrictions on the behavior of matter. They are restrictions on the behavior of physicists. If a physicist wants to build a model that applies at all times, places, and orientations, then he has no choice. That model will automatically contain the three conservation principles.

While the standard model of elementary particle physics is a long way from Noether's original work, it confirmed the general idea that the most important of what we call the laws of physics are simply logical requirements placed on our models in order to make them objective, that is, independent of the point of view of any particular observer. In my 2006 book *The Comprehensible Cosmos* I called this principle *point-of-view invariance* and showed that virtually all of classical and quantum mechanics can be derived from it.[14]

The book's subtitle is: *Where Do the Laws of Physics Come From?* The answer: They didn't come from anything. They are either *meta-laws* that are the necessary requirements of symmetries that preserve point-of-view invariance or *bylaws* that are accidents that happen when some symmetries are spontaneously broken under certain conditions. Note that if there are multiple universes, they should all have the same metalaws but probably different bylaws.

Although it is not widely recognized, Noether's connection between symmetries and laws can be extended beyond space-time to the abstract internal space used in quantum field theory. Theories based on this idea are called *gauge theories*. Early in the twentieth century it was shown that charge conservation and Maxwell's equations of electromagnetism can be directly derived from gauge symmetry.

In the late 1940s, gauge theory was applied to quantum electrodynamics, the quantum theory of the electromagnetic field that was described earlier. The spectacular success of this approach, which produced the most accurate predictions in the history of science, suggested that the other forces might also be derived from symmetries. In the 1970s, Abdus Salam, Sheldon Glashow, and Steven Weinberg, working mostly independently (they must have read the same papers), discovered a gauge symmetry that enabled the unification of the electromagnetic and weak forces into a single *electroweak* force. This was the first step in the development of the theoretical side of standard model. The three would share the 1979 Nobel Prize for Physics.

Let me explain what is meant when we say two forces are unified. Before Newton, it was thought (expressed in a modern context) that there was one law of gravity for Earth and another law for the heavens. Newton unified the two, showing that they are the same force, which described the motions of apples and planets in terms of a single law of gravity. In the nineteenth century, electricity and magnetism were thought to be separate forces until Michael Faraday and James Clerk Maxwell unified them in a single force called electromagnetism.

However, electromagnetism and the weak nuclear force hardly looked like the same force at the accelerator energies that were available until just recently. The electromagnetic force reaches across the universe, as evidenced by the fact that we can see galaxies that were over thirteen billion years away when they emitted the observed light.

The weak force only reaches out to about a thousandth of the diameter of a nucleus. It took a mighty feat of imagination to think these might be the same force! I recall Feynman being particularly dubious.

In the Feynman diagram scheme, a force is mediated by the exchange of a particle whose mass is inversely proportional to the range of the force. Since the range of electromagnetism seems to be unlimited, its mediator, the photon, must have very close to zero mass. Indeed, the mass of a photon is identically zero according to gauge invariance. On the other hand, the particles that mediate the weak force must have masses of 80.4 and 90.8 GeV. That is, they are almost two orders of magnitude more massive than the proton (0.938 GeV).

According to the Salam-Glashow-Weinberg model, somewhere above about 100 GeV (now known to be 173 GeV), the electromagnetic and weak forces are unified. Below that energy, the symmetry is broken spontaneously, that is, by accident, into two different symmetries—one corresponding to the electromagnetic force and the other to the weak force. The photon remains massless while the three "weak" bosons—the W^+ and W^-, which have electric charge $+e$ and $-e$, respectively, and the neutral Z—have the masses implied by the short range of the weak force.

In the electroweak symmetry-breaking process, the weak bosons, as well as leptons, gain their masses by the Higgs mechanism that was first proposed in 1964 by six authors in three independent papers, published well before the standard model was developed: Peter Higgs of Edinburgh University, Robert Brout (now deceased) and François Englert of the University Libre de Bruxelles, Gerry Guralnik of Brown University, Dick Hagen of the University of Rochester, and Tom Kibble of Imperial College, London.[15] To his embarrassment, the process was named after just one of the six, the unassuming British physicist Peter Higgs.

In the Higgs mechanism, massless particles obtain mass by scattering off spin-zero particles called Higgs bosons, and this became an intrinsic part of the standard model that was developed a decade later.

Basically you can think of it this way: The universe is a medium filled with massive Higgs particles that flit in and out of existence. As a zero-mass elementary particle tries to make its way through that medium at the speed of light, it bounces off the Higgs particles so that its progress through the medium is slowed down. This constitutes an effective increase in inertia, and mass is a measure of inertia.

The standard model predicted the exact masses of the weak bosons, 80.4 GeV for the Ws and 90.8 GeV for the Z. It also predicted the existence of weak neutral currents, mentioned in chapter 10 as playing a role in supernova explosions, which result from the exchange of the zero-charge Z boson. By 1983 these predictions had been spectacularly confirmed.

The full standard model containing both the strong and electroweak forces is based on a combined set of symmetries. The strong force is treated separately and is mediated, as mentioned, by eight massless gluons. The small range of the strong force, about 10^{-15} meter, results from a different source than the masses of gluons, which are zero; but we need not get into that.

By the end of the twentieth century, accelerator experiments had produced ample empirical confirmation of the standard model at energies below 100 GeV and had provided measurements of its twenty or so adjustable parameters, in some cases with exquisite precision. The model has agreed with all observations made in physics laboratories in the decades since it was first introduced.

On July 4, 2012, two billion-dollar experiments involving thousands of physicists working on the LHC at CERN reported with high statistical significance the independent detection of signals in the mass range 125–126 GeV that meets all the conditions required for the standard model Higgs boson.[16] Two of its six inventors, Peter Higgs and François Englert, shared the 2013 Nobel Prize for Physics.

Of course, as with all models, the standard model is not the final word. But with the verification of the Higgs boson and the availability of higher energies, we are now finally ready to move to the next level of understanding of the basic nature of matter—

and, as we will see, deeper into the big bang. The LHC is currently increasing its energy to 14 TeV, but we will have to wait a year or two to find what it reveals about physics at that energy.

At this writing, we now have both the data and a theory that describes the available data to inform us, with confidence, of the physics of the universe when its temperature was 1 TeV (10^{16} degrees), which occurred when it was only 10^{-12} second (one trillionth of a second) old.

PARTICLES OR FIELDS?

Relativity and quantum mechanics, and their offspring, quantum field theory and the standard model, compose the set of the most successful scientific theories of all time. They agree with all empirical data, in many cases to incredible precision. Yet if you follow the popular scientific media, you will get the impression that there is some giant crisis with these theories because no one has been able to satisfactorily explain what they "really mean."

Charlatans have exploited this perception of crisis, convincing large numbers of unsophisticated laypeople that the "new reality" of modern physics has deconstructed the old materialist, reductionist picture of the world. In its place we now perceive a holistic reality in which the fundamental stuff of the universe is mind — a universal cosmic consciousness. I call this *quantum mysticism*.[17]

Unfortunately, some theoretical physicists have unwittingly encouraged this new metaphysics by conjuring up their own mystical notion of reality. David Tong presents a typical example in the December 2012 *Scientific American*, writing,

> Physicists routinely teach that the building blocks of nature are discrete particles such as the electron or quark. That is a lie. The building blocks of our theories are not particles but fields: continuous, fluidlike objects spread throughout space.[18]

This is highly misleading. No one has ever observed a quantum field. However, we observe what can always simply and accurately be described as pointlike particles.

Quantum fields are purely abstract, mathematical constructs within quantum field theory. In the theory, every quantum field has associated with it a particle that is called the "quantum of the field." The photon is the quantum of the electromagnetic field. The electron is the quantum of the Dirac field. The Higgs boson is the quantum of the Higgs field. In other words, like love and marriage, you can't have one without the other. The building blocks of our theories are fields *and* particles.

But note that what Tong calls "a lie" is the notion that the building blocks of nature are discrete particles while it is fields that are the true building blocks of our theories. That is, he is equating ultimate reality to the mathematical abstractions of the currently most fashionable theory. That means it will change when the fashions change.

Tong is revealing his adoption of a common conception that exists among today's theoretical physicists. They believe that the symbols that appear in their mathematical equations represent true reality while our observations, which always look like localized particles, are just the way in which that reality manifests itself. In short, they are Platonists. It is important to note that twentieth-century greats Paul Dirac and Richard Feynman were not part of that school. And not all contemporary theorists are "field Platonists."

In his 2011 book *Hidden Reality*, physicist and bestselling author Brian Greene has this to say about particles and reality:

> I believe that a physical system is completely determined by the arrangement of its particles. Tell me how the particles making up the earth, the sun, the galaxy, and everything else are arranged, *and you've fully articulated reality*. The reductionist view is common among physicists, but there are certainly people who think otherwise.[19]

No one claims that the particles of the standard model must be thought of as classical billiard balls. But as philosopher Meinhard Kuhlmann points out, quantum fields also should not be thought of as classical fields such as the density of a gas. He writes:

> How can there be so much fundamental controversy about a theory that is as empirically successful as quantum field theory? The answer is straightforward. Although the theory tells us what we can measure, it speaks in riddles when it comes to the nature of whatever entities give rise to our observations. The theory accounts for our observations in terms of quarks, muons, photons and sundry quantum fields, but it does not tell us what a photon [field] or a quantum field really is. And it does not need to, because theories of physics can be empirically valid largely without settling such metaphysical questions.[20]

Although he does not claim this view for himself and presents all the alternatives, pro and con, Kuhlmann describes common positions found among hard-nosed experimentalists:

> For many physicists, that is enough. They adopt a so-called instrumentalist attitude: they deny that scientific theories are meant to represent the world in the first place. For them, theories are only instruments for making experimental predictions.

While others are a bit more flexible,

> Still, most scientists have the strong intuition that their theories do depict at least some aspects of nature as it is before we make a measurement. After all, why else do science, if not to understand the world?

I would just add that if some theory does not at least in principle imply some observable effect, it cannot be tested and we have little reason to think it correctly models reality. While such a theory may

228 GOD AND THE MULTIVERSE

be mathematically or philosophically interesting, its elements are not very good candidates for "aspects of nature."

While I cannot prove that particles are elements of ultimate reality, at least what we observe when we do experiments look a lot more like localized particles and are far simpler to comprehend than otherworldly quantum fields. After all, astronomical bodies look like particles when viewed from far enough away and we don't question their reality. So, for all practical purposes we can think of particles as real until we are told otherwise by the data.

Furthermore, as we saw in chapter 6, the wavelike phenomena that are associated with particles in quantum mechanics, and quantum field theory, are not properties of individual particles but of ensembles of particles. The expression "wave-particle duality" does not accurately describe observations. A lone particle is never a wave.

We also often hear that quantum mechanics has destroyed reductionism and replaced it with a new "holistic" worldview in which everything is connected to everything else. This is not true. Physicists, and indeed all scientists and, notably, physicians, continue to break the matter up into parts that can be studied independently. After a brief dalliance with holism in the 1960s, the success of the standard model returned physicists to the reductionist method that has served them so well throughout history of science, from Thales and Democritus to the present.

THE BIRTH OF PARTICLE ASTROPHYSICS

As we saw in chapter 10, by the 1990s nuclear astrophysicists had successfully described the formation of light nuclei when the universe was one second old by applying the model of big-bang nucleosynthesis. The calculated abundances agreed precisely with the data, including the very sensitive relationship between deuteron abundance and baryon density. Anyone seeing this result has to come away convinced that the big bang really happened.

While this was going on, particle astrophysicists (in most cases, the same people, with David Schramm leading the way) were applying the new standard model of elementary particles to describe what may have happened prior to 1 second. They adopted the idea of symmetry breaking, which had become fundamental in physics, to characterize a series of phase transitions starting from the first definable moment. At each critical temperature, the universe changed phase—like water freezing to ice—from higher symmetry to lower symmetry, with different sets of particles and forces emerging as described by the new symmetry.

Recall that prior to 1 second, when the temperature was above 1 MeV, the universe was a quasi-equilibrium mixture of electrons, neutrinos, antineutrinos, and photons in roughly equal numbers, along with a billion times fewer protons and neutrons that would form the light nuclei as the universe cooled and their equilibrium could no longer be maintained.

Let us now go back farther in time to 10^{-6} second, when the temperature was 1 GeV. This was still an era that we can describe with known physics, empirically as well as theoretically, so we are not just speculating. Just prior to this time, the universe was composed of the elementary particles listed in table 11.1 and there were no protons, neutrons, or composite hadrons of any type. The quarks were not free, however (they never are in quantum chromodynamics), but were confined along with gluons to a soup pervading the universe called a *quark-gluon plasma*. When the temperature dropped to around 1 GeV, a spontaneous phase transition occurred that generated the color-neutral hadrons my colleagues and I were studying in the 1960s with particle accelerators. Few hadrons except the protons and neutrons hung around in the early universe, however, because of their short lifetimes.

Although our physics measurements have so far extended to around 1 TeV, below which the strong, weak, and electromagnetic forces are distinct, the standard model makes the underlying assumption that the weak and electromagnetic forces were unified above that energy, prior to a trillionth of a second after the start of the big bang. The LHC allows us, for the first time, to empirically probe the region of higher symmetry and provide data on the physics of the universe earlier than 10^{-12} second.

MATTER-ANTIMATTER ASYMMETRY

Despite its success, the standard model does not explain a pretty important feature of our universe: the preponderance of matter over antimatter.

One of the principles contained in the standard model is *baryon-number conservation*. Each baryon has a baryon number $B = +1$; antibaryons have $B = -1$. Quarks have $B = +1/3$, antiquarks have $B = -1/3$. Leptons, gauge bosons (that is, the force mediators), and Higgs boson have zero baryon number. Baryon-number conservation says that the total baryon number of the particles engaged in an interaction is the same after the reaction has taken place as before. No reaction in particle physics, nuclear physics, and chemistry has ever been found that violates this rule.

If we reasonably assume that when the universe came into being its total baryon number was zero, then it should have equal numbers of baryons and antibaryons. By now they would have totally annihilated with one another and there wouldn't be any protons and neutrons to make atomic nuclei.

The standard model also contains a principle of lepton-number conservation. Leptons have $L = +1$, antileptons have $L = -1$. Baryons and gauge bosons have zero lepton number. So, by the same token, all the charged leptons and antileptons would have annihilated and there wouldn't be any electrons left. In short, the standard model says that the universe should be nothing but photons and neutrinos. This means no atoms, no chemistry, no biology, no you, me, or the family cat.

However, here we are. Protons and electrons outnumber antiprotons and positrons by a factor of a billion to one. At some point very early in the universe, before nuclei and atoms were formed, baryon-number conservation and lepton-number conservation were violated and the large asymmetry between matter and anti-matter was generated.

If baryon-number conservation is violated, then protons must be ultimately unstable. We need not worry about electrons being unstable, because they are so light there are no lighter charged particles they can decay into; charge conservation prevents them from decaying into photons and neutrinos. By contrast, there are many charged leptons that protons can decay into. The particle data tables, which list all the properties of elementary particles, provide dozens of possible decay modes.[21] Here's just one example:

$$p \rightarrow e^+ + \gamma$$

where e^+ is the positron. Note violation of lepton-number as well as baryon-number conservation.

Even before the standard model was completed in the 1970s, theorists were looking for ways it could be extended. One class of models considered was called GUTs, for *grand unified theories*. The standard model had unified the electromagnetic and weak forces into a single electroweak force, but the strong force was independent. GUTs attempted to unify the strong force with the others.

Most GUTs provided for *baryogenesis*, the generation of baryon asymmetry, along with *leptogenesis*, the generation of lepton asymmetry. A possible mechanism based on the original suggestion made in 1967 by the famous Russian physicist and political dissident Andrei Sakharov[22] is shown in figure 11.4. It involves a new gauge boson called X and its antiparticle.

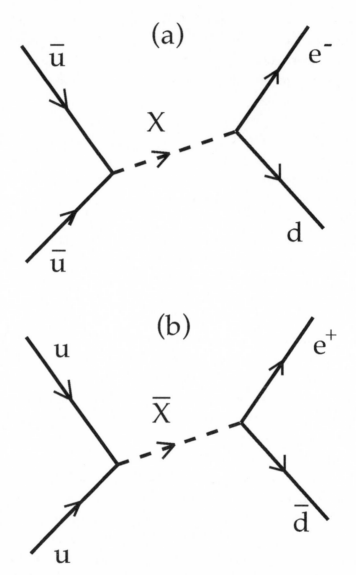

Figure 11.4. A mechanism for *B* and *L* violation. In figure 11.4 (a) two antiquarks annihilate into an X boson producing a quark and electron for a net gain of one unit each in baryon and lepton numbers. In (b) the same reaction is shown with particles replaced by their antiparticles, with the corresponding decrease in *B* and *L* by one unit each. CP-violation results in a greater rate for (a) than (b), generating a baryon and lepton excess. Image by the author.

The simplest of the early GUTs was proposed by Howard Georgi and Sheldon Glashow in 1974.[23] I will call it GG-GUT. (Technically it was called *minimal* SU(5).) It had the virtue of enabling a prediction of 10^{32} years for the proton lifetime.[24]

THE SEARCH FOR PROTON DECAY

This predicted proton-decay lifetime from GG-GUT was well within the range of existing technology, and four experiments were quickly mounted to seek proton decay. These experiments were conducted deep underground in mines in order to reduce background radiation, especially from high-energy cosmic-ray muons that can penetrate deep into the earth. At least two of the experiments were fully capable of seeing proton decay if the lifetime was 10^{32} years or less. One was in a salt mine near Cleveland, called IMB for the three primary institutions involved (University of California–Irvine, University of Michigan, and Brookhaven National Laboratory). Colleagues of mine at the University of Hawaii were also involved. The other sensitive experiment was in a zinc mine in Kamioka, Japan, and called KAMIOKANDE (Kamioka Nucleon Decay Experiment).

By 1982, all four experiments reported seeing no decays at the GG-GUT expected rate, thereby falsifying the model. (Despite some philosophers' assertions otherwise, falsification does happen in science.) Unfortunately, none of the other proposed GUTs predicted measurable proton lifetimes or other viable empirical tests.

Experiments continued to improve, with the most sensitive being an upgraded experiment in Kamioka called Super-Kamiokande. I played a small role in Super-K before retiring from research in 2000. Super-K has provided the best lower limit on proton decay so far, 1.01×10^{34} years as of 2011, two orders of magnitude higher than the GG-GUT prediction.[25]

Sometimes negative results are as important as positive ones. Knowing the lower limits for proton-decay lifetimes in various decay channels provides invaluable input for theorists as they look for physics beyond the standard model. They can rule out models that predict violations of that limit. When and if proton decay is observed, the rates by which it decays into various channels will help determine the structure of the physics beyond the standard model.

The underground experiments also had some positive ancillary discoveries that were almost as significant as the failure to see proton decay. In 1987, the experiments in Cleveland and Kamioka detected neutrinos from the supernova SN 1987A in the Large Magellanic Cloud, as mentioned in chapter 10.[26] This was the first observation of neutrinos from outside the solar system.

THE GUT PHASE TRANSITION

Given the success of the standard model, we can reasonably assume that prior to the electroweak phase transition at a presently estimated 173 GeV in temperature, which occurred at about 10^{-11} second, the universe can be described by the standard model with electroweak unification. That is, the strong force is still distinct but electromagnetism and the weak force are unified. The universe at this time still consisted of the quarks, leptons, and gauge bosons in table 11.1, but they were all massless and there were no Higgs bosons. Particles still outnumbered antiparticles by a billion to one. At some higher energy and earlier time, there almost certainly had to be a phase transition from a state of higher symmetry that, itself, was the result of a phase transition from an even higher symmetric state.

The best candidate remains some kind of GUT in which the strong and electroweak forces are united and both baryon- and lepton-number conservation are violated. This GUT, in turn, arose from another symmetry at higher energy in which B and L were conserved.

Most GUTs that have been proposed exhibit these properties. Here the symmetry is manifested by no distinction being made between quarks and leptons so that reactions such as those shown in figure 11.4 can take place. The X particle that is exchanged in the figure can be thought of as a "leptoquark," a combination of quark and lepton. The breaking of B and L come about, according to Sakharov, by differences in reaction rates caused by CP violation.

In the more symmetric state prior to the GUT phase transition, CP is invariant and B and L conservation are each restored. So the universe starts out with all the symmetries and equal numbers of particles and antiparticles. The asymmetry of matter and antimatter is generated after the phase transition from this earlier state to the GUT state.

So all we have to do is keep building more and more powerful colliding-beam accelerators so that we can keep probing farther and farther back in time until we reach the GUT regime. The trouble is, we aren't even close to having enough energy. The GUT phase transition is estimated to occur at about 10^{25} eV, twelve orders of magnitude above the energy of the LHC. A vast "desert" may exist between the GUT and electroweak phase transitions, during which time the universe remains in the unbroken phase of the electroweak-unified state.

At least the LHC will enable us to explore the unbroken phase. But will we ever be able to probe beyond this state? It is highly unlikely with accelerators, at least in any foreseeable future. However, we do have another window on the very early universe, and that is proton decay. Super-K may be nearing the point where proton decay is observed. Several GUTs predict decay modes that are within the range of Super-K or a larger detector.

SUPERSYMMETRY (SUSY)

A promising approach to physics beyond the standard model that has attracted the attention of a generation of theoretical particle physicists is *supersymmetry*, referred to as SUSY. This is a symmetry principle in which physical models do not distinguish between fermions and bosons. Recall that fermions have half-integer spins while bosons have integer or zero spins.

Under SUSY, every elementary particle is accompanied by a *sparticle* of the opposite spin property. Thus the spin-1/2 electron is accompanied by a spin-zero *selectron*, the spin-1 photon by a *photino* of spin 1/2, the spin-1/2 quark by a spin-zero *squark*.

If SUSY were a perfect symmetry, the sparticles would have the same masses as their partners and not only would we see them but they would not obey rules such as the Pauli exclusion principle that distinguishes bosons from fermions. In that case, there would be no chemistry. Since no sparticle has yet been seen and we have chemistry, the symmetry is broken at the low energies (low compared to those in the early universe) where we are able to live. If sparticles exist, they must have very high masses.

As we will see in chapter 13, the Large Hadron Collider has so far failed to detect any of the expected sparticles, leaving the whole notion in grave doubt.

M-THEORY

Supersymmetry offered the possibility of finding a theory that unified gravity with the other forces of nature described in this chapter, what is often referred to as the *theory of everything* (TOE). Originally called *string theory*, it was conjectured that the universe has more than three dimensions of space, with these extra dimensions curled up so tightly that that are undetectable. String Theory replaced zero-dimensional particles with one-dimensional strings.

Eventually a further generalization was introduced called M-*theory* in which objects of higher dimension were included, called *branes*. A two-dimension brane is called a *membrane*. A p-dimensional brane is called, appropriately enough, a *p-brane*. A particle is a 0-brane, a string a 1-brane, and a membrane a 2-brane. *M*-theory allows up to $p = 9$.[27] Although M-theorists have made many significant mathematical discoveries in their quest for the theory of everything, they have not yet managed to come up with empirical prediction that can be tested by experiment. Furthermore, since as just mentioned, supersymmetry has not yet been confirmed, as it was expected to be at the LHC. Although it gets a lot of publicity, leading the layperson to think the TOE is just around the corner, M-theory remains far from being verified and may soon be falsified.

12.
INFLATION

BIG-BANG BUGS

We have seen how, in the 1970s, the big bang became established with virtual certainty. Very often in science, however, a model that may agree beautifully with all the data and have no other viable competitor in sight is still beset with theoretical or philosophical problems. After all, a flat Earth once agreed with all the data available to primitive people. And look how long the geocentric model of the solar system lasted — not just as myth, but as an accurate predictor of the motions of planets.

As of 1980, the following theoretical problems with the big-bang model were common knowledge among cosmologists:

The Flatness Problem

Recall that cosmologists define a density parameter $\Omega = \rho / \rho_c$, where ρ is the average mass density of some component in the universe and ρ_c is the critical density at which the universe is exactly balanced on the edge between collapsing from gravity and expanding forever. If ρ refers to the average density of all the components, then $\Omega = 1$ and the geometry of the universe is flat, that is, Euclidean.

Now, here's the problem: According to Friedmann's equations, the density of the universe determines its rate of expansion. Consider the Planck time, $t = 10^{-43}$ second. If, at that time, Ω had been greater than 1 by just one part in 10^{60}, the universe would have immediately collapsed. If Ω was less than by just one part in 10^{60}, the universe would have expanded so rapidly that the visible universe would have quickly become so dilute that no life would have been possible. In the big-bang model, life can only exist if $\Omega = 1$ to great precision and the universe is extremely flat.

This is one of the parameters that Christian apologists claim had to have been fine-tuned by a creator God in order to make life possible.[1] In 2009 book *Life after Death: The Evidence*,[2] Dinesh D'Souza quotes Stephen Hawking from *A Brief History of Time*: "If the rate of expansion one second after the big bang had been smaller by even one part in a hundred thousand million million, the universe would have recollapsed before it ever reached its present size."[3] William Lane Craig has also referred to this statement in numerous debates.[4]

The Horizon Problem

When we look in opposite directions in the sky, we see the same temperature and spectrum for the CMB in both regions. This implies that they came from two sources that were in *causal contact* with one another at some earlier time so that they could interact with one another and achieve equilibrium. Two points in space can only be in causal contact when there is sufficient time for a signal to travel between them. Using the latest numbers, those points are now ninety-three billion light-years apart.

In chapter 10, we saw that the photons in the CMB were emitted when the universe became transparent at an age of 380,000 years. If you apply the standard big-bang model with linear Hubble expansion, you find that the distance between two points on the opposite side of the universe when the age of the universe was 380,000 years would have been about eighty-four million light-years, as shown in figure 12.1. This is much greater than the distance light could have traveled from the big bang, so the sources of the disturbances at A and B could never have been in causal contact so they could reach thermal equilibrium.

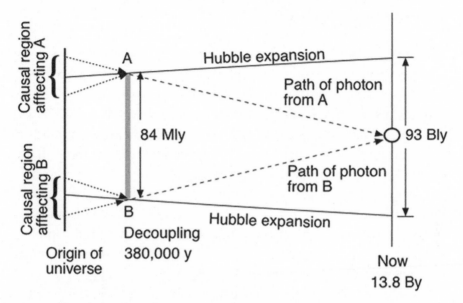

Figure 12.1. Illustration of the horizon problem. Time runs along the horizontal axis, and the distance between two points in the universe is plotted along the vertical axis. Points A and B are on opposite sides of the universe at decoupling, about eighty-four million light-years apart, emitting photons, shown by the dashed lines, that are detected from two opposite directions on the sky by a CMB observer today. Today they are ninety-three billion light-years apart because of the expansion of the universe. The dotted arrows are light rays showing that the regions that can affect A and B were never in causal contact. The drawing is not to scale. Image by the author.

The Structure Problem

We saw in chapter 10 that cosmologists had for many years struggled to try to explain how the complex structures in the visible universe came to be assembled. The problem was bad enough for a static universe. It was even worse for an expanding universe, which dispersed matter over greater distances making it even less likely for isolated clouds to collapse under gravity.

The Monopole Problem

In classical electromagnetic theory, the simplest electric charge is a point particle for which the electric field is visualized as lines of force radiating outward from the center. Two opposite point charges, positive and negative, constitute an *electric dipole*. So we can call a point charge an *electric monopole*. There are also quadrupoles, octopoles, and so on. If we take an electric dipole and pull the charges apart, we get two electric monopoles.

A bar magnet is an example of a *magnetic dipole*, with a north and south pole However, if you cut a magnetic dipole in two, you don't get two *magnetic monopoles*; you just get two more dipoles. No magnetic monopoles exist in the classical theory, and none have ever been observed empirically.

As French physicist Pierre Curie remarked in 1894, the absence of magnetic monopoles is the only difference between electricity and magnetism. In 1931, Paul Dirac showed that magnetic monopoles are consistent with quantum mechanics and restore the symmetry between electricity and magnetism.[5]

In 1974, Dutch physicist Gerard 't Hooft[6] and Russian physicist Alexander Polyakov[7] independently proved that unified gauge theories that include electromagnetism will contain magnetic monopoles. In 1976, British physicist Thomas Kibble (one of the six authors who proposed the Higgs mechanism in 1964, see chapter 11) showed that in a gauge symmetry–breaking phase transition, the new phase need not be uniform but could exhibit so-called *topological defects* such as appear in a ferromagnet. These defects include domain walls, strings, and monopoles.[8]

In 1979, Harvard graduate student John Preskill calculated that monopoles 10^{16} times as massive as a proton should have been produced during the GUT phase transition in numbers comparable to that of protons.[9] If that had been the case, the mass of the universe at the time would have been so great that it would have collapsed in less than 1,200 years.[10]

During the 1980s, numerous searches for magnetic monopoles were conducted and none found.[11] In 1987, I spent a six-month sabbatical in Italy working on MACRO (Monopole, Astrophysics, and Cosmic Ray Observatory) in the Grand Sasso Underground Laboratory. The lab is in a highway tunnel through the mountains near L'Aquila, scene of a 2009 earthquake (which did not affect the lab). MACRO was built primarily to search for magnetic monopoles and was the most sensitive monopole search ever performed. It detected none and by 2002 had set a very stringent upper limit on the monopole flux, well below what had been estimated should exist based on the effects they should have had on galactic magnetic fields.[12]

Nevertheless, at the worst, the failure to detect magnetic monopoles was a problem for GUTs, not for the big bang. I mention it mainly for historical reasons, since monopoles provided a major impetus for elementary particle physicists to become involved in early-universe cosmology.

INFLATION, OLD AND NEW

In 1980, several physicists and astrophysicists independently began to develop a scenario for an early universe that would ultimately provide viable solutions to the problems with the conventional big bang. On January 11, Russian physicist Alexsei Starobinsky, working with Stephen Hawking at Cambridge, submitted a paper to *Physics Letters* showing that quantum effects in the early universe could lead to a de Sitter universe and thus an exponential expansion we now call *inflation*.

In 1970, Hawking and Roger Penrose had used general relativity to prove that our universe began as a singularity, an infinitesimal point of infinite density.[13] Since then this result has been used by theologians to argue that our universe must have had a beginning and, although it doesn't follow, that there had to have been a personal Creator.[14] Starobinsky showed, and both Hawking and

Penrose agreed, that quantum effects in the early universe eliminate the singularity. Not being a quantum theory, general relativity breaks down at distances less than the Planck length, 10^{-35} meter.[15]

On May 5, 1980, an acquaintance of mine at the time, astrophysicist Demosthenes Kazanas of the NASA Goddard Space Flight Center, submitted a paper to the *Astrophysical Journal* with the title "Dynamics of the Universe and Spontaneous Symmetry Breaking."[16] There he argued that a phase transition in the early universe associated with spontaneous symmetry breaking would result in an exponential expansion that could account for the observed isotropy of the universe. I believe this was the first published paper explicitly recognizing exponential expansion as the solution to one of the major problems with the conventional big bang, namely the horizon problem.

On September 9, 1980, Japanese physicist Katsuhiko Sato submitted a paper to the *Monthly Notices of the Royal Astronomical Society* also showing that a first-order phase transition can lead to an exponentially expanding universe.[17] He suggested that fluctuations could account for the origin of galaxies but did not mention the other big-bang problems.

The seminal paper on inflation, however, was submitted to *Physical Review D* on August 1, 1980, by physicist Alan Guth, then working as a postdoctoral researcher at the Stanford Linear Accelerator.[18] Guth grasped the full significance of an early period of exponential inflation, emphasizing how it solves both the flatness and horizon problems and also suggested a possible solution to the monopole problem.

The flatness and horizon problems are by far the most significant, as Guth soon realized. Either was capable of falsifying the big-bang scenario had no solution proved possible. On the other hand, the magnetic-monopole problem was not critical. No magnetic monopoles exist in either classical or quantum electrodynamics, and none have been seen in nature. At best they add some symmetry to electromagnetism, but they appear to be required only in GUTs.

In Guth's 1997 excellent popular book *The Inflationary Universe*,[19] he relates how he was drawn to the notion of inflation by the monopole problem and admits he knew little about cosmology at the time. He had never heard of the horizon problem until December 1979. But he learned fast, and by the time he wrote the paper he fully appreciated the deep significance of both the flatness and horizon problems.

Guth nicely explains his original model in his book, but he and others soon realized it had to be modified. Rather than reviewing that history, which is highly technical, I will argue that exponential inflation is a natural consequence of general relativity.

If you write down Friedmann's equations for a de Sitter universe with a positive cosmological constant, it takes only college freshman math to prove that the solution is an exponential expansion. Regardless of the specific model, inflation solves the flatness, horizon, and monopole problems and sets the stage for solving the structure problem.

The Flatness Problem Solved

Recall that in chapter 8 I described expanding three-dimensional space in terms of the common analogy of the expanding two-dimensional surface of an inflating (!) balloon. Imagine a balloon starts out small and then expands by many orders of magnitude. A small patch on the surface will become very flat. The universe within our light horizon is like that small patch, which inflation has made very flat indeed.

Now this is usually interpreted to mean that the universe has $\Omega = 1$, that is, the density ρ is exactly equal to the critical value ρ_c for which the geometry of the universe is Euclidean. Recall that in this case the cosmological curvature parameter $k = 0$. The current empirical value is $\Omega = 1.002 \pm 0.011$. If it should turn out that ρ is just ever so slightly less than ρ_c, say by one part in 10^{100}, then we have a universe with slightly negative curvature, $k = -1$, which will also expand indefinitely.

On the other hand, if ρ is just ever so slightly greater than ρ_c, say by one part in 10^{100}, then we have a universe with slightly positive curvature, $k = +1$. In classical cosmology, when the cosmological constant was assumed to be zero, $k = +1$ referred to a "closed universe" that would someday collapse in a "big crunch." However, as we will see, even a "closed universe" dominated by a positive cosmological constant will continue to expand.

Later we will see that a universe with $k = +1$ provides a plausible mechanism for the origin of our universe that is completely consistent with existing knowledge and has been worked out fully mathematically.

Earlier in this chapter I mentioned that Christian apologists Dinesh D'Souza, William Lane Craig, and others have quoted Stephen Hawking that the expansion rate is fined-tuned to "one part in a hundred thousand million million." That was taken from pages 121–122 of *A Brief History of Time*. However, they conveniently ignored Hawking's explanation a few pages later, on page 128:

> The rate of expansion of the universe would automatically become very close to the critical rate determined by the energy density of the universe. This could then explain why the rate of expansion is still so close to the critical rate, without having to assume that the initial rate of expansion of the universe was very carefully chosen.[20]

In other words, inflation accounts for the fact that the expansion of the universe is equal to the critical rate out to sixty decimal places.

The Horizon Problem Solved

The horizon problem results from the observed fact that the CMB is highly uniform across the sky, with the same blackbody spectrum and temperature. As seen in figure 12.1, photons observed from opposite sides of the sky would never have been in causal contact according the big-bang model in which the Hubble expansion extrapolated back to the origin of the universe.

Inflation solves the problem, as illustrated in figure 12.2. In the time after the origin of the universe, but before inflation, the points A and B were close together and so came to quasi-thermal equilibrium with one another. Inflation stretched the distance between them by many orders of magnitude so that photons from them appear today as correlated signals from the opposite sides of the sky.

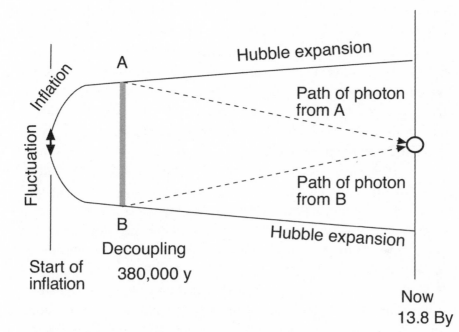

Figure 12.2. How inflation solves the horizon problem. A fluctuation in a small region of space sends photons in opposite directions. Inflation stretches the distance between them so that they provide correlated signals from opposite sides of the sky. Image by the author.

The Monopole Problem Solved

As mentioned, Guth was led to inflation while trying to solve the monopole problem in GUTs. However, since monopoles have never been seen anyway, this is only a theoretical problem for grand unified theories. The failure to see any monopoles might have falsified GUTs if no solution had been found but would not have killed the big bang.

Guth did not claim to solve the monopole problem in his original paper but suggested a solution. He and collaborator Henry Tye had come up with the idea that *supercooling* occurred, delaying the completion of the GUT phase transition and the nucleation of monopoles. Supercooling and superheating are familiar phenomena in thermodynamics as well as everyday life, and actually constitute the most common types of phase transitions, called *first-order* phase transitions. In a familiar example, when you boil water, it does not instantaneously convert into steam but forms bubbles. It takes a long time for all the water to convert to steam. If you put a cup of very pure water in your microwave, you will be able to heat it to above the boiling point without it boiling; it is superheated. Then, when perturbed, as by someone touching the cup, the water will all simultaneously turn to steam (be careful, people have been scalded by this phenomenon). Similarly, if you cool water below its freezing point, ice crystals will nucleate if some seed is present. But if the water is very pure, homogeneous nucleation takes place and the ice formed is uniform like glass.

It already had been established that the GUT phase transition was strongly first order. The bubbles of GUT phase that are formed in the phase transition will not immediately form monopoles since the fields remain scrambled until the temperature drops sufficiently. Guth and Tye proposed that the bubbles will expand sufficiently during the supercooled phase so that when the monopoles finally form they will be highly diluted.

Guth at first conjectured that our universe formed when the bubbles collided and their energies, concentrated in the bubble walls, were converted into particles. But calculations by Guth and Erik Weinberg showed that because of the continued expansion of the space between bubbles, they would never collect into a uniform mass but form finite clusters.[21] They considered the possibility that the universe was within a single bubble but concluded (prematurely) that it would be too empty to resemble any real universe.[22]

But that was for the Guth-Weinberg model. Without any empirical data, choosing a model is a matter of educated guesswork. At the same time, Russian physicist Andrei Linde[23] and American physicists Andreas Albrecht and Paul Steinhardt[24] had come up with their own models. These showed that it was possible for our universe to have been formed from just one of the bubbles. These models were termed *new inflation*. I will not discuss these either, since Linde soon came up with a better idea.

CHAOTIC INFLATION

Andrei Linde is one of the most original and productive of the many original and productive inflationary cosmologists who quickly became noticed after Guth's paper appeared. Guth graciously acknowledges that Linde independently invented much of the inflationary universe theory in the late 1970s, although Linde admitted that he did not immediately realize its significance.[25]

In 1983, Linde conceived another model called *chaotic inflation* that is so simple and natural that, while it is probably not exactly correct in detail, it is probably not far wrong and enables us to understand the process with minimal assumptions and a minimum of technical details. It is standing up well with the most recent observations.

Unlike other inflation models, chaotic inflation does not rely on trying to guess the shape of the inflating potential from GUTs or some other dynamical theory for which we have no observational guidance and no fundamental principle on which to base it. It simply starts out with almost nothing and lets quantum mechanics and statistics do their jobs.

I will follow recent convention and refer to the field responsible for inflation as the *inflaton* field, so we are not tied to GUTs or any other unnecessarily specific model. We just assume that any field that arises is simply a scalar field, equivalent to a cosmological constant in de Sitter space, which we have seen will produce an exponential expansion.

Once again, let us go back to the Planck time, 10^{-43} second and worry later what might have gone on before. Let me assume that the universe at that time was as small as it can be operationally defined, which is a sphere whose radius was equal to the Planck length, 10^{-35} meter (orders of magnitude are good enough at this level). The sphere would be empty except for a vacuum energy that would have a random value following a normal (Gaussian or bell-curve) distribution with a statistical standard deviation equal to the Planck energy, 10^{28} electron-volts. Note this is not a small number; it is equivalent to a temperature of 10^{32} degrees and a rest energy about thirty times that of a particle of dust.

A positive energy fluctuation equivalent to a positive cosmological constant would produce a de Sitter universe that expands exponentially. A negative fluctuation would result in exponential collapse, which we need not consider. Because of the constant energy density of the de Sitter vacuum, the universe gains internal energy as it expands. This is equivalent to a mass that we can call the "seed" of inflation. Energy is conserved and the internal energy or mass comes from the loss of gravitational energy as the universe "falls up" because of the negative pressure of the vacuum. The mass of the seed must exceed a certain limit sufficient to sustain inflation, or else the normal attractive gravity of the mass will cause it to quickly collapse.

In both classical and quantum field theory, fields have the mathematical properties of a one-dimensional simple harmonic oscillator such as a simple pendulum. The field value ϕ is analogous to the displacement of a pendulum from its equilibrium position. Because of the uncertainty principle, the quantum harmonic oscillator is never at rest but oscillates around its equilibrium point with a minimum energy called the *zero-point energy*. Thus any variation in ϕ can properly be called a "quantum fluctuation."

As illustrated in figure 12.3, we can visualize the oscillator metaphorically as a ball rolling up and down the side of a bowl. In this case, if the shape of the bowl is parabolic, the ball will exhibit simple harmonic motion, so this serves as a good model for the behavior of ϕ. The mathematics will be the same.

Normally the ball will roll quickly back down. However, the equation of motion for the oscillator in an expanding universe predicts that, because of the expansion of space, the return to equilibrium will be slowed by friction. This is analogous to the jar being filled with molasses. Actually, the contents of the bowl are more like molasses riding on water riding on air.[26] So, for small displacements, the ball just oscillates back and forth near the bottom of the bowl. However, Linde noted that occasionally, when the displacement is large, the molasses will slow the ball and it will stay for a while with a large displacement from equilibrium.

This so-called "slow roll" is a necessary feature of most inflation models and is artificially built into the "new inflation" models mentioned earlier. Here it is natural. Slow roll allows time for the original seed to expand by many orders of magnitude before the ball finally reaches bottom. Once at the bottom, it oscillates back and forth with ever-decreasing amplitude, never coming completely to rest. The energy lost to friction creates the elementary particles that then form our universe.

We can make this all quantitative, at least for illustration. For the inflaton field ϕ, we can write the potential energy density in harmonic oscillator form, $u(\phi) = m^2\phi^2/2$, where m is the mass of the quantum of the field, which we can think of as the inflaton particle. The value of m is unknown and so is treated as an adjustable parameter, the only one in the model. If we now put u into the equation of motion, we can use numerical methods to calculate ϕ, H, and the cosmological scale factor a as a function of time. You can find this all worked out in *The Comprehensible Cosmos*, including derivations of all the equations used at an undergraduate mathematics level.[27] Here I will just show the results.

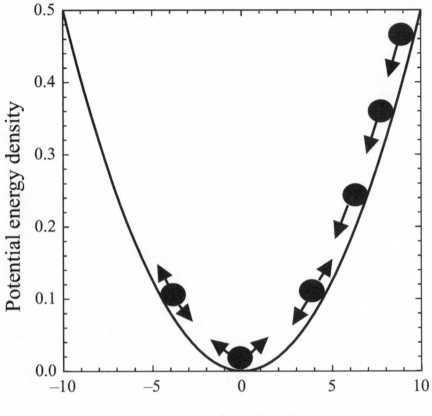

Figure 12.3. An illustration of chaotic inflation. The potential energy density is given by $u(\phi) = m^2\phi^2/2$, where ϕ is the scalar field and m is the mass of the inflaton. It is assumed to start at $\phi = 10$ Planck units. The evolution of the inflaton field is equivalent to a unit mass ball rolling down a parabolic well such as that of a damped, simple pendulum. Image by the author.

Let us work in "Planck units" where $\hbar = h/2\pi = c = G$ (Newton's gravitational constant) $= 1$. For illustration, I have chosen an initial fluctuation in ϕ of 10 Planck units and $m = 10^{-7}$ in Planck units (10^{11} GeV). The motion of the ball as it rolls down the hill from there is shown in figure 12.3. As the ball slowly descends, the volume of the universe increases exponentially. Its motion is damped by the expansion of space so the ball loses energy as it rolls down and then oscillates back and forth around the bottom of the potential with decreasing amplitude.

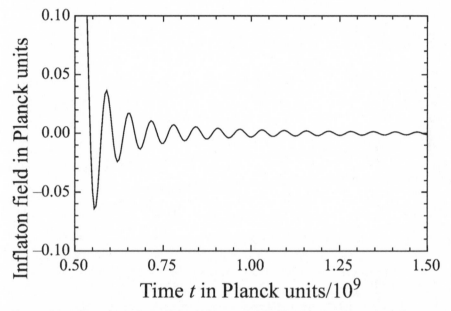

Figure 12.4. The evolution of the inflaton field $\phi(t)$ with time for chaotic inflation. The time scale is approximately seconds × 10^{-34}. The region $t < 0.5$ is suppressed. Image by the author.

Figure 12.4 illustrates how the field evolves with time t in units of the Planck time. The region $t < 0.5$ is not shown in order to illustrate the damped oscillation of the field. During the time $t < 0.6$, the field drops from 10 units (off scale) to zero and then oscillates about zero with ever-decreasing amplitude.

Figure 12.5 shows the evolution of the scale factor of the universe, a, which for our purposes we can think of as the universe's radius. Exponential inflation of 214 orders of magnitude is followed by a smooth transfer to the conventional Hubble expansion. This is illustrative only and not meant to exactly model our universe.

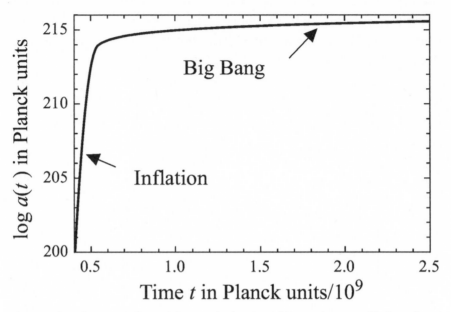

Figure 12.5. The evolution of the scale factor of the universe with time for chaotic inflation with $m = 10^{-7}$ an initial inflaton field of 10 Planck units. The time scale is approximately seconds × 10^{34}. The zero of the time axis is suppressed for illustrative purposes. The part of the curve labeled "Big Bang" refers to the normal Hubble expansion. Image by the author.

LARGE-SCALE STRUCTURE

While particle astrophysicists in the eighties were toying with the incredible idea that the universe expanded by many orders of magnitude during its first tiny fraction of a second, observational astronomers were discovering that the cosmos revealed with their latest telescopes was incredible in its own right.

In the 1970s, the general picture of the universe had been one of a more-or-less homogeneous distribution of clusters of galaxies moving away from each other uniformly with the Hubble flow. But by the 1980s, evidence began to accumulate that thousands of galaxies over a region of space millions of light-years across show small but measurable deviations from the purely radial recessional velocities they were expected to have from of the universe's expansion. The motions of the galaxies in our local cluster seemed to be directed toward a region of the sky that was about two hundred million light-years away at the center of the Hydra-Centaurus supercluster. This was termed the Great Attractor.[28]

Within a few years, other unexpected structures in the distributions of clusters, superclusters (clusters of clusters), and complexes (clusters of superclusters) would make their presence known. In 1987, my University of Hawaii colleague Brent Tully observed a filamentary structure one billion light-years long and 150 million light-years wide he called the Pisces-Cetus Supercluster Complex. It consists of five superclusters with a total mass of 10^{18} solar masses, including the Virgo supercluster of 10^{15} solar masses of which we are a part.

As we have seen, measuring distances have always provided a major challenge to astronomers. Over the years they have developed what they call the *cosmic distance ladder*, a series of methods where each applies out to a specific distance, at which point a new method then takes over with sufficient overlap so that each method can help calibrate the neighboring method.

I need not go into all the methods used. I have already discussed parallax, which works for nearby stars out to about one hundred light-years, and Cepheids, which works for galaxies out to thirteen million light-years. In 1977, Tully and his collaborator Richard Fisher had published a new method for determining distances to spiral galaxies that related their intrinsic luminosities to their rotational speeds.[29] As with other methods, you determine the distance by measuring the observed luminosity at Earth and assume it falls off as distance squared to give the observed value. Tully and Fisher used this method and others to provide an atlas of what they called "nearby" galaxies.[30]

But basically, redshift remains the most precise measurement that astronomers can perform, and Hubble's law can still give an approximate distance. The latest period in astronomical history has been marked by extensive redshift surveys of galaxies that have revealed a remarkable weblike structure to the visible universe.

The first extensive redshift survey was started in 1977 by the Harvard-Smithsonian Center for Astrophysics (CfA), which was completed in 1982. It was followed by another CfA survey from 1985 to 1995. Using these data, in 1989, Margaret Geller and John Huchra discovered a filament of galaxies whose redshifts implied it was roughly two hundred million light-years away with a length of five hundred million light-years, a width of three hundred million light-years, and a thickness of sixteen million light-years. It was dubbed the Great Wall.[31] As we will see in the next chapter, a large number of extensive redshift surveys have been conducted since 2000.

Basically, the visible universe is composed of one hundred billion to perhaps as much as a trillion galaxies that astronomers organize into groups, clusters, superclusters, sheets, filaments, and walls. These are separated by *voids* with diameters from thirty to five hundred million light-years that contain few galaxies. In 2013, Tully and collaborators produced a remarkable video that dramatically illustrates this structure.[32]

Nevertheless, the vastness, beauty, and complexity we see with our eyes and telescopes give us a false impression that the cosmos is highly complex and therefore intricately designed. In fact, the universe on the whole is simple and almost random. The invisible 99.5 percent of the mass of the universe has hardly any structure. The 69 percent that is dark energy has none while the structure of the 26 percent that is dark matter is not as finely detailed as the visible matter that is clumped around it. Furthermore, by a factor of a billion the universe is numerically composed mostly of photons and neutrinos in motion that is random to one part in 100,000. Far from looking as if it were designed by a supreme being of infinite intelligence, our universe looks as if it were the product of chance alone.

STRUCTURE AND INFLATION

At first, it was thought that inflation made the structure problem even worse. After all, one of the triumphs of inflation was to explain the extraordinary smoothness of the cosmic microwave background. So how could it be expected to explain the obvious lack of smoothness of all the visible matter around us—galaxies, stars, planets, the Rocky Mountains?

I have noted that prior to the introduction of inflationary cosmology several authors had proposed that structure formation in the universe was the result of primordial density fluctuations in the early universe. But without any knowledge of the nature of the primordial matter, they could do little more than speculate.

Inflationary cosmologists got the idea that small density perturbations caused by quantum mechanical zero-point fluctuations in the inflaton field were multiplied by many orders of magnitude during inflation and could have provided the variations in density needed for gravitational clumping and the formation of galaxies.

Applying their different models of inflation, cosmologists in the 1980s tried to calculate the density variation resulting from quantum fluctuations in the inflaton field. Guth describes the three-week workshop on the early universe held in Cambridge June 21 to July 9, 1982, organized by Stephen Hawking and Gary Gibbons, and how everyone was coming up with different estimates, most orders of magnitude too small to produce galaxies.[33] However, shortly after, a 1983 paper by James Bardeen using a novel calculational technique that was model-independent argued that a density fluctuation on the order of 10^{-3} to 10^{-4} from inflation was not implausible.[34]

Despite the uncertainty in magnitude, inflation was expected to give, at least approximately, the scale-invariant fluctuations that, as we saw in chapter 11, are deemed to be necessary for structure formation. In the simplest model where the inflaton field is the uniform scalar field in a de Sitter universe, scale invariance follows from the time translation invariance of the exponential solution.

Specific models of inflation are more complicated, but they all give something very close to scale invariance. Actually, we will see that they, including chaotic inflation, predict a slight but significant variation from scale invariance that serves as yet another risky test for the inflationary model.

During the inflationary period, tiny quantum fluctuations in the density of the inflaton field expanded by many orders of magnitude. When inflation ceased, the universe was a hot, highly dense gas of elementary particles, which then followed the more sedate Hubble expansion. The fluctuations caused the expanding sphere of gas to vibrate and produce sound waves that propagated in all directions with the speed of sound. Since the medium vibrating was mostly photons, the sound speed was essentially the speed of light divided by $\sqrt{3}$. In the sophisticated models we will discuss later, the speed of sound is allowed to vary as the baryon-to-photon ratio varies, thus providing a tool to measure their relative contributions.

As the universe continued to expand and cool, the various processes described in chapter 10 took place. During all this time, the particles remained tightly coupled in quasi equilibrium, with a well-defined temperature that grew less as the universe expanded, from 10^{27} degrees at the end of inflation following the straight line on the log-log plot shown in figure 10.2.

Recall the artificial distinction made by astronomers and cosmologists between radiation and matter. They are both composed of material particles, but radiation is extreme relativistic ($v \gg c$) while matter is nonrelativistic ($v \ll c$). Radiation composed of photons remained the dominant ingredient of the universe for fifty-seven thousand years, so that period is called radiation dominance. However, because of the redshift caused by the universe's expansion, the energy density of radiation falls off faster than the energy density of matter, and the universe passed from radiation dominance to matter dominance. As we will see, matter dominance ended about five billion years ago. Since then, the universe has been increasingly dominated by what is called *dark energy* that behaves very much like a cosmological constant with an accelerating expansion of the universe.

Now we are talking here about mass/energy density, not number density. Zero-mass photons and low-mass neutrinos continued to outnumber everything else during matter domination. Then, at 380,000 years, the temperature dropped to the point where atoms could form. This is referred to as "recombination."[35] Recombination cleared the universe of most of its charged particles as positively charged nuclei and negatively charged electrons neutralized each other in atoms. And, most significantly, because photons no longer had charged particles with which to collide, they "decoupled" from the rest of the universe, which then became transparent. Over the next 13.8 billion years, these photons would cool to 2.725 K and become the cosmic microwave background we observe today.

At present, photons still outnumber atoms by a billion to one. Neutrinos decoupled much earlier, at two seconds, to form a relic background of their own at 1.95 K. Although there are hundreds in every cubic centimeter, these neutrinos produce no measurable effects, at least with current technology, and are generally ignored. As we will see, they are unlikely to be the ingredient of dark matter.

Decoupling also meant that the photon pressure in matter that served to counteract gravity and prevent it from collapsing was released and the process of structure formation by gravity infall could begin. This was aided by dark matter, which was there all along but had not participated in the electromagnetic interactions that had previously kept the photons and charged particles in equilibrium. And so whatever pattern of vibration the sphere of photons had at decoupling was frozen for all time. The regions of higher density were also hotter and the less dense regions colder, so temperature fluctuations in the gas traced its density fluctuations, and that pattern appears today in a CMB temperature variation across the sky.

LOOKING BACK TO THE BEGINNING

By the 1980s, the realization that the CMB carried information on the earliest moments of the universe spurred many efforts to make more-precise measurements of any possible deviations from the smooth distribution over the entire sky that had so far been observed. Inflation explained that smoothness but also suggested that variations or "wrinkles" should appear at about the one-in-a-hundred-thousand level. These anisotropies would prove to be a crucial test that would make or break the theory.

One of the leaders of this observational effort was a University of California at Berkeley particle physicist George Smoot. Working with Nobel laureate Luis Alvarez and others at the Lawrence Berkeley National Laboratory, which I frequently visited on other projects at the time, Smoot and colleagues had developed a *differential microwave radiometer* that measured temperature differences of the CMB from two directions in the sky.

In 1976, the Berkeley instrument made a series of flights aboard the Lockheed U-2 "spy plane." It detected the difference in temperature caused by the motion of the Milky Way, which includes our sun and Earth, at six hundred kilometers per second through the background radiation field. This is the so-called *dipole anisotropy* with the frequency of the CMB being blueshifted on the side we are moving toward and redshifted on the side we are moving away from because of the Doppler effect.[36]

In the mid-1970s, Smoot and colleagues proposed to NASA that they develop a satellite to be named COBE, the Cosmic Microwave Background Explorer. The spacecraft was to carry three instruments:

- DMR, a Differential Microwave Radiometer that was an improved version of Smoot's previous instrument for measuring temperature variations of the CMB across the sky at three wavelengths: 3.3 mm, 5.7 mm, and 95 mm.

- FIRAS, the Far-Infrared Absolute Spectrophotometer that would measure the wavelength spectrum from 0.1 to 10 mm.
- DIRBE, the Diffuse Infrared Background Experiment that would search for the cosmic infrared background in the wavelength range 1.25 to 240 microns.

After considerable delay caused by the *Challenger* disaster and other problems, COBE was launched on November 18, 1989, on the back of a Delta rocket. And thus began the next amazing chapter in humanity's comprehension of the cosmos.

13.

FALLING UP

WRINKLES IN TIME

As the decade of the nineties began, skeptics who had their own favorite theories were still questioning both the big bang and inflation.[1] In January 1990, at a meeting of the American Astronomical Society held in Arlington, Virginia, the first results from COBE were presented by John Mather, the principle investigator of FIRAS, the Far Infra-Red Absolute Spectrophotometer, one of the three onboard instruments. As described by George Smoot and Keay Davidson in *Wrinkles in Time*, when Mather showed the graph in figure 13.1, "A moment's silence hung in the air as the projection illuminated the screen. Then the audience rose and burst into loud applause."[2] The blackbody nature of the CMB had been conclusively confirmed.[3]

At this point it seemed that the big bang could no longer be held in doubt. No other proposed alternative could explain this result without ad hoc assumptions. However, inflation was still not out of the woods.

Bitter opponents such as the prominent astronomers Fred Hoyle and Geoffrey Burbidge, whose great contributions to stellar nucleosynthesis are not to be diminished, continued to speak out and even claim that inflation was already falsified by the lack of empirical confirmation.

But they were a bit premature.

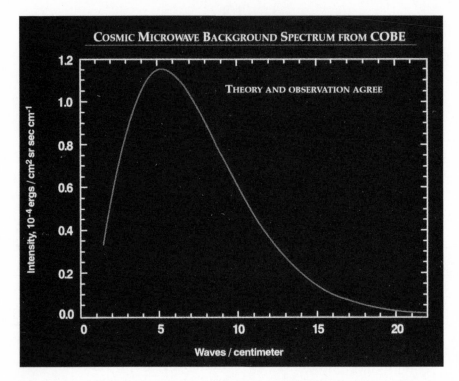

Figure 13.1. The CMB spectrum measured by FIRAS. The bottom scale is the reciprocal wavelength, which is proportional to frequency. The curve is a fit to the Planck blackbody spectrum for a temperature of 2.75 K. Image courtesy of NASA/GSFC.

On April 23, 1992, Smoot showed a series of maps of the CMB sky before a packed audience at an American Physical Society meeting in Washington, DC. To another standing ovation like the one given to his colleague John Mather two years earlier, Smoot demonstrated what he called "wrinkles in time" that fully confirmed the predictions of inflation.

Stephen Hawking, with some hyperbole, called it "The scientific discovery of the century, if not all time." Smoot said it was "like seeing God."[4] The *National Enquirer* (or a similar publication) showed its version of the data—the face of Jesus in the sky.

The temperature variations of the CMB predicted by inflationary cosmology were, after a decade of intense effort, finally confirmed.[5] Mather and Smoot would share the Nobel Prize for Physics in 2006.

In *The Inflationary Universe*, Alan Guth presents a graph of the COBE results redrawn in less-technical fashion, which is reproduced with his permission in figure 13.2.[6] The graph presents the CMB temperature difference squared averaged over all directions in the sky as a function of the separation angle of the two directions measured by the differential radiometer, from 0 to 180 degrees. The data agree perfectly with the shape predicted by inflation, although in the book Guth does not make any claim about the magnitude of the effect, which was adjusted in the figure to fit the data.

However, it was soon realized that if the anisotropy had not come in at about a magnitude of one part in one hundred thousand, inflation would have been in big trouble indeed, if not falsified.

In science, a model that is not falsifiable is not science. But when a model passes a very risky, falsifiable test such as this was, it earns the right to be taken seriously. Still, a note of caution must be added based on the history of science. Even when a model passes a test that could have falsified it, this does not mean that the model has been proved conclusively and will not someday be superseded by a better model. As we will see, however, when included as a part of a full-blown cosmological model, inflation still has more to offer than any alternative can match.

In a paper interpreting their results, the COBE team compared its observations with many proposed models and basically found that its measured anisotropies were very large compared to the inhomogeneities observed in galaxy surveys and so must be primordial. They concluded that COBE has provided "the earliest observational information about the origin of the universe, going back to 10^{-35} second after the big bang."[7]

At first it was thought that primordial fluctuations at the time of inflation on the order of 10^{-5} were insufficient to produce the galaxies given the observed density of matter. However, the answer was quickly found (someone in Smoot's audience had actually shouted it out): "Dark matter!"

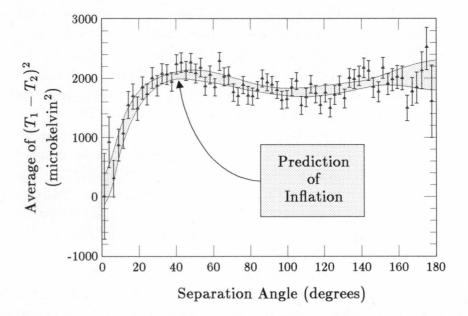

Figure 13.2. CMB temperature difference squared averaged across the sky as a function of the separation angle of the COBE differential microwave radiometer compared with the prediction of inflation. The magnitude was adjusted to fit the data. Image courtesy of Alan H. Guth.

How much was needed? As we will see, exactly the amount that seemed to be present—about five times as much mass as is carried by visible matter.[8]

As shown in figure 13.2, for separation angles above about thirty degrees the distribution is basically flat, which confirms the prediction of scale invariance. We need not get into the more-detailed shape since succeeding experiments, yet to be discussed, would provide greatly improved data and smaller angles would reveal remarkable structure that defined the early universe in great detail.

NEW WINDOWS ON THE UNIVERSE

CMB observations were not the only significant achievements of the 1990s, in what was another banner decade for astronomy and cosmology. I will only briefly mention a few selected examples that bear most directly on cosmology.

In 1990, the Hubble Space Telescope (HST) was carried into orbit by the space shuttle *Discovery*. Unfortunately, the primary mirror was severely flawed and had to be repaired in orbit. This was accomplished by shuttle astronauts in 1994, a spectacular achievement that in my mind was the most important contribution to science made by the whole shuttle program. Other Hubble-servicing missions were carried out in 1997, 1999, 2002, and 2009.

Covering the near-ultraviolet, visible, and near-infrared regions of the spectrum, Hubble provided the most detailed images ever recorded in astronomical history and mapped the universe out to the horizon. Still operating at this writing after twenty-three years, HST contributed to cosmology by providing greatly improved estimates of the Hubble constant and other key parameters. In its deep field observations, HST would show that the farthest, oldest galaxies were smaller and more chaotic than the closer, younger spirals, putting yet another nail in the coffin of the already-so-dead steady-state model.

The year 1990 also saw the Keck I multi-mirror ten-meter optical telescope installed on Mauna Kea. It was joined by Keck II in 1998. Recall that Mauna Kea is the best observational site on Earth and capable of peering into the infrared as well as visible part of the spectrum. One of the most productive ground-based astronomy projects of recent years, the Keck telescopes were among the first to find evidence for planets around stars (besides the sun). By determining the orbital speeds of stars near the center of our galaxy, the Keck telescopes helped establish that a four-million solar mass black hole resides at the center of the Milky Way.

In 1993, the Very Long Baseline Array (VLBA) of ten radio

telescopes was distributed from Hawaii to the Virgin Islands and operated remotely from New Mexico. By utilizing long-baseline interferometry, the array was able to achieve angular resolutions from 0.17 milliarcseconds to 22 milliarcseconds over a band of ten wavelengths from 0.7 centimeters to 90 centimeters. The array discovered two giant black holes, each 150 million times more massive than the sun and separated by only twenty-four light-years! They are located at the center of the galaxy 0402+379 that is 750 million light-years from Earth.

Indeed, it is now clear that most, if not all, large galaxies have supermassive black holes at their centers.

In 1995, the Infrared Space Observatory (ISO) was placed in orbit. It was designed to study the region of the spectrum from 1.5 to 196.8 microns. ISO made twenty-six thousand successful observations before failing in 1998.

Moving to the gamma-ray region of the spectrum, back in 1967, the Vela satellites whose purpose was to detect nuclear-weapons tests on Earth had serendipitously detected nonrepeating bursts of gamma rays located randomly in the sky. It was generally assumed that they were from inside our galaxy because of their brightness.

In 1991, the Compton Gamma Ray Observatory was launched into space. One of the instruments onboard, BATSE (Burst and Transient Source Experiment) was designed to detect and analyze gamma-ray bursts. It found an average of one a day, for a total of 2,700 detections. From these observations, the gamma-ray bursts were determined to originate from distant galaxies and thus to be emitting enormous amounts of energy.

Recently, NASA announced that the Hubble Space Telescope has detected such a burst and calculations have identified it as the collision of two neutron stars.[9]

VERY HIGH-ENERGY ASTROPHYSICS

Although the signals measured by radio telescopes are usually described as radio waves, they are composed of photon particles just like any other electromagnetic wave. The energy E of a photon in the beam of photons that make up an electromagnetic wave is given by $E = hc/\lambda$, where λ is the wavelength. You can write this $E = 1.97 \times 10^{-7}/\lambda$ eV when λ is in meters. Since the longest wavelength for the VLBA is about one meter, the photon energy in that case is less than a millionth of an electron-volt.

At the other end of the spectrum, the EGRET (Energetic Gamma Ray Experiment Telescope) detector on the Compton spacecraft had a maximum detectable photon energy of 30 GeV = 3×10^{10} eV, corresponding to a wavelength on the order of 10^{-17} meter.

At this time, there were a few, including myself, who sought to go even further, both in higher energy and in type of particles to look for. In the midseventies, I became involved in a project that proposed to place a large detector on the bottom of the ocean at a depth of 4.8 kilometers off the southern (Kona) coast of the Big Island of Hawaii. It was called DUMAND, for Deep Undersea Muon and Neutrino Detector. The purpose was to open up a whole new window ("nu window") on the universe by searching for very high-energy neutrinos from the cosmos, those with energies greater than 1 TeV (10^{12} electron volts). The original leader of the project was Fred Reines, who would share the 1995 Nobel Prize for Physics for his codiscovery of the neutrino with Clyde Cowan in 1956.

It was theorized that very high-energy neutrinos would be produced by the enormous energy sources that exist at the centers of active galaxies. (See discussion of active galaxies in chapter 9.) Since they would likely come from deeper in the heart of the galaxy than photons, we hoped they would provide information about these great sources of energy. In 1984 I published a paper in *Astrophysical Journal* showing that active galaxies could produce detectable very high-energy neutrinos under certain conditions.[10]

The proposed technique is still basic to all the experiments in very high-energy astrophysics that are still being conducted, along with proton-decay experiments. When a charged particle is moving faster than the speed of light in a transparent medium (but still less than c) such as water or air, it emits an electromagnetic shock wave called *Cherenkov radiation*, which is a bluish light that is detectable with very highly sensitive photodetectors called *photomultiplier tubes*.

DUMAND was designed to deploy a large array of these photodetectors at the bottom of the ocean where the background light from cosmic-ray muons was minimal. The proton-decay experiments Kamiokande and IMB mentioned in chapter 11 also used this technique, installing photomultiplier tubes inside large tanks of very pure water in mines deep underground.

While working on DUMAND in the 1980s, I became involved in another experiment that I thought would provide complementary information useful for DUMAND. A team led by physicist Trevor Weekes of the Harvard-Smithsonian Center for Astrophysics had built a very inexpensive mirror ten meters in diameter made of flat plates forming a spherical reflecting surface at the Whipple Observatory on Mount Hopkins in Arizona. At its focus were mounted several small photomultiplier tubes.

When a very high-energy gamma-ray photon hits the top of the atmosphere, it generates a shower of thousands of electrons and other charged particles that cascade toward Earth. The Whipple telescope was designed to detect the Cherenkov light produced by this air shower.

In 1989, after I had left the Mount Hopkins project to work on a similar experiment closer to home, on Mount Haleakala on Maui, Weekes and coauthors reported a highly statistically significant signal from the Crab Nebula.[11] The Crab is the remnant of a supernova that was recorded by Arab, Chinese, Indian, and Japanese astronomers in 1054.

The Crab always had been our most likely candidate and we watched it closely. In 1968, a rotating pulsar had been discovered in the center of the nebula and identified as a neutron star. With its rapid rate of rotation, once every 33.5 milliseconds, the neutron star's magnetic field can accelerate electrons to very high energy. When they collide with the surrounding gas, they generate gamma-ray photons, and, I hoped, also neutrinos.

The Crab is within our galaxy. In 1992, Weekes and company reported a detection of an extragalactic source, the blazar Markarian 421. My graduate assistant Peter Gorham and I had also regarded blazars as promising sources since their beams point toward Earth.

In the meantime, a German group installed another gamma-ray telescope called HEGRA (High Energy Gamma Ray Astronomy) in the Canary Islands. It confirmed the Whipple sources in 1996 and in April 1997 reported the detection of another blazar, Markarian 501.[12]

So the upper limit of the energy spectrum of observed signals from the cosmos was moved up another order of magnitude above EGRET, at a monetary cost many orders of magnitude less, I might add.

Yet these trillion-electron-volt photons, eighteen orders of magnitude more energetic than the radio photons detected by the VLBA, are not the highest-energy objects observed in the universe. With arrays of particle detectors spread over large areas on Earth, showers of particles produced by cosmic rays hitting the atmosphere, primary cosmic rays up to 1 ZeV = 10^{21} eV have been observed.

But, there is a limit, called the Greisen-Zatsepin-Kuzmin limit, of 0.5 ZeV for cosmic-rays particles to make their way across the universe. Above that limit they will lose energy by collisions with the CMB. Thus, these ZeV particles likely come from fairly close to Earth. One possibility is Messier 87 in the constellation Virgo, which is "only" fifty-three million light-years away and has an active nucleus that is believed to contain a supermassive black hole.

On the other hand, note that very high-energy neutrinos are not subject to this limit and are the only known method by which we can observe such energies at great distances.

At this writing, the nu window on the universe has been opened with the 1987 supernova and now some exciting new results at much higher energies. However, after a huge effort involving a number of very challenging and expensive ocean deployments of test instruments, DUMAND was determined to be technically too difficult and in 1995 was canceled by its funding agency, the US Department of Energy. Nevertheless, DUMAND did serve as a proving ground for the concept of very high-energy neutrino astronomy, and a number of other similar projects have been developed that built on what we learned. As we will see in the next chapter, these are beginning to bear fruit with the report in 2013 of the observation of twenty-eight neutrinos above 30 TeV in an experiment at the South Pole.

NEUTRINO MASS

Neutrinos from the sky still made headlines in 1998 when Super-Kamiokande provided the first conclusive evidence that neutrinos have mass. I played a small role in this experiment, which was my last research project before retiring in 2000. However, I had worked on neutrino physics and astrophysics for over two decades and the technique used for this discovery was one I suggested at a neutrino-mass workshop in 1980 and published in the proceedings.[13]

Neutrinos with nonzero mass were predicted to have a property known as *neutrino oscillation*. The three types of neutrinos listed in table 11.1 and their antiparticles are produced by weak decay processes, such as beta decay,

$$n \rightarrow p + e^- + \bar{v}_e$$

where $\bar{v}_e$ is an antielectron neutrino. However, these neutrinos do not have definite masses. The quantum state of each is a mixture of three other neutrino states that have definite mass, denoted v_1, v_2, and v_3. The masses (rest energies) are different so that the wave function describing a beam of each will have a different frequency.

Because of this difference, the mixture changes as time progresses. Suppose we start with a pure beam of v_μ's. The mixture changes with time so that when we detect a neutrino there is some probability that it will be one of the other types, v_e or v_τ. Neutrino oscillation does not occur when the masses are zero, so our observation of neutrino oscillation was direct evidence for neutrino mass.

High-energy cosmic-ray protons and other nuclei hitting the upper atmosphere produce copious amounts of short-lived pions and kaons. Their decay products include substantial numbers of muon and electron neutrinos, plus fewer tauon neutrinos. To reach the Super-K detector underground, a neutrino coming straight down from the top of the atmosphere travels about fifteen kilometers. By contrast, a neutrino passing straight up from the opposite side of Earth travels about thirteen thousand kilometers, so it has more time to oscillate.

Super-K observed an up-down asymmetry of muon neutrinos that was almost 50 percent at the highest energy of 15 GeV. Applying the theory of neutrino oscillations, this implied a difference in the mass-squares of two neutrino species to be in the range 5×10^{-4} to 8×10^{-3} eV².[14]

Additional experiments have pinned down the mass differences between neutrinos and determined that at least one has a mass on the order of 0.1 eV. This is to be compared with previously lightest known particle with nonzero mass, the electron, whose mass is 5.11 $\times 10^5$ eV, ten million times greater.[15]

Also in 1998, Super-K provided a neutrino photograph of the sun, shown in figure 13.3, taken at night through the earth — the first glimpse anyone has ever had of the core of a star.[16] For those who think that the sun disappears when it drops below the horizon at night, the picture proves it's still there.

Masatoshi Koshiba would share the 2002 Nobel Prize for Physics for his leadership of the Kamioka experiments.

Figure 13.3. The sun at night, as seen through Earth in neutrinos by Super-Kamiokande. Image from R. Svoboda, UC Davis (Super-Kamiokande Collaboration).

DARK MATTER

As we have seen, one of the biggest problems with the original big-bang model was that if at the earliest operationally definable moment in our universe the average mass density of the universe were greater than the critical density by more than one part in 10^{60}, it would have immediately collapsed. If it had been lower by that amount, it would have expanded so rapidly that the current universe would be essentially empty. This was called the flatness problem since it required that the geometry of the universe be almost exactly Euclidean. Inflation solves the flatness problem by expanding space by many orders of magnitude so it became flat and the density became critical.

However, astronomers have long known that the density of

visible matter of the universe, most of which is in luminous stars and dust, was far less than critical. Although reasonably good evidence for a large invisible component to the universe called dark matter goes back to the 1930s, most astronomers were slow to accept its reality for the very reasonable reason that they didn't see it directly with their telescopes. Missing mass was inferred by applying Newton's laws of motion and gravity to the observed orbital motions of stars in galaxies.

Maybe, some thought, these laws had to be modified on astronomical scales and some specific models were proposed. Ockham's razor, however, favors not rushing in to replace an existing theory — especially one so firmly ensconced as Newton's law of gravity — unless there is no other choice. And, so far, dark matter is the most parsimonious choice.[17] Note that, while Newton's law of gravity is superseded by general relativity, this does not change the conclusions about missing mass since Newton's law still applies in this case.

Still, there were problems to be solved if dark matter and inflation were going to survive. As described in chapter 9, the remarkable success that big-bang nucleosynthesis had in calculating the exact abundances of the light nuclei, especially deuterium, showed that the density of baryons, that is, familiar atomic matter, is at most 5 percent of the critical density. This includes not only luminous matter, which (galaxies and all) is a miniscule 0.5 percent, but also all the bodies made of atoms (planets, brown dwarfs, black holes) that do not emit detectible light. Dark matter is not just dark — it is not matter as we know it.

Since dark matter must be electrically neutral, stable, and weakly interacting in order to have remained undetected, only neutrinos among the known elementary particles are possible candidates for the dark-matter particles. They are not baryons.

Two types of dark-matter models have been generally considered: *hot dark matter* where the particles are relativistic, that is, travel at speeds close enough to light that they must be described by relativistic kinematics; and *cold dark matter* where the particles are nonrelativistic. However, we should still keep open an in-between possibility: *warm dark matter*. In the case of hot dark matter, the gravitational mass of the dark-matter particles essentially equals the particle's kinetic energy since the rest energy is negligible. For cold dark matter, the gravitational mass essentially equals the particle's inertial mass since the kinetic energy is negligible. The temperature, that is, average kinetic energy of dark matter, should be the same as that of the CMB, since they are in equilibrium and generate no heat of their own, although the relic cosmic neutrino background (CvB) is slightly cooler at 1.95 K. For warm dark matter, neither mass contribution can be neglected. However, since the temperature of the universe changes so rapidly, on the cosmic timescale, there is usually a rapid transition from hot to cold for any given particle.

Neutrinos were early candidates for hot dark matter. As we saw in the previous section, the mass of at least one neutrino is no more than 0.1 eV, the rest are even lower. So whether cosmic neutrinos are either hot or cold depends on their temperature. The transition from hot to cold occurred at about a million years after the big bang. Earlier a neutrino of this mass is hot; later it is cold.

However, the number of neutrinos with this mass needed to provide an appreciable fraction of the critical density at that mass is something like 10^{90}, an unlikely number. By comparison, there are "only" 10^{88} relic neutrinos, about the same number as photons in the CMB. The number of atoms is a billion times less. Thus, dark matter composed of familiar light neutrinos is pretty much ruled out from the latest CMB data, so we need to look for other dark-matter candidates. And, the proper sequence to follow is to examine first those possibilities that require the fewest new hypotheses.

While the standard model contains no candidates, there are two that do not require a complete overhaul of the theory but require only minor extensions: *sterile neutrinos* and *axions*.

Once it was discovered that the known neutrinos had mass, it was understood that they must be accompanied by another set of neutrinos not yet observed. It is conjectured that the additional neutrinos are *sterile*, that is, interact only gravitationally or at best very weakly. If they have appreciable mass, say more than a few hundred electron volts, then they are dark-matter candidates that still fit within the existing physics of the standard model, slightly extended to include parameters describing these states.[18]

While this book was in production, a variety of new observations had suddenly pushed sterile neutrinos to the forefront of the search for dark matter. This will be discussed in chapter 14.

Another hypothetical dark-matter candidate that still fits within the basic framework of the standard model is the axion, which was introduced back in 1977 to solve some technical problems with quantum chromodynamics. They are estimated to have masses less than 1 eV.

WIMPS AND SUSY

No other cold dark-matter candidate exists within reach of only minor changes in the standard model. If not a sterile neutrino or axion, it then has to be something totally new. Generically dubbed WIMP for Weakly Interacting Massive Particle, it most likely would be massive and nonrelativistic.

For a long time, the favorite candidate among physicists has been one of the particles predicted by an extension of the standard model that includes supersymmetry (SUSY), described in chapter 11. A generic term used for the WIMP that would arise from supersymmetry is *neutralino*. Four possible neutralinos are proposed that are the fermionic spartners of the gauge bosons of the standard model.

Evidence for SUSY was fully expected to be uncovered during the first runs of the LHC. It was not. Much of the theoretical effort over the last forty years has been based on supersymmetry, in particular, most theories of quantum gravity (super gravity) and *M*-theory. These could come crashing down if SUSY is not confirmed during the next run starting in 2015.

If this happens, many physicists will be disappointed, but not all — including me. Major discoveries in physics usually lead to simpler theories with fewer adjustable parameters. SUSY roughly doubles the number of adjustable parameters, and M-theory has 10^{500} different variations.[19] Despite their mathematical beauty, that makes them ugly in my empirical mind.

Now, these were not the only problems that confronted cosmologists as the second millennium of the Common Era drew to a close. By 1998 it had been established that dark matter, whatever its nature, can at most contribute about 25 percent toward making the density of the universe critical. Three-quarters of the mass of the universe required by inflation was still missing. Once again, inflation was on the verge of being falsified. And, once again, nature came to its rescue.

DARK ENERGY

Since Hubble first plotted the recessional speeds of galaxies against their distances in 1929, astronomers steadily improved their measurements, but their data continued to fit a straight line. That is, the slope of the line, H, which gives the rate of expansion of the universe, appeared constant. Indeed, it was called the Hubble constant.

However, there was no reason for H, the expansion rate, to be constant. At some point the graph was expected to start curving downward as the expansion was slowed by mutual gravitational attraction. That is, the expansion should decelerate.

On the other hand, in 1995 cosmologists Lawrence Krauss and Michael Turner pointed out that existing data at the time suggested that the universe had a positive cosmological constant that, in fact, contributes the bulk of the critical density. They noted that this should produce an accelerated expansion that shows up as an increase in recessional speed at greater distances, that is, the graph should curve up.[20]

The possibility of a positive cosmological constant had been mentioned earlier, in 1982, by the distinguished French astronomer Gérard de Vaucouleurs.[21] He noted that the spatial distribution of quasars indicated a small positive curvature that could result from a positive cosmological constant.

As we will now see, later observations confirmed this effect, but until then the conclusions of these papers were not widely appreciated or accepted.

The recessional speeds of galaxies are easy to measure from their redshifts. But, as we have seen, measuring distances has always been a challenging problem for astronomers. In the 1990s, two research groups applied a new standard candle based on a special kind of supernova that involves a white dwarf. This method greatly improved the accuracy of estimated distances to galaxies at the greatest distances.

White dwarfs are the remnants of relatively typical stars (such as our sun) that have burned all their nuclear fuel. What dim light they still give off results from stored thermal energy. If a white dwarf has mass less than 1.38 solar masses, it will remain relatively stable. However, if it is part of a binary system, it can accrete matter from its companion and gain sufficient mass to explode as a supernova. This is called a *Type Ia supernova*. Because the explosion occurs when a unique mass is reached, the peak luminosity of the explosion will be about the same for all such events.

From energy conservation, light from the supernova will fall off with the square of distance. So, by measuring the observed brightness of a Type Ia supernova, identified by its light curve (variation of brightness with time), its distance can be determined with unprecedented precision.

One of the research groups was called the High-Z Supernova Search Team, led by Brian Schmidt of the Australian National University and Adam Riess of the NASA Space Telescope Science Institute. It included twenty-five astronomers from Australia, Chile, and the United States who analyzed observations from the European Southern Observatory in La Silla, Chile.

The other group, led by Saul Perlmutter of the Center for Particle Astrophysics at the University of California, Berkeley, was called the Supernova Cosmology Project and included thirty-one scientists from Australia, Chile, France, Spain, Sweden, the United Kingdom, and the United States. They analyzed data from the Calán/Tololo Supernova Survey performed at the Cerro Tololo Inter-American Observatory, also in Chile.[22]

In September 1998, the High-Z group published evidence that the Hubble curve was turning up at large distances.[23] On June 1, 1999, the Supernova Cosmology Project published its results that it interpreted as evidence for a positive cosmological constant with 99 percent confidence.

The High-Z results are shown in figure 13.4. The top panel shows a measure of distance used by astronomers called the *distance modulus*, which is based on the measured and expected luminosities, plotted against the redshift z, which measures the recessional velocity. The data are compared with three models assuming different values of Ω_M, the energy density of matter, and Ω_Λ, the energy density of the vacuum, each as a fraction of the critical density. The lower panel gives the difference between the measured distance modulus and its expected value a model with $\Omega_M = 0.2$ and $\Omega_\Lambda = 0$. While the error bars on individual points are large, the data as a whole clearly favor a model dominated by vacuum energy rather than matter. Matter domination would make the data curve downward, as the mutual gravitational attraction of galaxies at great distances would cause them to recede more slowly. Instead we witness a speeding up, indicating gravitational repulsion. That is, the expansion of the universe is accelerating. The source of that repulsion was been termed *dark energy*, with an energy density estimated by these data to be about 70 percent of critical density.

Now, this would seem to violate conservation of energy. However, it does not. Recall from chapter 5 that the first law of thermodynamics is basically a generalized form of conservation of energy that applies to any material system — gas, liquid, solid, or plasma. Familiarly, an expanding gas does work, as in the cylinders of your car (assuming it has an internal-combustion engine). That's because most gases have positive pressure caused by the motions of its molecules and their collisions with the walls of the container.

However, according to general relativity the pressure induced by a positive cosmological constant is negative. This means that the work done by this pressure as the volume expands is negative. Unlike the familiar expanding gas, an expanding universe with negative pressure does work on itself. Since the amount of work equals the increase in internal energy, conservation of energy is preserved.

Perlmutter, Riess, and Schmidt would share the 2011 Nobel Prize for Physics for the dramatic proof that the universe is falling up.

As mentioned, this discovery was not totally surprising. It was well known to cosmologists that a positive cosmological constant, introduced by Einstein in his general theory, provides for gravitational repulsion. Indeed, we have seen that a de Sitter universe, which has no matter or radiation and just a positive cosmological constant, expands exponentially and offers a simple model for inflation in the early universe. Now we seem to have inflation going on today as well, albeit considerably slower.

Let us look briefly at the physics involved. The cosmological constant (see chapter 6) is equivalent to a scalar field of constant energy density uniformly filling the universe. Thus, as the universe expands, its total internal energy increases as its volume increases.

While the accelerating expansion of the universe could be the result of a cosmological constant, it need not be. Alternatively, the universe might be filled with a quantum field that has negative pressure. This field has been termed *quintessence*. Negative pressure is not unheard-of in other branches of physics. For certain ranges of pressure and temperature, a Van der Waals gas has negative pressure when its molecules are so close to one another that their electron clouds push away from each other and the molecules experience a net attraction.

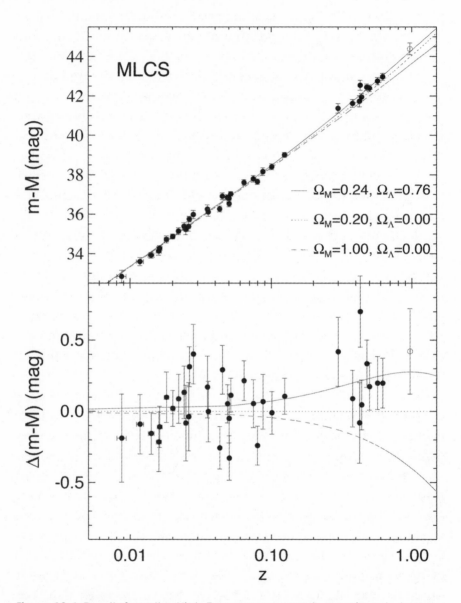

Figure 13.4. Results from the High-Z supernova search experiments. Image from Adam G. Riess et al., "Observational Evidence from Supernovae for an Accelerating Universe and a Cosmological Constant," *Astronomical Journal* 116, no. 3 (1998): 1009. © AAS. Reproduced with permission.

The quanta of the quintessence field would be bosonic, most likely spin-zero. It would be expected to have negative pressure by virtue of the quantum mechanical tendency of bosons to collect in a single state. The possibility of quintessence is included in the most advanced cosmological models by not automatically assuming a cosmological constant is the source of the acceleration.

THE COSMOLOGICAL-CONSTANT PROBLEM

In 1989, Steven Weinberg pointed out what he called "the cosmological-constant problem."[24] Because of the uncertainty principle, a quantum harmonic oscillator has a minimum energy that is not zero because it can never be brought exactly to rest. The lowest energy level has a *zero-point energy*.

A quantum field is mathematically equivalent to a quantum harmonic oscillator. So, if you take a quantum electromagnetic field, for example, and remove all its quanta (photons), you will still have energy left over even though no photons are present. Weinberg associated the energy density implied by a cosmological constant with the quantum zero point energy of the vacuum. When he made the calculation, it turned out to be 120 orders of magnitude greater than the highest possible value it could have and be consistent with observations.

Actually, Weinberg only considered photons, which are bosons. Fermions have a negative zero-point energy, so that will cancel some of the positive energy from bosons. That cancellation would be perfect if the universe were supersymmetric. But it is not—at least at low energies. So, the discrepancy is still fifty orders of magnitude. This is the cosmological-constant problem.

Any calculation that is so far from what is observed is surely wrong, and there are many proposed solutions. I discussed several in *The Fallacy of Fine-Tuning*,[25] but none have satisfied a consensus of physicists. Nevertheless, it is clear to me why the calculation is wrong.

The calculation of the energy density of the vacuum involved a sum over all the quantum states in a volume of space. However, the maximum number of states in a volume is equal to the number of states of a black hole of the same volume. And it is easy to show that the number of states of a black hole is proportional to the surface area of the black hole, not its volume. When you do the calculation by summing the states on the surface rather than the volume, you get a value consistent with observation.

BACK TO THE SOURCE

As we have seen, the CMB we observe today was produced when atoms were formed 380,000 years after the big bang and photons decoupled from the rest of matter. At that time, that surface contained the ripples in density that evolved from the original source over that period. Since then the universe has expanded by a factor of 1,100 and the radiation has cooled from 3,000 K to 2.725 K.

In CMB anisotropy observations, investigators measure temperature differences between two directions in the sky separated by an angle θ. When they look at the CMB in two regions of the sky separated by $\theta = 180°$ and find a temperature difference, this is called a *dipole anisotropy*. Recall that this particular anisotropy, which results from our motion relative to the CMB, was observed by Smoot and his team when they flew their original differential microwave radiometer onboard the U-2 spy plane in 1976. This effect is subtracted out for the analysis of the early universe.

When investigators look at four regions separated by 90° and see different temperatures, they have a *quadrupole anisotropy*. This has a background contribution from the Milky Way and is also ignored. In general, for a separation angle θ_ℓ in degrees we have a *multipole index* $\ell = 180/\theta_\ell$ and, as we will see, the higher poles, that is, smaller angles, are the most important.

When the square of the fractional temperature difference is

plotted against ℓ we have what is called the *angular power spectrum*. Theoretical analyses and computer simulations are able to use these measurements to reconstruct the *acoustic power spectrum* produced by the primordial fluctuations. With its limited angular resolution of 7°, COBE was only able to explore out to about $\ell = 20$. However, this was sufficient to verify that the fluctuations had at least the approximate scale invariance predicted by inflation. Calculations predicted that at angles less than 1° or ℓ greater than about 200 one should start observing peaks in the angular spectra that correspond to harmonics in the original acoustic oscillations (see chapter 11).

JUMPING ON THE BANDWAGON

Even before the COBE results were announced, research groups from all over the world rushed to hop on the bandwagon of what was recognized as a one of the greatest scientific opportunities ever presented — the ability to look back to the first moments of universe. On its Lambda website devoted to CMB science, NASA lists twenty experiments in operation during the 1990s using either ground-based telescopes or high-altitude balloons designed specifically to measure the anisotropy.[26]

Most had angular resolutions much better than COBE's 7 degrees, although they were unable to collect the amount of data that a satellite in orbit can. The Canadian SK telescope in Saskatoon, Saskatchewan, had an angular resolution from 0.2° to 2° in six frequency bands between 26 and 46 GHz enabling it to cover the range $\ell = 54$ to 404.[27]

Even more impressive was ACTA, the Australia Telescope Compact Array composed of five twenty-two-meter-diameter antennas 30.6 meters apart in an East–West line. Its angular resolution was a remarkable 2 arcminutes (0.03 degrees) at a frequency of 8.7 GHz and covered $\ell = 3350$ to 6050.[28] These experiments gave the first hints that much more was to be learned from the CMB, in particular, that the spectrum was not flat at smaller angles.

Although it would be quicker for me at this point to jump to the latest results, in this chapter and the next I am going to present a chronological series of plots of increasing precision both to demonstrate how science works and to give deserved credit to those who pioneered in this remarkable advance.

Figure 13.5 shows the status as of 1998 of the angular spectrum up to $\ell = 1000$ obtained by seventeen experiments.[29] There the first signs of the first (fundamental) acoustic peak can be seen.

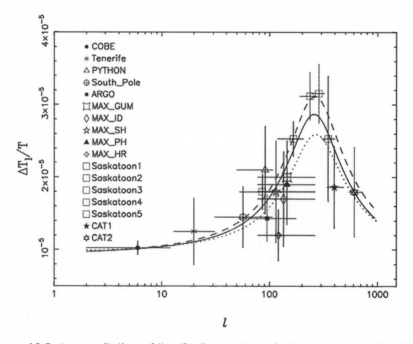

Figure 13.5. A compilation of the CMB angular anisotropy data as of 1998. Image used by permission of Oxford University Press, from S. Hancock et al., "Constraints on Cosmological Parameters from Recent Measurements of Cosmic Microwave Background Anisotropy," *Monthly Notices of the Royal Astronomical Society* 294, no. 1 (February 11, 1998): L1–L6.

At the time, two exceptional experiments using high-altitude balloon flights, called BOOMERANG and MAXIMA were gathering data that greatly improved the spectrum. We will hear about their results and those of the even more spectacular Wilkinson Microwave Anisotropy Probe (WMAP) and then the Planck space telescope in the next chapter.

And so, the second millennium of the Common Era ended with convincing evidence that our universe began with a rapid exponential expansion called inflation that ended after about 10^{-32} second. After a few billion years of more sedate expansion, our universe is now once again inflating exponentially, albeit at a much slower rate that is likely to continue forever. At some point in the far future, any inhabitants on a planet still heated by a sun will be unaware of anything else in the universe beyond the Milky Way and the Andromeda halo from when the two merged, since the rest will be beyond their horizon.

14.

MODELING THE UNIVERSE

SURVEYING THE SKY

I n chapter 12, I described how galaxy-redshift surveys revealed a visible universe marked by an incredible weblike structure of clusters of galaxies organized in filaments separated by almost empty voids. Since 2000 there have been dozens of new surveys that have greatly enhanced the database.[1]

The most extensive, the Sloan Digital Sky Survey (SDSS), uses the 2.5-meter-wide-angle optical telescope at the Apache Point Observatory in New Mexico. Starting in 2000 and still gathering data, SDSS has collected observations of five hundred million objects, including the spectra for five hundred thousand new objects for which light left seven billion years ago.

Of special cosmological significance, included in the SDSS survey is a baryon oscillation spectrographic survey (BOSS) that maps the spatial distribution of highly redshifted luminous red galaxies and quasars in order to obtain the acoustic signal of the baryons (atomic matter) in the early universe.[2] Just as the sound waves produced by primordial fluctuations have left an imprint on the CMB, they also left an imprint on the distribution of early galaxies. While these fluctuations produced inhomogeneities in the dark matter as well as in atomic matter, the dark matter offers no resistance to the gravitational collapse of high-density regions while the atomic matter has radiation pressure that opposes gravity. The two opposing forces result in an oscillation that affects the distribution of galaxies.

Using data for 46,748 luminous red galaxies in the redshift range 0.16 to 0.47, in 2005 a team led by Daniel Eisenstein of the Harvard-Smithsonian Center for Astrophysics reported a slight excess of galaxies separated by five hundred million light-years, matching the shape and location of the expected imprint of the acoustic oscillation at the time of recombination as predicted by the standard cosmological model.[3]

HEARING THE BANG

In the previous chapter, we closed out our review of the final decade of the second millennium of the Common Era by showing the angular spectra from COBE along with those from sixteen ground-based or balloon CMB experiments having better angular resolution but less statistical precision that followed shortly after (see figure 13.5). They provided the first indication of the expected fundamental acoustic peak that was inaccessible to COBE. The first year of the new decade would witness the strong confirmation of that peak and the observations of two more peaks by two spectacular high-altitude balloon flights and two more magnificent space telescopes.

The balloon experiments BOOMERANG (Balloon Observations of Millimetric Extragalactic Radiation and Geophysics) and MAXIMA (Millimeter Anisotropy Experiment Imaging Array) were each the product of large international collaborations. BOOMERANG flew over the South Pole in 1998 and 2003 at altitudes over forty-two thousand meters. MAXIMA made flights at forty thousand meters over Palestine, Texas, in 1998 and 1999. Their combined results, shown in figure 14.1, were published on a joint paper in 2001.[4] These confirm not only the fundamental peak at ℓ = 220 but also the smaller secondary peaks at ℓ = 500 and 750.

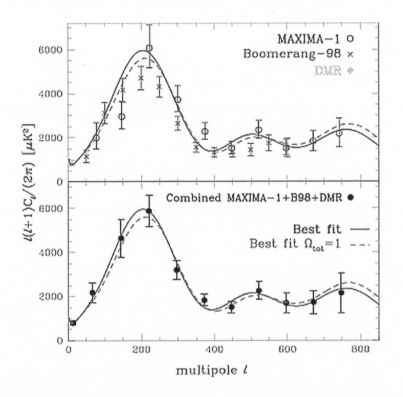

Figure 14.1. CMB angular spectra from BOOMERANG and MAXIMA. Reprinted figure with permission from Andrew H. Jaffe, P. A. R. Ade, A. Balbi, J. J. Bock, J. R. Bond, J. Borrill, A. Boscaleri, et al., "Cosmology from MAXIMA-1, BOOMERANG, and COBE DMR Cosmic Microwave Background Observations," *Physical Review Letters* 86, no. 16 (2001): 3475–79. © 2001 by the American Physical Society.

The data were fit to two models, the better fit having 70 percent dark energy and 20 percent cold dark matter and 10 percent baryons with the total density of the universe equal to the critical density within 4 percent.

But there's nothing like space (including the cost). On December 30, 2001, the NASA Microwave Anisotropy Probe spacecraft was launched from Cape Canaveral. It would later be renamed the Wilkinson Microwave Anisotropy Probe (WMAP) in honor of the microwave-astronomy pioneer David Wilkinson, who died in 2002.

WMAP gathered data for nine years and published its final results in 2013.[5]

Figure 14.2 shows the angular power spectrum from the first seven years of data.[6] The secondary acoustic peaks are evident. The curve is a fit to these data alone for a six-parameter model that will be described shortly.[7]

Just like light from the sun is polarized, so are microwaves. Polarization results, also shown, were also published by WMAP and other experiments.

Now, it is important not to expect that the sound spectrum plotted here will resemble exactly that of a musical instrument. Indeed, if the frequencies and intensities are shifted to ranges audible to the human ear, they are indistinguishable from noise. Watch, and listen to, the *Great Courses* lectures 15 and 16 by Mark Whittle for a delightful demonstration and his attempts to disentangle the various harmonics to make the "music of the spheres" more musical.[8] Also, see his webpage "Cosmic Acoustics."[9]

A number of "distortions" are present in the expanding ball of photons and other particles whose vibrations are producing the sound. These fill in the gaps and reduce some of the power in the higher harmonics in the angular spectrum. But what is wonderful is that these distortions provide us with information about the nature of that medium that we would not have from the pure spectrum alone.

A program that simulates the big bang called CMBFAST, written by Uros Seljak and Matias Zaldarriaga, is widely used to fit the CMB anisotropies and polarizations to various models.[10] Let us look at this model, which continues to describe all the data remarkably well although more sophisticated models will probe deeper as the data get better.

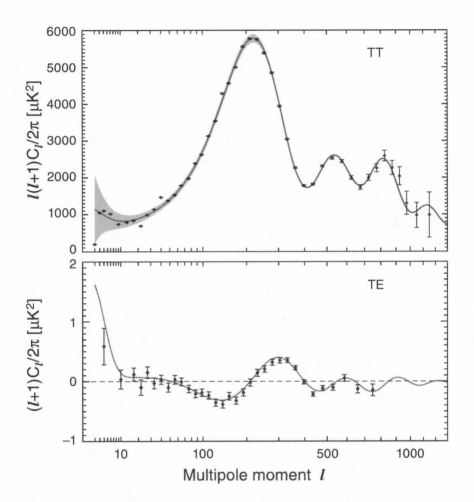

Figure 14.2. The WMAP seven-year temperature and temperature-polarization power spectra. The curves are fits to the data with the six-parameter ΛCDM model described in the text. Image from N. Jarosik et al., "Seven-Year Wilkinson Microwave Anisotrophy Probe (WMAP) Observations: Sky Maps, Systemic Errors, and Basic Results," *Astrophysical Journal Supplement Series* 192, no. 2 (2011): 14. © AAS. Reproduced with permission.

ΛCDM

As the increasingly precise measurements of the CMB angular power spectra and polarizations, along with other remarkable astronomical observations such as the accelerated expansion and the web of galaxies, were being produced by large collaborations of observational astronomers and astrophysicists, theoretical physicists and cosmologists were developing models to describe the data in terms of basic physics.

A comparatively simple model that was applied to the 2005 WMAP data is called the six-parameter ΛCDM model. It assumes a universe composed of baryonic (atomic) matter, cold dark matter (CDM), and dark energy (Λ) resulting from the cosmological constant. The model parameters are:

- Ω_b: the density of baryonic matter relative to the critical density
- Ω_c: the density of cold dark matter relative to the critical density
- Ω_Λ: the density of dark energy relative to the critical density
- n: the spectral index characterizing the original fluctuation power spectrum (see chapter 11)
- A: the amplitude of the original fluctuation
- τ: the reionization optical depth

Reionization has not been previously mentioned. To describe it I need to elaborate on the evolution of the universe from decoupling to the formation of the first stars, which is an important part of the story.

THE FIRST STARS

Just after photon decoupling at 380,000 years, the universe was a ball of thermal atomic gas, mostly hydrogen and helium along with the now-noninteracting photon gas, all at the same temperature of 3,000 K. This temperature corresponds to a peak wavelength of about one micron, which is in the near-infrared region of the black-body thermal spectrum. However, since the spectrum is broad, there is still plenty of visible light and the sky is bright orange.

As the ball of gas expanded, both components cooled accordingly, their spectra peaking at longer and longer wavelengths and the sky became increasingly redder until at about six million years the universe emitted little visible light. The period that followed, called the "Dark Age," lasted a few hundred million years until the first stars were formed and finally there once again was visible light.

Also expanding was the dark matter, which began to clump as it cooled, causing the much less massive atomic matter to clump along with it. Because of its weak interaction with other matter, dark-matter clumping did not result in any dissipation of energy. The atoms, on the other hand, collided with one another more frequently and so dissipated energy, cooling more rapidly than they would have from the expansion alone. This made it even easier for the atoms' own gravity and that of the dark matter to compress the atomic matter even further so that hot, dense cores formed inside a cooler surrounding medium. These cores eventually reached a high-enough temperature and pressure to ignite nuclear fusion so that the first stars could form.

Now these were not much like the stars in our current universe. The earliest stars were some hundred times more massive than today's stars and, furthermore, almost entirely composed of just hydrogen and helium. As a result, they burned at very high temperatures and emitted ultraviolet radiation that ionized the surrounding medium. This is called "reionization."

The first galaxies that formed as these stars clumped were

quasars and other active galaxies with supermassive black holes at the center that emitted intense radiation, also contributing to reionization.

And so, the previously dark, electrically neutral universe now once again contained charged particles. While their density was much lower than it was prior to decoupling, it was sufficient to cause space to lose some of the transparency it gained when the universe became neutral. The fog produced by reionization served to reduce the intensity of the CMB that we eventually would observe on Earth. This is parameterized in ΛCDM by the reionization depth parameter τ, which measures the thickness of the fog. From this, the time at which reionization occurred was estimated to be about four hundred million years after the big bang.

THE PLANCK SATELLITE

On May 14, 2009, the European Space Agency Planck satellite was launched from the Guiana Space Centre in French Guiana. It started taking data in February 2010 and published its initial results in March 2013. The CMB angular power spectrum is shown in figure 14.3.[11]

The media played up the fact that several of the numbers were different from the previous consensus, notably a somewhat-greater age of the universe. However there were no statistically significant disagreements for these numbers.

Much has been made of a so-called tension between the Planck results and those of the Hubble Space Telescope and other instruments that study galaxies that formed long after the CMB was emitted at decoupling. In particular, the model described above that fits the Planck data predicts a mass for galactic clusters that is about 80 percent of that measured from an all-sky survey of clusters.[12] We will discuss this further in a moment.

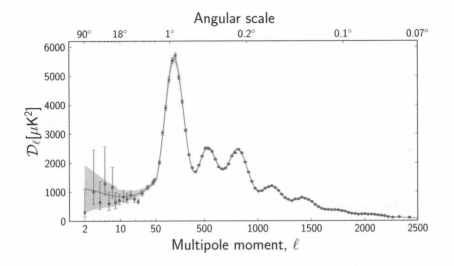

Figure 14.3. Angular power spectrum from Planck satellite published in 2013. Here we see the full harmonic structure determined with spectacular precision. Note that we have seven significant bumps. Table 14.1 lists a selection of parameters determined from fitting both the Planck and WMAP data along with other observations that I need not detail. The six-parameter ΛCDM described above fits the data well, although extended models containing more adjustable parameters have been tested. Image from Planck Collaboration, P. A. R. Ade, et al., "Planck 2013 Results. I. Overview of Products and Scientific Results," arXiv preprint arXiv:1303.5062 (2013).

The media often highlights reports of new theories that do away with the expansion of the universe, the big bang, or inflation. Pay no attention until one of those theories is able to reproduce the observations in figures 10.4 and 14.3 as accurately and parsimoniously as the inflationary big-bang model described here. And, as we will shortly see, recent observations have made inflation almost as certain as the big bang itself.

t_0	Age of the universe in billions of years	13.798 ± 0.037
H_0	Hubble's constant in km/sec/Mpc	67.80 ± 0.77
Ω_b	Baryon density as a fraction of critical density	0.04816 ± 0.00052
Ω_c	Density of cold dark matter as a fraction of critical density	0.2582 ± 0.0037
Ω_Λ	Density of dark energy as a fraction of critical density	0.692 ± 0.010
n	Spectral index of primordial fluctuations	0.9608 ± 0.0054
τ	Reionization optical depth	0.092 ± 0.013
Ω_k	Energy density of spatial curvature, 95% confidence level	-0.0005 ± 0.0066
Σm_v	Sum of masses of neutrinos in eV, 95% confidence level	< 0.230
N_{eff}	Effective number of neutrinos, 95% confidence level	3.30 ± 0.53
Y_P	Fraction of baryon mass in helium, 95% confidence level	0.267 ± 0.039
w	Dark energy equation of state factor, 95% confidence level	-1.13 ± 0.24

Table 14.1. Parameters determined by a fit to 2013 data from Planck, WMAP, and other experiments. Please note that these are just the best numbers as of the date of this publication and are sure to be modified as time goes on and the data improve.

The spectral index of primordial fluctuations is now slightly less than 1 with high statistical significance, confirming the expectation from inflation. Thus inflation has passed yet another falsifiable test. Of particular note is that the dark-energy equation-of-state factor is still consistent with $w = -1$, continuing to support the hypothesis that the cosmological constant is the source of dark energy. However the 21 percent error is still large, and some form of quintessence remains a possibility. Indeed, if the dark energy is quintessence with w very close to -1, it will be very difficult to distinguish any differences from a cosmological constant and to determine its exact nature.

Theoretical cosmologists continue to propose other models, one or more of which may turn out superior to what I have described. However, at this writing none is anyway near replacing ΛCDM.[13]

The Planck team has compared its data with a number of different models for the potential energy function that gave rise to inflation.[14] While several are ruled out, others remain viable. In particular, Linde's chaotic model, which I have been using as a specific example since it is the most natural and simple, is still in the running and even somewhat supported by the most recent results on primordial gravitational waves, which will be discussed shortly.

After exhausting its supply of liquid helium used as a coolant, the Planck Observatory shut down on October 23, 2013.

GRAVITATIONAL WAVES

One of the predictions of general relativity is the existence of *gravitational waves*. Just as an electromagnetic wave is produced by an oscillating charge and causes another charge at a distance to oscillate, a gravitational wave is produced by an oscillating mass and causes another mass at a distance to oscillate. But the gravitational effect is much smaller.

For years, attempts have been made to directly observe the oscillations of mass caused by gravitational waves. The most recent is the Laser Interferometer Gravitational Wave Observatory (LIGO). It is located at two sites 3,002 kilometers apart in Hanford, Washington, and Livingston, Louisiana, enabling source location by triangulation. Each observatory is composed of an L-shaped, high-vacuum pipe five kilometers on each side with a laser beam bouncing back and forth between mirrors at the ends of each pipe. A gravitational wave would be detected when it produces a slight change in the length of one beam relative to the other, resulting in an interference between the two laser beams, similar to a Michelson Interferometer. Since 2002, the results have been negative. Currently the observatories are being upgraded.

Once again, the cosmic background radiation provides another avenue of approach to a fundamental phenomenon. Not only has it, at this writing, presented the first significant evidence for gravitational waves, it has also provided the strongest confirmatory evidence yet for inflation.

Recall our discussion of B-Mode polarization in the section on gravitational lensing in chapter 11. B-Mode polarization cannot be produced by normal scalar inflaton field perturbations. However, its detection in the CMB power spectrum in the multipole range 30 $< \ell < 150$ is an almost-sure sign that fluctuations in the tensor gravitational field in the early universe have been greatly amplified by inflation. In their book *Endless Universe: Beyond the Big Bang*, Paul Steinhardt and Neil Turok called the detection of B-Mode polarization the "sixth milestone test of the inflationary scenario."[15] In 2001, Steinhardt, Turok, and two other coauthors proposed an alternative to inflationary cosmology called the *Ekpyrotic universe* in which the universe is produced by colliding "branes."[16] Branes are two-dimensional objects that occur in M-theory (see chapter 11). *Ekpyrotic* comes from the Greek *ekpyrosis* (ἐκπύρωσις) which was used by the Stoics to mean "conversion into fire."

The discovery of significant B-Mode polarization in the CMB with the expected power spectrum, peaking at $\ell \sim 80$, was announced with great fanfare on March 17, 2014, by another South Pole experiment called BICEP2 (Background Imaging of Cosmic Extragalactic Polarization).[17] The null hypothesis was ruled out at a statistical significance level of at least one part in 3.5 million. The data, shown in figure 14.5, fit the ΛCDM model with a tensor/scalar ratio of 0.20 ± 0.06. Cosmologists have urged caution, awaiting independent replication and the complete ruling out of all other possible sources of the effect. But if that occurs, we will have witnessed a discovery ranking among the most important in scientific history.

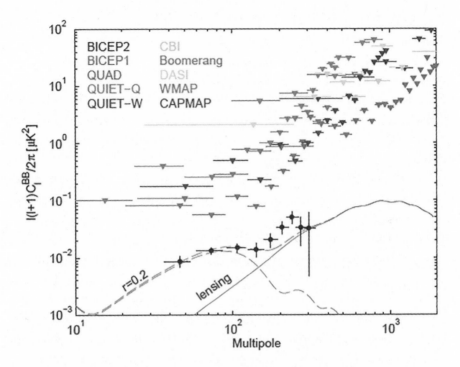

Figure 14.4. BICEP2 results for the B-mode polarization power spectrum compared to previous lower limits from a variety of observations. The circles are the data points at different multipoles ℓ, with error bars indicated. The dashed curve labeled $r = 0.2$ is the prediction of the ΛCDM model for a tensor/scalar ratio of 0.2. The solid curve is the expectation from gravitational lensing. Image from P. A. R. Ade et al., "Detection of B-Mode Polarization at Degree Angular Scales by BICEP2," *Physical Review Letters* 112, no. 24 (2014): 241101.

Note that the lensing effect, observed earlier by others, has a very small contribution at the lower multipoles.

The detection of this form of CMB polarization rules out most models that attempt to address the flatness, horizon, and homogeneity problems in cosmology without inflation, including the Ekpyrotic universe — as Steinhardt has admitted.[18]

SEARCHING FOR DARK MATTER

In chapter 11, I mentioned the phenomenon of gravitational lensing in which a large mass such as a cluster of galaxies can bend the light from a source so that multiple images of the same source are seen. Gravitational lensing has been used to great effect in confirming the existence of dark matter and mapping out its distribution, and we can expect a lot more examples in the future.[19]

Multiple images are produced when the mass of the lens is very high. This is called *strong lensing*. When the lens mass is lower we have *weak lensing* in which we do not get multiple images but simply a distortion of the image of the source. It can be stretched or magnified, or both. While a single galaxy can be elongated in shape, when a set of elongated galaxies line up in some way, we have a good sign that some invisible mass is acting as a distorting lens. From the amount of distortion and the distribution of the distorted galaxies, the mass of the lens and its distribution can be estimated. When this is done, there is clear evidence for dark matter.[20]

However, gravitational lensing does not tell us what the dark-matter particles may be. As of this writing, for two decades some thirty or so experiments have been conducted or are in current operation to detect and identify dark matter.[21] Recent results are tantalizing but not yet confirmed.

Basically two techniques are employed: *direct searches* that attempt to detect the passage of dark-matter particles through detectors, and *indirect searches* in which one looks for the secondary particles produced when dark-matter particles annihilate with one another. Until now, both techniques have been largely directed toward WIMPS, weakly interacting massive particles, and, in particular, the neutralinos predicted to exist from supersymmetry, with a few direct searches specially designed to detect axions. However, with the failure so far to see SUSY particles at the LHC, a lot of attention is now being given to sterile neutrinos as a possible alternative, as mentioned in chapter 13.

Most direct searches are deep underground to reduce backgrounds from cosmic rays, while indirect searches are done with high-altitude balloons or spacecraft. On October 30, 2013, the first results were reported from what is called "the world's most sensitive dark matter detector," LUX (Large Underground Xenon experiment) in Lead, South Dakota.[22] It failed to confirm earlier reports of signal "hints" from several less-sensitive experiments, pretty much ruling out WIMPs in the mass range 5–20 GeV.

Hints of dark-matter signals have also been reported from several indirect searches. Again the emphasis is on WIMPs, especially those suggested by supersymmetry.

Neutralinos are expected to be their own antiparticles, so they should annihilate into high-energy gamma rays, electron-positron pairs, or proton-antiproton pairs. The three experiments I will describe are indirect experiments that look for the products of neutralino annihilation.

PAMELA (Payload for Antimatter Exploration and Light-nuclei Astrophysics) is an experiment assembled by a collaboration from Russia, Italy, Germany, and Sweden. It is mounted on the Russian satellite Resure-DK1 that was launched on a Soyuz rocket on June 15, 2006. It is still operating. In August 2008, the collaboration announced that it had detected an excess of positrons in cosmic rays above 10 GeV.[23]

In November 2008, the high-altitude balloon experiment ATIC (Advanced Thin Ionization Calorimeter) launched from Antarctica, reported an excess of electrons in the energy range 300–800 GeV, although it could not distinguish positrons from electrons.

The Fermi Gamma-Ray Space Telescope is a collaborative effort of NASA and agencies in France, Germany, Italy, Japan, and Sweden. It was launched on a Delta rocket from Cape Canaveral on June 11, 2006. In 2009, the collaboration reported an excess of positrons that agreed with the PAMELA results.[24] More significant, in February 2014, PAMELA reported "a compelling case for annihilating dark matter" with the observation of an excess of 1–3 GeV gamma rays from a region around ten degrees from the center of the Milky Way. The signal is well fit by a 31–40 GeV dark-matter particle.[25] This may be the best indication yet for neutralino dark matter. Dark-matter masses around a few TeV implied by previous reports are higher than expected.[26]

Another major project to search for dark matter is the NASA Alpha Magnetic Spectrometer (AMS-02) onboard the International Space Station. The principle investigator of this large multinational collaboration is Nobel laureate physicist Samuel Ting of the Massachusetts Institute of Technology. AMS-02 was deployed by the space shuttle *Endeavor* on May 19, 2011.

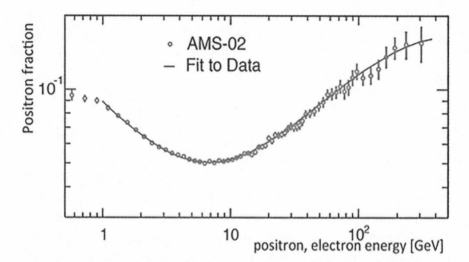

Figure 14.5. The positron fraction measured by AMS-02. The data are fit to a model that parameterizes the position and electron fluxes as the sum of individual diffuse power law spectra and the contribution of a single source. Reprinted figure with permission from M. Aguilar et al., "First Result from the Alpha Magnetic Spectrometer on the International Space Station: Precision Measurement of the Positron Fraction in Primary Cosmic Rays of 0.5–350 GeV," *Physical Review Letters* 110, no. 14 (2013). © 2013 by the American Physical Society.

In a paper published in 2013, Ting and collaborators presented the results shown in figure 14.5, showing a dramatic increase in the fraction of positrons from 10 GeV to 250 GeV based on 6.8×10^6 positron and electron events.[27] The positron fraction shows no fine structure, and no significant variations with time or preferred direction were observed, as expected if their source was dark matter. The result is consistent with the results from PAMELA but not the most recent Fermi result mentioned previously.

Note that the peak seems to be flattening out at high energy. If with more data the peak turns over, this would be strong evidence for the annihilation of a particle with mass of several hundred GeV. However, the result is still preliminary and we will have to wait and see. In any case, the authors note, the features of their data imply the existence of a new phenomenon of some type.

RECENT HINTS OF STERILE NEUTRINO DARK MATTER

In chapter 13 I mentioned the possibility that dark matter is composed of sterile neutrinos. These are three species of neutrinos that interact very weakly with the rest of matter, even more weakly than the three familiar neutrinos species, v_e, v_μ, and v_τ. They are expected to exist with a minimal extension of the standard model. The original form of the model assumed but did not require massless neutrinos and has been modified to allow for neutrino mass since its discovery in 1998.[28]

While this book was in production, two independent groups looking at satellite observations from superimposed clusters of galaxies reported a signal of 3.5 keV slightly above background.[29] It is conjectured that these results for the decay of 7 keV sterile neutrino into two photons, although this remains far from being a confirmed discovery. Galactic clusters are particularly concentrated centers of dark matter, and this mass is sufficient to constitute dark matter. These neutrinos would be "cold" by decoupling time.

Since a sterile neutrino is likely to be accompanied by two

others, sterile neutrinos in the mass range of a few eV might also solve the empirical tension mentioned earlier that exists between the predictions of the model used to fit the CMB data and telescopic observations of galactic clusters.[30] At decoupling, these neutrinos would still be "hot," in which case they would not clump so readily. Since they would still be a component of dark matter, this would result in lower cluster formation when galaxies later formed.

HIGHEST-ENERGY NEUTRINOS EVER

In the last chapter I mentioned that I worked for many years on a project called DUMAND, which sought to place a large neutrino detector on the bottom of the ocean near the Big Island of Hawaii to look for very high-energy neutrinos from extraterrestrial sources. A number of other experiments of the same nature have also been underway at other locations such as Lake Baikal in Siberia and the Mediterranean Sea.[31] DUMAND was eventually terminated because it was deemed too difficult and expensive to work in the deep ocean. A more hospitable environment than the ocean off Hawaii was found by a different team of scientists headquartered at the University of Wisconsin—the South Pole.

Once again, the technique involves detecting the Cherenkov light from the charged particles produced when very high-energy neutrinos collide with nuclei in a transparent medium, in this case Antarctic ice. In the 1990s, the AMANDA project (Antarctic Muon and Neutrino Detector Array) deployed strings of photomultiplier tubes deep in the ice at the Amundsen-Scott South Pole Station. Starting in 2005, it was expanded to encompass a cubic kilometer at a depth between 1,450 meters and 2,450 meters and renamed IceCube. Construction was completed in December 2010. IceCube is by far the most sensitive of any of the existing experiments.

On November 21, 2013, the IceCube collaboration announced that it had detected twenty-eight neutrinos above 30 TeV, two of which had energies exceeding 1 PeV (10^{15} eV).[32] If future data enables it to pinpoint sources, IceCube will have finally opened up the nu window on the universe. While this book was in production, a third PeV neutrino was announced.

Astrophysicist Floyd Stecker of the Goddard Space Flight Center, with whom I have worked in the past, has shown that the PeV neutrinos are consistent with a prediction he and three collaborators made in 1991 that ultra-high-energy neutrinos could be produced in the cores of active galaxies.[33]

A WORD OF CAUTION

As in the previous section, several of the new results I describe in this chapter (also in chapter 11) were inserted into the manuscript after it had already been submitted to the publisher. Some of the references still exist only in preprint form. Obviously, particle physics and cosmology are very rapidly developing areas of study, so nothing here should be taken as final. I can only present a snapshot of the situation as it existed when this book was published.

15.

THE ETERNAL MULTIVERSE

FROM THE BIG BANG TO NOW

Let us use a timeline to summarize what we know as of this writing, about the history of our universe from the big bang to the present. By *our* universe I mean what we can see with our telescopes today and what we can infer from those observations about the past of that universe. We will consider the future in the following section and the possibility of other universes in the next chapter.

Since our current physics knowledge, based on accelerator experiments, "only" goes back to 10^{-12} second after the big bang, the best we can do is speculate about earlier times. However, our speculations need not be merely wild guesses if we base them on solid existing knowledge.

The reader should keep in mind that I do not claim to depict what actually exists in some ultimate, metaphysical reality. As I have emphasized, my philosophical position is that that we cannot access precise knowledge of that reality. All we can do is make observations, as quantitative as possible, and describe those with mathematical models. These are based on our own human notions, operationally defined, such as time, space, and temperature. The following is such a description, with temperatures given in eV. Keep in mind it is a simplified model and certainly not the final word.

10^{-43} second, 10^{28} eV. The Planck Time.

At the Planck time, our universe was an empty sphere of Planck dimensions, 10^{-35} meter. Any model seeking to describe an empty universe in terms of the human-contrived notions of space and time is maximally symmetric. It will automatically contain the implied conservation principles people call "laws" that I have termed "metalaws." No lawgiver was involved, natural or supernatural. These models automatically obey the rules of quantum mechanics and relativity, which also follow from symmetries. This includes, in particular, the uncertainty principle and the Friedmann equations, which we use to model what happened next.

While empty of particles, quantum fluctuations in vacuum energy density or, if you prefer, the curvature of space-time, caused the sphere to expand and contract. At some unknown time, usually taken to be about 10^{-35} second, a positive fluctuation was large enough to trigger inflation. Friction in the expanding universe, predicted by the model, sufficiently slowed the return to equilibrium and our universe expanded exponentially by many orders of magnitude.

~10^{-34} second, 10^{24} eV. The End of Inflation (exact times unknown).

Like a damped harmonic oscillator, the same friction brought inflation to a halt and the almost-linear Hubble expansion took over. The first particles were generated from the energy lost. They were all massless.

Being maximally symmetric, the forces by which the particles interacted were unified as a single force. Maximal symmetry includes supersymmetry, with each particle and its spartner being identical with zero mass.

As the expanding universe cooled, a series of spontaneous (accidental, uncreated, undesigned) symmetry-breaking phase transitions took place that differentiated the types of particles and forces from one another. This was first step of the development of asymmetric structure in the universe. Particles were distinguished from antiparticles and given a slight preference. Or, more precisely, we label those given preference "particles" and the others "antiparticles." Bosons were separated from fermions, as supersymmetry was broken. Leptons were differentiated from baryons (quarks). Gravity was distinguished from the forces described by a grand unified theory (GUT) model. The particles remained massless.

Then, with further cooling, the GUT symmetry was broken and the strong force was separated from the electroweak force. Gluons were separated from the other bosons. The universe was now composed of all the particles and antiparticles of the standard model, with the quarks and gluons forming a strongly interacting *quark-gluon plasma* while the other particles moved around freely, scattering off one another in what was a very dense medium.

Note: Most timelines you will see in earlier books and articles place symmetry-breaking phase transitions prior to inflation, since they assume the inflaton field was the GUT Higgs field (GUTH?). However, this need not have been the case and the GUT and phase transitions could have occurred after inflation when there were particles present to work with.

10^{-10} second, 100 GeV. Electroweak Symmetry Breaking.

At this point we can begin to make more definite statements because as a result of accelerator experiments we already understand the physics involved. About this time, the electroweak unity of the standard model was broken and the electromagnetic forces and weak nuclear forces separated, giving us the four distinct forces we experience today — gravity, electromagnetism, and the weak and strong forces. The Higgs boson appeared and gave mass to the weak bosons and leptons while the photon and gluon masses remained zero. The weak force thus became limited to a range of about 10^{-18} meter while the electromagnetic force remained infinite in range. Quarks also gained mass, however only partially by the Higgs mechanism and mainly from their strong interactions with gluons, which themselves remained massless and buried in the quark-gluon plasma.

The sparticles left behind by the breaking of supersymmetry gained large masses with the lightest neutral one possibly becoming the dark matter. But, as we have seen, this scenario has not yet produced the SUSY particles expected at the LHC, so this remains uncertain. Perhaps the sparticles are too high in mass to see at that collider. We will have to wait and see.

10^{-6} second, 1 GeV. Quark Confinement.

When the temperature of the universe dropped to about a billion electron volts, quarks and gluons fashioned nucleons and the many composite hadrons that were discovered in particle accelerators in the 1960s and 1970s. Most are very short-lived and so decayed away with only protons and neutrons and their antiparticles remaining, along with electrons, positrons, and photons. All were in quasi equilibrium.

About this same time, antinucleons and nucleons annihilated, leaving behind a residue of one in a billion protons and neutrons. Photons and leptons dominated.

1 second, 1 MeV. Synthesis of Light Nuclei.

Neutrinos decoupled, forming the cosmic neutrino background. Light nuclei began to form. Free neutrons were all absorbed into nuclei or decayed away into protons, antineutrinos, and electrons.

10 seconds, 100 keV. Positron Annihilation.

Positrons and electrons annihilated, leaving a residue of one in a billion electrons.

3 minutes, 25 keV. Radiation Dominance.

Nucleosynthesis terminated. The energy density of photons exceeded that of nuclei, giving radiation dominance. The universe was opaque since the photons were swimming around in a charged plasma of nuclei and electrons with which they interacted as they do in a dense fog.

60,000 years, 1 keV. Matter Dominance.

The density of nuclei exceeded the photon density, moving from radiation dominance to matter dominance. The universe remained opaque.

380,000 years, 700 eV. Photon Decoupling.

Atoms form ("recombination"), photons decouple, and the universe becomes transparent. The sky is bright orange and gets redder as the universe cools. Atomic matter starts to clump along with the dark matter.

5 million years, 0.01 eV. The Dark Age Begins.

The universe has cooled so that the background radiation is now well outside the visible spectrum and the sky becomes dark.

200 million years, 0.002 eV. Star Formation Begins.

First stars form and the Dark Age ends. The stars are much larger than the sun and contain no heavy elements, so they burn out fast and we have many supernovae that synthesize heavier elements. Supernova radiation reionizes space and makes it slightly foggy, but nowhere near as much as during the Dark Age. Light can still get through, though somewhat dimmed. Active galaxies such as quasars may also start forming, adding to the ionizing radiation.

~1 billion years, 0.001 eV. Galaxy formation.

Galaxies form. They have frequent collisions and contain super-novae that continue spreading heavy elements throughout space, which then become ingredients of the next generation of stars. These stars are less massive and more slowly burning, like the stars we have today. Active galaxy formation slows down.

~6 billion years, 4×10^{-4} eV. Cluster Formation.

Denser regions begin collapsing and forming the various structures of galaxy clusters and superclusters.

~7 billion years, 4×10^{-4} eV. Acceleration Begins.

Up until this time, the expansion of the universe was decelerating because of the dominance of matter and radiation, which have attractive gravity. However, their densities have been dropping while the dark-energy density has remained constant. That density now exceeds the other densities, and since it has negative pressure, the expansion of the universe slowly begins to accelerate.

~8 billion years, 3 × 10⁻⁴ eV. The Familiar Universe Appears.

First spiral galaxies are formed.

9.1 billion years, 3.2 × 10⁻⁴ eV. Solar System Forms.

Our sun and planets form.

13.8 billion years, 2.6 × 10⁻⁴ eV. Now.

THE FUTURE

While predicting the future is always a risky task, we can still ask what timeline into the future is implied by current knowledge.

5 billion years. Bye, Bye Earth.

Our sun uses its last hydrogen fuel and becomes a red giant, incinerating Earth. In the following billion years, the sun contracts to a white dwarf.

17 billion years. Merging.

Milky Way and Andromeda merge.

~40 billion years. Structure Formation Ends.

The exponential expansion caused by dark energy overcomes any remaining gravitational clumping and structure formation ceases.

~100 billion years. Bye, Bye, Other Galaxies.

Other galaxies have all moved behind the horizon of Milky Way/ Andromeda, leaving it alone in the universe. Eventually everything is beyond the horizon of everything else.

~1 trillion years. Bye, Bye, Stars.

Remaining stars begin to die out, leaving behind black holes, neutron stars, brown dwarfs, and planets.

10^{33}–10^{37} years. Bye, Bye, Matter.

Protons and other heavy particles decay, leaving behind a gas of photons, electrons, and neutrinos that continues to thin forever.

No Heat Death

In chapter 5, we learned that physicists in the nineteenth century developed the concept of "heat death" in which the universe ultimately must reach an equilibrium state of maximum entropy. However, they still labored under the assumption that the universe was a firmament in which bodies maintained a constant average separation from one another. In this situation, the entropy of the universe has a maximum, which we now know (they didn't) is equal to the entropy of a black hole of the same size.

But heat death never happens. Rather, the universe continues to expand indefinitely and evolves to a pure de Sitter space where each causal patch never reaches maximum entropy. Now, the possibility remains that the universe is ultimately closed, that is, the curvature parameter $k = 1$. Of course, inflation implies that the universe is still very flat, better than one part in 10^{60}, but this can still happen with $k = 1$ where the universe has a tiny positive curvature.

Nevertheless, we can't look forward to a "big crunch" in which the expansion stops, the universe contracts back down to a Planck sphere, and everything begins all over again. This *oscillatory universe* was once very popular, but that was before the discovery of dark energy. As indicated above, about seven billion years ago (halfway back to the big bang) the energy densities of matter and radiation dropped below that of dark energy and the universe's expansion began to accelerate. If the source of dark energy is a cosmological constant, or something that looks very much like it, the energy density will remain constant while the radiation and matter die away and the universe continues to expand forever.

DID THE UNIVERSE HAVE A BEGINNING?

In 1970, Stephen Hawking and Roger Penrose showed that, within the framework of general relativity, our universe began as an infinitesimal point of infinite energy density in space called a *singularity*.[1] Theologians, notably William Lane Craig, have lifted this out of context and used it to claim that space and time must have been created at that instant, thus showing that the universe had a beginning. They then proceed to argue, without theoretical proof or empirical evidence, that anything that begins has a cause and, in the case of the universe, in a huge leap of incongruity, that cause must be the personal God of Christianity.[2]

However, as Hawking pointed out in his 1988 bestseller *A Brief History of Time*, "There was in fact no singularity at the beginning of the universe . . . it can disappear [from the theory] once quantum effects are taken into account."[3] Penrose agreed. General relativity is not a quantum theory and so cannot be applied wherever quantum effects are important. And this, in particular, includes the beginning of the universe.

But Craig and his fellow faithful never give up. In a debate with philosopher Alex Rosenberg held at Purdue University on February 1, 2013, and in other debates and writings, Craig referred to a 2003 paper by Arvind Borde, Alan Guth, and Alexander Vilenkin (BGV), which Craig says proves that the universe had to have a beginning.[4] It is to be noted that this paper is also based purely on general relativity and does not take quantum mechanics into account.

I have asked Vilenkin, whom I have known professionally for many years, "Does your theorem prove that the universe must have had a beginning?" He replied, "No. But it proves that the expansion of the universe must have had a beginning. You can evade the theorem by postulating that the universe was contracting prior to some time."[5] The same point has been made by Anthony Aguirre and Steven Gratton,[6] and Sean Carroll and Jennifer Chen.[7] Carroll skillfully engaged Craig in a debate on cosmology in New Orleans on February 21, 2014.[8]

In figure 15.1 we have a space-time diagram showing *worldlines*, which are paths in space-time, of particles emerging from the origin of an expanding universe.

BGV proved that all the worldlines had to originate from a point, which we can interpret as the beginning of *our* universe, that is, the one we live in. The other authors showed that the worldlines can be continued through the origin to the negative side of the time axis.

In short, *our* universe had a beginning, but it need not have been the beginning of *everything*.

Craig has come up with yet another argument for everything having a beginning and thus a creator. If it did not, Craig says, then it would have begun an infinite time ago, in which case we would never have reached the present.[9]

Recall that the same argument was made in the sixth century by John Philoponus and in the ninth century by Ya'qūb ibn Ishāq al-Kindī (see chapter 2). Theological arguments never die. Nor do they fade away.

In this regard, it is to be noted that most physicists and cosmologists (and even many mathematicians) use the term *infinity* when they really mean "endless" or "unbounded," or often just "a very big number." However, as brilliant work by mathematician Georg Cantor (1845–1919) showed, infinity is not a real number.[10]

time

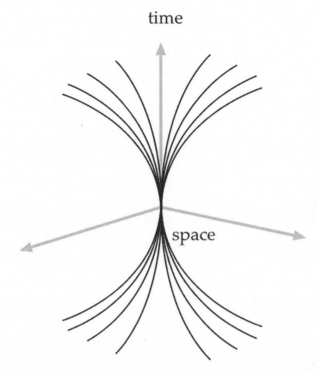

Figure 15.1. A space-time diagram showing worldlines of particles emerging from the origin of our universe. They can be extended in the negative time direction. Image by the author.

In 1925, the renowned mathematician David Hilbert (1862–1943) gave a talk in which he said, "The infinite is nowhere to be found in reality. It neither exists in nature nor provides a legitimate basis for rational thought."[11]

I assume that by "reality" he was referring to the measurements of observed phenomena we make in the world with our scientific instruments. Also, I am not sure what Hilbert meant in saying that the concept of infinity provides no legitimate basis for rational thought. Certainly his and Cantor's work with infinity was eminently rational. And theoretical physicists are not being irrational when they use the well-defined mathematical concept of infinity in their theories, so long as they are careful not to apply it to any quantities measured by us experimental physicists or to give it ontological, that is, Platonic, meaning.

Craig quotes Hilbert to bolster his argument for a beginning to everything. If there is no actual infinity, then everything cannot be eternal, that is, could not have begun an infinite time ago. Otherwise it would have taken an infinite time to reach the present. However, as I pointed out in chapter 2 when discussing the same argument made by Philoponus, an eternal universe would not have had a beginning an infinite time ago. It had no beginning. Running a clock backward and counting the ticks: -1, -2, -3, . . . , will never get it to -∞. The time from now to any moment in the past is a finite number of ticks.

Another simple way to see why everything need not have had a beginning is based on the fact that there is no reason for it to have an end. Cosmology indicates that our current universe will simply go on expanding forever. Then, since time is perfectly symmetric (with the conventional direction of time just being a definition based on common experience), even if we can identify a point in the past as the beginning of the big bang, that need not have been the beginning of everything. If there is no end, there must be no beginning.

AT THE PLANCK TIME

Figure 15.1 shows worldlines passing through a point at the origin of a space-time diagram. However, it should be recalled that quantum mechanics requires that this be a finite region of space and time and not an infinitesimal point. If this region is as small as possible, it should be of Planck dimensions, that is, a four-dimensional sphere with a spatial radius on the order of the Planck length and temporal dimension on the order of the Planck time — small, but not an infinitesimal point. Because of the quantum uncertainty principle, such a sphere would have a mass equal to the Planck mass, 10^{28} eV, and thus meets the main criterion for a black hole described in chapter 6.

Since we cannot see inside a black hole, we have zero information about what went on before the Planck time. And so the earliest possible time that has any instrumental meaning in our universe is the Planck time, 10^{-43} second. Here we must again make a distinction between theory and observations. While we can always write equations assuming time is a continuous variable and having any possible value no matter how small, we cannot even in principle measure a shorter time interval than the Planck time. This is why I do not get excited when some theory based on continuous space-time has problems with infinities. They are not empirical physics problems; they are theoretical physicist problems. The theory is simply wrong and has to be corrected.

As mentioned, the maximum entropy of a spherical body is equal to the entropy of a black hole of the same radius. So, at the Planck time, the universe was in a state of complete disorder or maximum entropy. This implies that anything that may have gone on earlier has left no imprint on our universe—other than utter and complete randomness. So, even if there were a creation, divine or natural, our universe possesses no memory of it. Not only is the god of most religions ruled out, so is the Enlightenment deist creator god (see chapter 2). The only possible god is a "quantum deist" god who started up the universe by throwing dice and then teleporting itself away to another reality. But then, who needs such a god—one who has no effect on anything?

Now, you might inquire, if the universe began with maximum entropy, does that not violate the second law of thermodynamics, which says that the entropy has to be lower in the past than it is now? No, because the entropy was lower in the past. But then, you ask, how could it have been lower in the past if it was maximal in the past?

Easy. At the Planck time the entropy was maximal for a sphere of Planck dimensions. As the universe expanded in the big bang, so did its maximum allowable entropy. And so, since the Planck time there has been room for local structures to form as the loss of local entropy during the formation of structure is offset by an increase in entropy of the environment, namely, the rest of the universe.

QUANTUM GRAVITY

The common wisdom is that we will not be able to describe even theoretically what may have gone on prior to the Planck time until we develop a quantum theory of gravity that encompasses all the features of both quantum mechanics and general relativity. Currently, quantum field theory includes the special theory of relativity but not the general theory.

Early attempts at quantum gravity, by Richard Feynman and others in the 1950s, were patterned after highly successful quantum electrodynamics. In place of the spin-1 photon, which is the mediating boson of the electromagnetic force, a massless spin-2 *graviton* was introduced to carry the gravitational force between two masses.

But the mathematics simply did not work out, requiring other avenues of approach that have yet to succeed.[12] Most depend heavily on supersymmetry and could come crashing down if SUSY fails to be verified at the Large Hadron Collider. This includes string theory, to which almost an entire generation of theoretical physicists has devoted their professional lives.

Gravity is clearly a different kind of force from the other two. You will often hear it said, even by prominent physicists, "Gravity is 10^{39} times weaker than the electric force." But that number is just the ratio of strengths of the forces between a proton and an electron and is not true in all cases. If instead of these you have two particles with the same electric charges but masses equal to, say, the Planck mass (a more natural mass than that of a proton or an electron), gravity is 137 times *stronger* than the electric force! There simply is no way to define an absolute strength of gravity the way you can for the other forces.

However, I can suggest a simple explanation for why gravity

on the scale of elementary particles is so weak compared to electro-magnetism. Their masses are intrinsically low, starting out zero and gaining a small mass by the Higgs mechanism described in chapter 11. I will elaborate further in chapter 16.

It also should be remembered that in Einstein's general rela-tivity, the phenomenon of gravitation results from the curvature of space and no explicit gravitational force enters the equations. In this model, Earth doesn't stay in orbit around the sun because the sun's force of gravity is pulling on it by exchanging gravitons or what-ever; it follows the natural path it should follow in the absence of any forces—a geodesic through space-time, which happens to curve around the sun. More recently, it has been suggested that gravita-tion can be described as an "emergent" phenomenon that arises from the tendency of systems to move in the direction of higher entropy (see chapter 5 for a discussion of emergence).[13]

Even if we had a quantum theory of gravity, we would seem to have no way of checking its predictions in the Planck regime where it supposedly applies. Still, quantum gravity may possibly have measurable effects. For example, at distances near the Planck scale, space-time is expected to be "lumpy," that is, instead of a smooth continuum we should see the famous "quantum foam" proposed by John Wheeler in 1955.[14]

Amazingly, it turns out that this foaminess is in principle detect-able. Recall the discussion of gamma ray bursts in chapter 13. They take place in distant galaxies, billions of light years away and emit photons of high energies. They are believed to result from the col-lision of two neutron stars. These photons could interact with the quantum foam and be delayed in reaching Earth. This effect might be observed by measuring the arrival times of two or more photons that were emitted simultaneously from the same burst.

Using data on gamma-ray bursts taken by the Fermi Gamma-Ray Space Telescope, astronomer Robert Nemiroff and collabora-tors compared the arrival times of gamma-ray photons of different energies emanating from the same sources. In the case of gamma-ray burst GRB 090510 recorded in May 2009, which is seven billion light-years away, three photons were observed to arrive within a millisecond.[15] The result places a limit on the size of the bubbles of space-time foam of 525 times less than the Planck length. Although the result needs to be independently confirmed, it would appear that space-time is smooth in the observable universe.

It might be asked how this result complies with the assertion I made in chapter 6 that no distance smaller than the Planck length can be measured. The answer has to do with the difference between theory and experiment, to which I have alluded time and again in this book. The actual measurement was not that of a distance shorter than the Planck length. However, that measurement was inserted into a theoretical model that then proceeded to predict a limit on bubble size much less than the Planck length. But that quantity itself exists only within the model and is not a direct observation.

THE BIVERSE

Let us now ask what might have existed on the negative side of our time axis, that is, prior to $t = 0$ in our past. Where did this primordial sphere of total chaos come from? While we possess no observable information on what may have gone on before the Planck time, we can continue to apply our best theoretical knowledge, that is, general relativity and quantum theory, which were both founded on empirical evidence from later times.

In a 2006 book and published article, I described a scenario that provides for a natural origin of our universe that follows from well-established physics and cosmology.[16] It was based on a model proposed in 1982 by David Atkatz and Heinz Pagels.[17] I worked out the scenario fully mathematically at a level accessible to undergraduate physics majors, relying heavily on a very nice tutorial published in 1994 by Atkatz, *Quantum Cosmology for Pedestrians*.[18] Here I will just outline the procedure.

In 1982, Atkatz and Pagels had shown how our universe could appear as a quantum-tunneling event. This mechanism was also proposed by Vilenkin in 1982[19] and by James Hartle and Stephen Hawking in 1983.[20]

We start with the Friedmann equations for an empty, homogeneous, isotropic universe with positive curvature, that is, with curvature parameter $k = +1$. Although our universe is very close to being flat, this does not necessarily require that the global curvature parameter $k = 0$; it could have $k = +1$ or $k = -1$ and still be very, very flat after inflation. Atkatz and Pagels showed that tunneling only works for $k = +1$.

With this equation in hand, we then follow the usual rules by which one goes from a classical equation to a quantum mechanical one in which real numbers are replaced by mathematical operators.[21] The result is surprisingly simple. You obtain the quantum mechanical time-independent Schrödinger equation for a nonrelativistic particle of mass equal to half the Planck mass and zero total energy, with a single dimension that is just the cosmological scale factor of the universe, which we can take to be the universe's radius. Note this is just a mathematical equivalence and does not imply that such a particle exists.

The derived equation is a simplified form of the Wheeler-DeWitt equation, whose solution is grandly referred to as "the wave function of the universe."[22] In the usual quantum mechanical way of interpreting wave functions, squaring the amplitude of the wave function of the universe gives the probability of finding a particular universe in an ensemble of many similar universes.

The resulting scenario is illustrated with the space-time diagram in figure 15.2. Time, t, is plotted vertically and two of the three dimensions of space, x and y, are shown in perspective. At a given time, the expanding spherical universe is projected as a circle perpendicular to the time axis. It emerges from a sphere of Planck dimension located at the origin, $t = 0$.

Also shown is a mirror universe on the opposite side the time axis from our universe. If we look at the Friedmann equations and other equations of cosmology, nothing prevents negative time. The mirror universe emerges from the same Planck sphere in the opposite time direction. Note this is consistent with the worldline picture shown in figure 15.1.

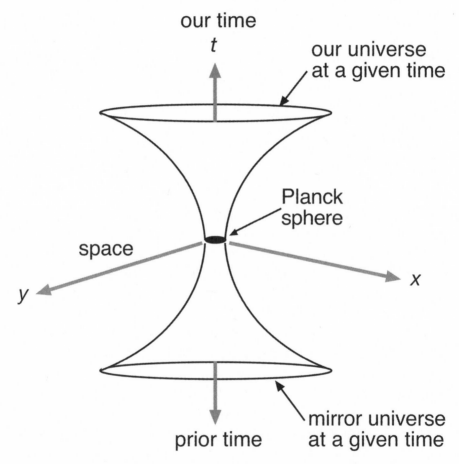

Figure 15.2. A space-time diagram showing our universe and a mirror universe emerging from the same Planck sphere in opposite time directions. One space dimension is suppressed, so the spherical universes and Planck sphere are projected as circles. Image by the author.

Now here is a widely unrecognized implication: While the mirror universe exists in our past, its arrow of time points opposite to ours. As we saw in chapter 5, no direction of time can be found in the equations of physics. The familiar, everyday arrow of time is a purely statistical definition in which forward time is the direction of most probable occurrences, as specified by the increasing entropy of our universe.[23] And so, because the mirror universe is expanding in the direction opposite to ours, the direction of the arrow of time that would be defined by an observer in that universe points opposite to ours.

Since the Planck sphere contains no information or structure, it is functionally indistinguishable from "nothing" and we have two universes tunneling from nothing. I will call our universe and its mirror partner the *biverse*.

I should add that we do not expect the mirror universe to be identical to ours, since it is just as likely to be as dominated by randomness as ours is.

While I cannot prove this is how our universe came to be, no one has proved it did not. That is, we have a plausible scenario for the natural, uncreated origin of the universe based on established physics and cosmology. It is just one of several possible scenarios that have been proposed, but this strikes me as by far the simplest since it requires no new assumptions — just established physics and cosmology. If nothing else, the biverse serves to refute any claim that our universe could have originated only by means of a supernatural creation.

At this point, the reader might ask where the "established physics and cosmology" came from. Please be patient. We will shortly get to that.

THE MULTIVERSE

If two universes are possible, then why not three or four—or any number? Since the 1980s, scientists have recognized the possibility that our universe may be just one of an unlimited number of similar but not identical universes dubbed the *multiverse*. As we saw in chapter 12, soon after physicists came up with the idea of inflation it was realized that the bubble-formation process that produced our universe would likely generate many more.

From the beginning, theists objected vehemently to the whole notion of multiple universes. Virtually all world religions teach the divine creation of a single universe with a central place for humanity. The multiverse challenges that belief.

On July 7, 2005, Christoph Schönborn, Cardinal Archbishop of Vienna wrote in the *New York Times*, "The multiverse hypothesis in cosmology [was] invented to avoid the overwhelming evidence for purpose and design found in modern science."[24]

Of course the Church also objected when Copernicus moved the Earth from the center of the solar system. And when Giordano Bruno said that ours was just one of many planets orbiting many suns, they burned him at the stake. Theists talk about humility, but they don't like it when science gives them a reason to be humble.

William Lane Craig has expressed similar sentiments as Schönborn. In the Purdue debate Craig said, "The proponents of chance have been forced to postulate the existence of a World Ensemble of other universes, preferably infinite in number and randomly ordered, so that life-permitting universes will appear by chance somewhere in the Ensemble."[25]

These statements are not only wrong but are insults to serious scholars. The "World Ensemble" or multiverse was motivated by established science—with no thought whatsoever to theology. It is the conclusion of our best current models of cosmology based on the extremely precise observations of modern astronomy and our best knowledge of fundamental physics.

Besides, suffice it to say, none of the claimed scientific evidence for purpose and design in the universe can be honestly classified as overwhelming.

A common objection to the notion of multiple universes is that it introduces additional entities when only a single entity, a lone universe, is needed. In 1986 astronomer Edward Harrison wrote: "Take your choice: blind chance that requires multitudes of universes, or design that requires only one."[26]

To help us decide on the best choice, we can apply the test of Ockham's razor, which favors the simplest hypothesis when there are several to choose from. At first glance, it might seem that a single universe is more parsimonious than multiple universes. However, Ockham's razor does not apply to the number of objects in a theory but rather to the number of hypotheses. The atomic theory of matter multiplied the number of objects physicists had to deal with by trillions of trillions. Yet it was simpler and more powerful than macroscopic thermodynamics, which preceded it and which can be completely derived from atomic theory. Similarly, since current science based on observations implies multiple universes, to postulate that only a single universe exists requires an additional hypothesis not required by the data. That is, the single universe hypothesis is the one that violates Ockham's razor.

In another objection, many nonbelieving scientists have joined theists in arguing that that the multiverse is "nonscientific" because we have no way of observing a universe outside our own. In fact, this is wrong. The multiverse is a legitimate scientific hypothesis since it seems to be an unavoidable consequence of eternal inflation, the current model of the early universe, to be discussed in the next section, which is based on our best observational data.

Our theories often contain unobservables such as quarks and black holes. And, as we will see shortly, empirical evidence for other universes is not beyond the realm of possibility. Any phenomenon forming part of a viable theory that is in principle detectable is a legitimate part of science.

The whole idea that many universes may exist is of such staggering consequence that it has been the subject of a huge literature in science, philosophy, and theology over the last thirty years or more.[27] Rather than attempt to review all that has been written on the subject, I will restrict my discussion to the latest conclusions we can draw from the simplest assumptions.

ETERNAL INFLATION

In 1983, cosmologist Alexander Vilenkin, with some trepidation, suggested what is now called *eternal inflation*.[28] According to eternal inflation, once expansion starts it never ends, with new universes being created all the time. In 1986, Andrei Linde elaborated the idea, showing how it was possible that the universe reproduces itself indefinitely and "may have no beginning or end."[29]

Eternal inflation, as conceived by Vilenkin and Linde, results in the continual production of universes inside of other universes in a fractal-like structure.[30] Basically, while a bubble universe is exponentially inflating to a much larger size, other bubbles can nucleate in an ever-growing empty de Sitter space surrounding the original bubble. This process continues eternally into the future.

In a model proposed by Anthony Aguirre of the Institute for Advanced Studies, Princeton, and Steven Gratton of Princeton University, the bubble nuclei are biverses similar to what I describe earlier, each with opposite arrows of time.[31]

What about the past? William Lane Craig has continued his effort to find evidence for a creation in cosmology by denying that time extends indefinitely into the past as well as to the future. Finally recognizing that our universe may not be all there is, he now says, "Even if our universe is just a tiny part of a so-called 'multiverse' composed of many universes, the BGV theorem requires that the multiverse itself must have an absolute beginning."[32]

As we have seen, the BGV theorem requires only that inflation have a beginning. It says nothing about the beginnings of multiverses. Indeed, in the biverse scenario, two universes have the same beginning and expand in opposite directions of time. In the eternal inflation picture, new universes are continually formed in each of these expanding spaces with opposite arrows of time.

I suppose Craig can claim, without proof or reason, that the whole shebang, what we might call the *Big Shebang*, was created by God. But let us put off discussing the theological implications until the end of the book and first address some remaining scientific issues.

SOLVING THE ENTROPY PROBLEM

In *From Eternity to Here,* Sean Carroll addresses a common question: Why was the entropy of the universe so low to begin with? This is called the *entropy problem*. Well, my glib response is that it was low to begin with because that's how we define "begin"—when the entropy was minimum. However, the real question, as Carroll explains, is why did such a highly improbable state occur at any time at all? If the universe is the result of random processes, then it should have begun with much higher entropy. It was like tossing a billion dice and having them all come up boxcars.

Of course, if the multiverse is unlimited, then every possible combination will come up an unlimited number of times. But that's too easy an answer and about as uninformative as saying God did it.

After carefully considering all the options, Carroll shows that the eternal multiverse offers a plausible solution to the entropy problem. He asks the key question: What should the universe look like if it is perfectly natural? His answer: "A natural universe—one that didn't rely on finely-tuned low entropy boundary conditions at any point, past, present, or future—would basically look like empty space."[33]

Now, as we have seen, empty space is described by the de Sitter solution of Friedmann's equations and can have a positive cosmological constant Λ, which is equivalent to a constant vacuum energy density and results in exponential inflation. The entropy can be shown to be inversely proportional to the magnitude of Λ. So when that is large, the entropy will be small.

In chapter 12, I described how, as suggested by Linde, quantum fluctuations in de Sitter space can drive the inflaton field up in potential energy, like a dad pushing his daughter on a swing. This is equivalent to generating a cosmological constant. In this case, we can envisage a region of space experiencing such a fluctuation and becoming an expanding bubble. Usually that bubble will shrink back to nothing.

Occasionally, however, the fluctuation will be occasionally very large. Then, because of friction, one bubble will remain at a higher energy long enough to expand by many orders of magnitude. It then can pinch off from the original background space to become a separate bubble universe.

Since the bubble universe has a large cosmological constant, it will have low entropy. This solves the entropy problem: In order for an inflating bubble universe to emerge from a quantum fluctuation, it must have low entropy. This does not violate the second law of thermodynamics because we can view the bubble universe plus background space as a single system in which the total entropy still increases, the background gaining the entropy lost (or more) when producing the bubble universe.

Note that all this can be done without any underlying dynamics, that is, by way of purely random, chaotic inflation. Since everything is symmetric, no specially created laws of physics are needed, just metalaws and bylaws. The models we use to describe what happens are logically required to contain, within their formulations, those principles that are required by symmetries—metalaws. This includes all the conservation laws, special and general relativity, and quantum mechanics. They follow from the symmetries of emptiness. Spontaneously broken symmetries then provide the bylaws needed for complexity to evolve.

DISCOVERING OTHER UNIVERSES

A common argument against other universes is that we have no way of ever observing them. However, perhaps we can. Early in our universe another universe may have been sufficiently close for its gravity to affect the isotropy of the CMB. Or, the bubbles may have collided, leaving a bruise on each. A detection of a large-scale anisotropy in the CMB could provide evidence for a universe outside our own. The Planck space telescope has confirmed several unexplained anomalies of this nature that were hinted at in earlier observations by WMAP.[34]

Since the observation of another universe beside our own would be the greatest scientific discovery in history, don't expect any cosmologists to make such a claim until they have ruled out every other possibility to the highest level of confidence and have seen the data independently verified several times. In the case of Planck, the investigating team has not deemed the evidence sufficiently significant to make any published claim.

Put simply, the single-universe hypothesis requires that the universe be spherically symmetric. Any significant deviation from this would prove there is something outside the universe. At some point, our theories may be able to make a prediction of the quantitative deviation from spherical symmetry expected in the multiverse model. And, at some point, the CMB data from future experiments may become sufficiently precise to test that prediction. This would make the multiverse hypothesis falsifiable. This prospect alone should be sufficient to permit the notion of multiple universes to remain a part of legitimate scientific discourse.

THE MANY WORLDS
OF QUANTUM MECHANICS

Conventionally, the multiverse of modern cosmology has been considered as unconnected with the *many-worlds interpretation* (MWI) of quantum mechanics. Recently, however, some authors have suggested a connection. Let's see what that connection might be.

It is important for us to again make a clear distinction between the *mathematical model* that constitutes the quantum theory used in making calculations and the *ontological interpretation* of what that theory may be telling us about the real world. The first is physics. The second is metaphysics.

The quantum model, which was already largely in place by the 1930s, has been hugely successful in describing the behavior of matter at the extremes of small distances, low temperatures, and high densities. However, the radical manner by which quantum methodology deviates from that of classical physics, which still works very well in other domains, has created a never-ending debate over what it all means.

Unlike classical models such as Newtonian mechanics or Einsteinian relativity (considered "classical" in this context), the quantum model does not predict where a particle will be at some later time but rather only the probability that it will be found in a certain volume of space. The probability per unit volume is given by the square of the amplitude of a mathematical object called the *wave function* or, more generally, the *state vector* (see chapter 6).

In the 1920s, Niels Bohr and Werner Heisenberg formulated what is known as the *Copenhagen interpretation* of quantum mechanics. While undergoing countless changes and appearing in many different forms over the years, for a long time Copenhagen was the closest thing we had to a "conventional" philosophical view of the meaning of quantum mechanics. Because of its many variations, some of which have been given other identifications, I will not try to present a comprehensive review but instead will focus on a few basic ideas common to most interpretations that follow more or less along the same lines as Copenhagen.

The fundamental, underlying assumption of these interpretations is that individual physical events are not predetermined by the laws of physics, as in Newtonian mechanics, but occur spontaneously. However, the statistical behavior of ensembles of similar events is predetermined, and this is what the mathematical model provides.

For example, when an atom in an excited state drops down to a lower state and emits a photon, that specific event is not predetermined, which in practice means it is unpredictable. However, the intensity of that particular spectral line, which results from a large number of photons making the same transition, is precisely calculable.

Similarly, no existing theory can predict that a particular radioactive nucleus will decay at a certain moment, but the hypothesis that such decays are equally likely for all times in a given time interval leads to the exponential decay "law" that is obeyed to great precision. In fact, this result and that for atomic transitions provide strong empirical evidence that these processes are not predetermined. That is, it is not our ignorance that makes these events random. They really are random.

Crucial to our discussion of many worlds, Copenhagen treats measuring devices as classical systems and so the act of measurement constitutes a quantum-to-classical transition that is not described by the theory but is simply assumed to take place upon the act of measurement. Before the measurement of a particle's position is made, the wave function gives the probability for wherever the particle was known to be prior to that time. If nothing is known, the particle can be anywhere in the universe. After a measurement, the particle is known to be within the volume of the detector and so the wave function is said to instantaneously "collapse" to give the new probability. This is illustrated in Figure 15.3.

Einstein objected to the whole idea of instantaneous wave-function collapse, calling it a "spooky action at a distance."[35] The collapsing wave function must move faster than the speed of light, indeed, infinite speed.

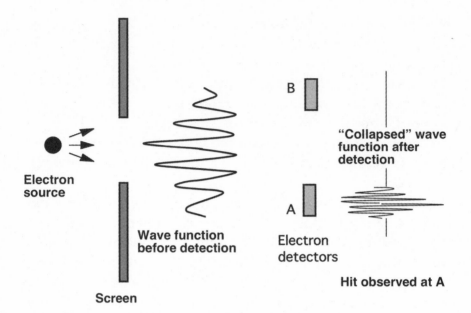

Figure 15.3. An illustration of the collapse of the wave function of an electron. After passing through a large hole in a screen, the wave function is spread out in a region of space about as big as the hole. After being detected, it collapses to the size of the detector. Note that, in this example, the electron had an equal prior likelihood of being detected by either A or B. Image by the author.

Most common interpretations of quantum mechanics usually are said to be "deterministic" in the sense that the statistical probabilities, given by the wave function or state vector, are predetermined in the theory—analogous to the way that particle motion is predetermined in Newtonian mechanics. However, here it's the collective behavior of an ensemble of identical systems that is determined statistically, not the behavior of an individual system. I prefer to call quantum mechanics "indeterministic."

Now let's look at some of the alternatives to this view. In the 1950s, David Bohm proposed an interpretation of quantum mechanics, based on a much earlier idea of Louis de Broglie, in which particle motion is predetermined. That motion is controlled by undetected subquantum forces, what de Broglie called "pilot waves."[36] While deterministic in principle, the model based on this interpretation does not predict individual particle motions. It still leads to the same statistical predictions as all other interpretations.[37]

In his 1957 doctoral dissertation at Princeton University, Hugh Everett III presented a brilliant new mathematical formulation of quantum mechanics. It removed the artificial separation between the quantum and classical realms that existed in the Copenhagen interpretation and it also did away with wave-function collapse.[38] Both were significant improvements. Everett's formalism included the detector along with the object being observed in the full quantum system and incorporated all possible outcomes of an experiment.

While Everett's mathematics was impeccable, other authors attempted to give the model a philosophical interpretation in which each time a measurement is performed the universe (really) splits into two separate, disconnected universes.[39] This became known as the many-worlds interpretation (MWI) of quantum mechanics.

Consider the situation shown in figure 15.3. The electron has a fifty-fifty chance of being detected at A or B. In the place of wave-function collapse, Everett's mathematical formalism has both possibilities take place. In the many-worlds *ontological* interpretation, the two events happen in separate universes or "worlds." In one world, the electron hits A. In the other world, it hits B. If the probability of hitting A is 3/4 and that of hitting B is 1/4, then we have four worlds, three of which have A hit and one of which has B hit. In this way the statistical nature of quantum mechanics is explained.

Indeed, each time an observation is made, the universe is envisaged as splitting into multiple universes, one for each possible outcome of the experiment. However, once again we have an ontological prescription that predicts the same empirical, statistical results as all the other interpretations of quantum mechanics.

While still a graduate student at Princeton in the 1940s, Richard Feynman proposed a mathematical formulation of quantum mechanics called *sum-over histories* that was quite different from the models that had become standard at that time. (Isn't it wonderful the way physics graduate students are able to solve problems that their more-experienced and much-higher-paid mentors cannot?) In Feynman's model, a wave function or state vector does not even appear, although they can be derived from it.

In the traditional classical procedure, the path of a particle is calculated by applying some equation of motion or, equivalently, a principle such as least action. In Feynman's quantum picture, the particle is assumed to take all possible paths between a source and detector. The probability for a particular observation is then constructed from all the possible histories that could have led to the observation. Although very useful and popular among particle physicists, sum-over histories again led to the same empirical results.

Feynman never attempted to give an ontological interpretation of his model. In *Timeless Reality* I showed how this notion can be reified using time reversal.[40] Although Everett does not reference Feynman, and their mathematics is different, Feynman's model was based on the same idea, namely, incorporate everything that can happen within the formulation. But please note: Just because we have empirically successful models in which everything happens that can happen, it does not follow that everything does happen in empirical reality.

Other interpretations of quantum mechanics exist, including variations on Feynman, that, like his, are not so extravagant as many worlds. The problem with the different interpretations is that some, notably many worlds, skirt very close to metaphysics. All are consistent with the data (those that are not are quickly discarded), but none makes a unique prediction. So we have no way of using the primary principle of scientific method—empirical testing—in order to judge one from the other. Since the calculational methodology we call quantum theory continues to agree with all observations, many of my fellow experimentalists are underwhelmed by the whole discussion and ask, "What else is new?"

Recent authors have seen a possible role for the many-worlds interpretation in cosmology. Recall my discussion of the biverse earlier in this chapter. I stated that what is called the wave function of the universe gives the probability of finding a universe with specific properties. Applying the many-worlds interpretation of quantum mechanics, the wave function of the universe represents an ensemble of parallel universes. Noted string theorist Leonard Susskind and others have suggested that this ensemble of universes can be associated with the many universes of the cosmological multiverse, which Susskind calls the *megaverse*.[41]

TIMELESS REALITY

One of the puzzles of the Wheeler-DeWitt equation, which gives the wave function of the universe, is its lack of a time variable. As described earlier in the section on the biverse, in its simplest form the Wheeler-DeWitt equation reduces mathematically to the quantum mechanical time-independent Schrödinger equation for a nonrelativistic particle of mass equal to half the Planck mass and zero total energy and with a single dimension that is just the cosmological scale factor. Solving it is analogous to perhaps the most familiar application of quantum mechanics, calculating the stationary states of the hydrogen atom from the Schrödinger equation. This gives the energy levels of the stable atom. Similarly, the wave function of the universe describes the stationary state of the universe, which has zero total energy.

So where, then, does time enter the picture? How can we have a model of the universe that does not contain that most basic of human conceptions—the passage of time?

In 1983, physicists Don Page and William Wootters wrote that a closed system such as the universe must be in a stationary state and so time is not a necessary element in the description of the world. They show that "the observed dynamical evolution of a system can be described entirely in terms of stationary observables."[42]

In a second paper published in 1984, Wootters further explained:

> Any statement we would ordinarily make regarding the time dependence of a system can without loss of observational content be cast in the form, "If the clock is found to be in the state . . . then the probability of finding the system in the state . . . is . . ."[43]

More recently a rash of papers have appeared from both physics and philosophical perspectives describing time as an "emergent phenomenon."[44] They are all highly technical, but I think I can present a simplified basic idea.

In my 2000 book *Timeless Reality*,[45] I showed how the quantum state of a particle at a given place in space is actually a linear superposition of two states, each going in opposite directions in time. The result is that the state of the particle is "timeless." A measurement of that particle breaks the coherence between the two states, and time "emerges," although I did not use the term. Note that the process is time-symmetric so the biverse is a natural outcome.

Note also that, as I explained in chapter 6, space and time are closely related. In fact, both are operationally defined in terms of what is read on a clock. So, if time is emergent, so is space. Wootters recognized this and suggested that a workable quantum theory of gravity might be built on the idea of timelessness (and spacelessness).[46] This possibility had been suggested even earlier by others, but the idea has not yet achieved fruition.

None of this should be given the mystical interpretation that the mind of a human observer creates time and space as some intrinsic reality. Time and space are names we give to quantities that are registered in the mechanism of a clock and have no meaning without a clock. No human or other sentient being need be involved in the actual measurement.

Once again we witness the futility of trying to attach metaphysical meaning to the quantities in physical models. If you can't read all these papers on emergent time (and emergent gravity), don't worry about it. This is just another example of model-builders seeking ways to logically build the operationally defined quantity we call time into the overarching model of a multiverse described by the timeless Wheeler-DeWitt equation.

ARE MANY WORLDS REAL? WHAT IS REAL?

Cosmologist Max Tegmark has made a strong claim that the many universes of both cosmology and quantum mechanics are part of an ultimate reality that is fundamentally mathematical. He has introduced what he calls the *Mathematical Universe Hypothesis* (MUH): "Our external physical reality is a mathematical structure."[47]

Tegmark defines mathematical structure as "abstract entities with relations between them." He argues that MUH provides for a description of external reality, a reality independent of us humans, involving "no baggage." By that he means, "Our descriptions of entities in the external reality and relations between them [are] completely abstract, forgoing any words or other symbols used to denote them to be mere labels with no preconceived meanings whatsoever."[48]

I find myself agreeing with Tegmark, with the exception of his calling this scheme "reality." Based on my different experience as an observer and experimenter, I consider the equations of theorists such as Tegmark to be no more than beautiful and useful descriptions of what we record with our eyes and instruments, analogous to a photograph of the sunset taken from Waikiki Beach. For all its spectacular beauty, the photograph remains a crude human construction that, at best, only images reality. I believe Tegmark's picture is consistent with that interpretation — describing the objects of theoretical physics as mathematical relations between observations. He calls them "reality." I call them "mathematical relations between observations."

I am hardly alone in this opinion. In their bestselling book *The Grand Design*, published in 2010, Stephen Hawking and Leonard Mlodinow say, "There is no picture- or theory-independent concept of reality." Instead, they endorse a philosophical doctrine they call *model-dependent realism*, which is "the idea that a physical theory or world-picture is a model (generally of a mathematical nature) and a set of rules that connect the elements of the model to observations."[49] They add, "It is pointless to ask whether a model is real, only whether it agrees with observations."[50]

I am not sure how model-dependent realism differs from *instrumentalism*, which is the view I have promoted in this and previous books. In both cases we concern ourselves only with observations and, while we do not deny that they are the consequence of some ultimate reality, we do not insist that the models describing those observations correspond exactly to that reality.

William Lane Craig seems to think that model-dependent realism is equivalent to *ontological pluralism*, which is the notion that there are many different, independently valid realities.[51] I can see how this ontological doctrine can result from taking our successful theories too seriously, as so many theoretical and experimental physicists do, and assuming they are the "true reality." But this is not rational. First, theories are always being replaced by better ones. Second, often, as in the case of quantum mechanics, we have multiple interpretations. In both of these cases the different theories and interpretations are often logically inconsistent, although they all fit the same data. If one is real, they all are.

However, I don't think that's what Hawking and Mlodinow are saying at all, as the above quotation makes amply clear. Still, their choice of terms, calling their doctrine model-dependent *realism*, serves to confuse their position rather than clarify it.

Getting back to the connection between many worlds and the multiverse, at this stage in the game, until some empirical consequence can be extracted from this discussion, the significance of an association between the two becomes a matter of one's philosophical perspective. Are the many worlds in Everett's quantum mechanics "real"? Is the quantum wave function "real"? From my empiricist's perspective, I am not at all disturbed that a collapsing wave function is "spooky." Casper the Ghost is spooky too, but no more and no less a human creation. All the worlds in Everett's model are likewise human creations—part of a model that successfully describes observations but provides no basis for assuming they all exist in whatever ultimate reality is out there. Similarly, Tegmark's mathematical reality and Hawking's and Mlodinow's model-dependent realism should not be mistaken for whatever ultimate reality may lie beyond our observational verification.

THE MORMON MULTIVERSE

We have seen that Christian theologians going back to Pope Pius XII in the 1950s and exemplified by William Lane Craig today have seized on the big bang as evidence for the creation of a finite universe a finite time ago in accord with the conventional Judeo-Christian-Islamic tradition. Even when the multiverse is brought into the picture, Craig argues that it had to be created *ex nihilo* as well. However, as we have seen in this chapter, the multiverse is now thought to be eternal.

No doubt the fine art of Christian apologetics will always find ways to reconcile Christian theology with whatever science comes up with, as they did with the teachings of Plato and Aristotle: Pick and choose what you like and ignore what you don't like. Other religions can do likewise.

However, I find it fascinating that Mormon scholars are saying that the multiverse is not only congenial to their particular cosmology but also inherent in it and helpful in resolving some of the difficulties that Mormon cosmology has with single-universe big-bang theory.[52]

According to Mormon scholar Hyrum Andrus,[53] "Revelations

to Smith concerning the nature and order of the cosmos indicate that there are innumerable worlds in space populated by organized beings in varying degrees of progressive attainment."[54]

And, the universe is eternal. Andrus adds:

> Joseph Smith had no part in the prevailing misconception of his day that matter was created out of nothing. Instead, he proclaimed that matter and intelligence are eternal and that God is in time and space as the great Organizer of self-existing matter and things. The Prophet instructed the brethren concerning the "Eternal Duration of Matter," stating that "the elements are eternal," that "earth, water, etc., had their existence in an elementary state from eternity," and that "no part or particle of the great universe could become annihilated or destroyed."

In any case, just as the cosmologies in the Torah, Old Testament, and Qur'an bear not the slightest resemblance to modern cosmology, neither does that found in the Mormon *Book of Abraham*, which like the *Book of Mormon* is regarded as a revelation from God to Joseph Smith. This describes a star or planet called Kolob at or near the throne of God that later Mormon scholars interpreted as the center of the universe.[55] Needless to say, there is no empirical evidence for Kolob.

THE MULTIVERSE IN PHILOSOPHY AND LITERATURE

The idea of multiple universes and alternate worlds has been a favorite topic of speculative philosophy and science fiction for many years.[56] In a *Star Trek* episode in the 1960s called "Mirror, Mirror," Captain Kirk meets his other self. In Philip K. Dick's 1963 novel *The Man in the High Castle*, America is a Nazi puppet state.[57] In Philip Roth's 2004 novel *The Plot against America*, Nazi sympathizer Charles Lindbergh defeats Franklin Roosevelt for president in

1940.[58] Robert Cowley has edited a collection of essays by eminent historians titled *What If?* In which they envisage what history would have been like had certain major events turned out differently.[59]

As we saw in chapter 1, multiple worlds were included in the cosmology of the ancient atomists. In *Academica*, the Roman historian Livy (59 BCE–17 CE), who was not an atomist, wrote:

> Would you believe that there exist innumerable worlds . . . and that just as we are at this moment close to Bauli and are looking towards Puteoli, so there are countless persons in exactly similar spots with our names, our honours, our achievements, our minds, our shapes, our ages, discussing the very same subject?[60]

In chapter 3, we saw that the great philosopher Leibniz recognized that many alternate worlds were possible. He justified the existence of a benevolent deity in the presence of all the evil and suffering in the world by proposing that this was still "the best of all possible worlds." He pictured an infinite pyramid with a room for each possible world with ours, the one true world, at the apex.

In *Candide*, Voltaire ridiculed Leibniz's optimism through the adventures of his title character, who experiences many of the evils and sufferings of human life including the Lisbon earthquake and tsunami of 1755 that killed an estimated 40,000–50,000 people in Portugal, Spain, and Morocco. Leibniz is represented in the tale by Dr. Pangloss, "the greatest philosopher in the Roman Empire," who continues to promote his optimistic view while suffering many misfortunes himself before being hanged by the Portuguese Inquisition to appease God.[61]

Finally, the "multiple worlds" mentioned in the previous sections and by the various historical figures described in the early chapters of this book should not be taken too literally as presaging the multiple universes of the scientific multiverse. These individuals were likely thinking in terms of other worlds like Earth, rather than other universes.

In the final chapter we will examine the theological implications of multiple planets and universes.

16.

LIFE AND GOD

PLANETS AND LIFE

Another recent gigantic leap forward in our knowledge of the universe, which will play a large role when we consider the theological implications of the cosmos as we now understand it, has been the growing evidence that the universe is teeming with planets. Furthermore, a huge number are likely to exist that are capable of sustaining some form of life based on chemical processes. This does not even include the possibility of other forms of life that may not be chemical.

There was never any theoretical reason to think that planets weren't plentiful. But finding actual evidence has been a daunting task because of the brightness of the stars they orbit. Nevertheless, modern technology has once again met the challenge.

Only a dozen or so planets have been detected by direct imagining, all except one, Fomalhaut b, being much larger than Jupiter. However, thousands of others have been discovered by indirect means in which the motion of the planet causes observable changes in the light seen from its parent star. As of November 6, 2013, 1,039 confirmed planets were listed in *The Extrasolar Planets Encyclopedia*.[1]

The Kepler Space Observatory was launched on March 7, 2009, with the express purpose of searching for Earth-sized planets. As of November 2013, Kepler had confirmed the detection of 167 planets with another 3,568 candidates awaiting confirmation.[2] On May 11,

2013, the failure of reaction wheels that are used to steer the spacecraft brought the main mission to a close. However, since the rest of the system is intact, Kepler is now operating in a new mode.

On November 4, 2013, astronomers from the Universities of California and Hawaii reported on their analysis of the measured brightnesses of forty-two thousand stars from the Kepler mission. They found 603 planets including ten that were Earth-sized, that is, with radii 1–2 times the Earth radius that also received comparable levels of stellar energy. That is, the planets orbited in a habitable zone where liquid surface water could exist.[3]

The authors concluded that 22 percent of sunlike stars harbor Earth-size planets orbiting in their habitable zones. Furthermore, they speculate that the nearest such planet may be as close as twelve light-years.

Seth Shostak, senior astronomer of the SETI (Search for Extraterrestrial Intelligence) Institute has crunched the data further.[4] Since stars like the sun make up about 20 percent of the approximately two hundred billion stars in the Milky Way, this gives nine billion planets in our galaxy able to support life. In addition, Shostak refers to another analysis of Kepler data that implies 16 percent of all red dwarf stars sport planets in the habitable zone, thus adding another twenty-four billion candidates giving a total of perhaps thirty-three billion habitable worlds within our own galaxy.

Since there are about 150 billion other galaxies in the visible cosmos, this multiplies out to 5,000,000,000,000,000,000,000 planets (five sextillion) within our horizon capable of supporting life of some form. However, this number could be off by orders of magnitude, so just remember — it is very, very big.

These are just the planets within forty-six billion light-years of Earth, which is the horizon beyond which light cannot reach us in the 13.8 billion-year age of the universe. If inflation is correct, beyond that limit exists a far vaster region of space that Alan Guth estimates contains a *minimum* of twenty-three orders of magnitude as many galaxies as those inside our horizon.[5] It is probably much greater.

This is not pure speculation. It is based on the empirically well-established cosmology. Also note that I am not talking about other universes here — just our universe — the result of a single big bang.

More speculative are estimates of the likelihood of life anywhere except Earth. As long as we do not have a proven theory of how life began, there will be those who will argue that it could only have been the result of a miraculous creation. As we will see later in the chapter, life on other planets, in particular, intelligent life, presents great difficulties for traditional theologies, which were formulated by people whose best knowledge of the universe was that Earth sat alone at the center of all there is.

However, based on our best current knowledge it is hard to imagine that in this immense universe there aren't countless planets with some form of life. That life is likely to be very different from our own, given the major role of random chance in the evolution of life. All we know about our kind of life based on carbon chemistry is that it developed very easily and quickly on Earth once the conditions were right. We also know that the chemicals needed for our kind of life, including amino acids, are very plentiful in space. We just have not found any evidence yet for extraterrestrial life because of the immense distances between stars.

THE FINE-TUNING QUESTION

In recent years, many theologians and Christian apologists have convinced themselves and their followers that they have a knock-down, drag-out scientific argument for the existence of God. They claim that the parameters of physics are so finely tuned that if any one of these parameters were just slightly different in value, life — and especially human life — would have not been possible anywhere in the universe.

Of course, like all design arguments this is a God-of-the-gaps argument that they cannot win in principle because they can never prove conclusively that the values of these parameters cannot be natural. But they keep trying.

Assuming, on no basis whatsoever, that those parameters are independent and could have taken on any value over a wide range, they conclude that the probability of a universe with our particular set of parameters is infinitesimally small. Further assuming, on no basis whatsoever, that the probability of a divine creator is not equally infinitesimally small, they conclude that such a creator existed who fine-tuned the universe for life, particularly human life. Note that there is also no basis whatsoever to assume that this creator was the personal God worshipped by Christians, Muslims, and Jews or the god of any major religion. A deist creator works equally well.

William Lane Craig summarized the argument this way in his 1998 debate with philosopher Massimo Pigliucci (and in other debates available on his website):

> During the last 30 years, scientists have discovered that the existence of intelligent life depends upon a complex and delicate balance of initial conditions given in the Big Bang itself. We now know that life-*prohibiting* universes are vastly more probable than any life-*permitting* universe like ours. How much more probable?
>
> The answer is that the chances that the universe should be life-permitting are so infinitesimal as to be incomprehensible and incalculable.
>
> John Barrow and Frank Tipler estimate that a change in the strength of gravity or of the weak force by only one part in 10^{100} would have prevented a life-permitting universe.[6] There are around 50 such quantities and constants present in the Big Bang which must be fine-tuned in this way if the universe is to permit life. And it's not just *each* quantity which must be exquisitely fine-tuned; their *ratios* to one another must be also finely-tuned. So improbability is multiplied by improbability by improbability until our minds are reeling in incomprehensible numbers.[7]

Just because Craig's mind reels and he personally can't comprehend the numbers, it does not follow that they are in fact incomprehensible to the rest of us.

I have written extensively on the subject of fine-tuning but need to cover some of it again here for completeness. I will try to bring the subject up to date without, I hope, being too repetitive. Recently I contributed a chapter arguing against fine-tuning for an Oxford University Press anthology *Debating Christian Theism* and will refer to some of that material.[8] Christian philosopher Robin Collins of Messiah College presented the case for fine-tuning in an accompanying chapter.[9] In it he criticizes a number of my previous arguments, to which I will respond here. These responses have not appeared elsewhere.

The multiverse provides a very simple, purely natural, solution to the fine-tuning problem. Suppose our universe is just one of an unlimited number of individual universes that extend for an unlimited distance in all directions and for an unlimited time in the past and future. If that's the case, we just happen to live in that universe that is suited for our kind of life. Our particular universe is not fine-tuned to us; we are fine-tuned to it.

The multiverse explanation is adequate to refute fine-tuning. Note the multiverse does not need to be proved to exist to refute fine-tuning claims. It just must be plausible. Those who dispute this have the burden of proving otherwise. This they have not done.

Nevertheless, the multiverse remains unverified so it behooves us to continue to examine the credibility of the divine fine-tuning hypothesis for our single, lone universe.

In a book published in 2011 titled *The Fallacy of Fine Tuning*, I provided purely natural explanations for the values of the so-called fine-tuned parameters that appear most frequently in the theistic literature.[10]

Many authors have written on fine-tuning, often misleadingly referred to as the *anthropic principle*, which suggests it has something to do with human beings. They insist certain parameters are fine-tuned to exquisite precision. And, by "exquisite precision," they don't just mean within an order of magnitude or 10 percent, or even 1 percent. Rather, they assert that some parameters must be tuned to one part in fifty to a hundred orders of magnitude for any life to be possible.

Before I get to the specific parameters that are supposedly so fine-tuned, let me say a word about my basic interpretation of their meaning. The models of physics are human constructions, and so it follows that the quantities, parameters, and "laws" that appear in these models are likewise constructed. It strikes me as somewhat-incongruous to think of them as fine-tuned by God or nature. Physicists in another galaxy might have their own models with a totally different set of parameters.

Thus the parameters that are supposedly fine-tuned need not have any specific ontological significance. Of course, the models must agree with observations and so, as I have emphasized, they must have something to do with whatever objective reality is out there. They are not arbitrary, just as a landscape painting is not a random splash of colors (unless it's by Jackson Pollock).

Let us now look at the specifics. Physicist and Christian apologist Hugh Ross lists twenty-nine "characteristics of the universe" and forty-five characteristics of the solar system "that must be fine-tuned for any kind of life to be possible."[11] Right off the bat this statement is incorrect. More than half of Ross's parameters do not address life in general but life on this planet alone and, in some cases, even more particularly to human life.

The most common fallacy made by Ross and others who agree with his position is to single out the carbon-based life we have on Earth and assume that it is the only possible type of life. According to Christian belief, humans are made in the image of God (Genesis 1:26), so it is not surprising that they find it difficult to imagine other life-forms. However, with only one example available, they simply do not have the data to allow them to conclude that all other forms of life are impossible, whether based on carbon chemistry or not.

Referring to the possibility that the parameters can vary randomly, Collins asks, "Why should they give rise to precisely the right set of laws required for life?"[12] Well, that's the whole point. They didn't have to be precise to lead to *some* form of life *somewhere* in this vast universe. In *The Fallacy of Fine Tuning* I showed that wide ranges of physical parameters could plausibly lead to conditions, such as long ages of stars, that could in principle allow for the evolution of life of one form or another.

TRIVIAL PARAMETERS

Two of the parameters that appear in most lists of fine-tuned quantities are

- the speed of light in a vacuum, c, and
- Planck's constant, h.

As basic as these parameters are to physics, their values are arbitrary. The fundamental unit of time in physics is the *second*. As we saw in chapter 6, the units for all other measurable quantities in physics, except for those that are dimensionless, are defined relative to the second. The value of c is chosen to define what units will be used to measure distance. To measure distance in meters you *choose* $c = 3 \times 10^8$. To measure distance in light-years you *choose* $c = 1$.

The value of Planck's constant h is chosen to define what units will be used to measure energy. To measure energy in joules you *choose* $h = 6.626 \times 10^{-34}$. To measure energy in electron-volts you *choose* $h = 4.136 \times 10^{-15}$. Physicists like to work in what they call "natural units," where $\hbar = h/2\pi = c = 1$. Other arbitrary quantities that are often claimed to be fine-tuned include Boltzmann's constant, k_B, which simply converts from units of absolute temperature, degrees Kelvin, to energy, and Newton's gravitational constant, G, which also depends on the choice of units. In Planck units, $G = 1$.

PARAMETERS NEEDED
FOR ANY FORM OF LIFE

Less trivially, let us look at five parameters that are claimed by theists to be so finely tuned that no form of life could exist in any universe in which any of the values differed by an infinitesimal amount from their existing values in our universe.[13] These are:

- The ratio of electrons to protons in the universe
- The ratio of the electromagnetic force to gravity
- The expansion rate and mass density of the universe
- The cosmological constant

The ratio of electrons to protons in the universe

Ross asserts that if this ratio were larger, there would be insufficient chemical binding. If smaller, electromagnetism would dominate gravity, preventing galaxy, star, and planet formation.

The fact that the ratio is exactly equal to one can be easily explained. The number of electrons in the universe should equal the number of protons from charge conservation, on the reasonable expectation that the total electric charge of the universe is zero. While there are other charged particles in the standard model, the proton and electron are the only ones that are stable.

The ratio of electromagnetic force to gravity

Ross says that if this ratio were larger, there would be no stars less than 1.4 solar masses and hence short and uneven stellar burning. If it were smaller, there would be no stars more than 0.8 solar masses and hence no heavy element production.

The ratio of the forces between two particles depends on their charges and masses. As I have already remarked, despite the statement often heard in most (if not all) physics classrooms—that gravity is much weaker than electromagnetism—there is no way one can state absolutely the relative strengths of gravity and any other force. Indeed, if one were to define the strength of gravity using the only natural mass, the Planck mass, you find that gravity is 137 times *stronger* than electromagnetism.

The reason gravity is so weak in atoms is the small masses of elementary particles. This can be understood to be a consequence of the standard model of elementary particles in which the bare particles all have zero masses and pick up small corrections by their interactions with other particles.

Collins misunderstands this point when he writes: "Stenger's attempt to explain away this apparent fine-tuning [the low mass of the proton and neutron] is like someone saying protons and neutron are made up of quarks and gluons, and since the latter masses are small, this explains the smallness of the former masses."[14]

This is a complete misrepresentation of my position. Nowhere have I used this argument. Collins provides no direct quotation or citation. In truth, I make the very reasonable assumption, based on the standard model, that all the elementary particles (the proton and neutron are not elementary) were massless when they were first generated in the early universe. All have low masses today, compared to the Planck mass, since those masses were just small corrections provided by the Higgs mechanism. And, before Collins complains that the Higgs mechanism is another arbitrary assumption, recall that it is part of the standard model, which emerged undesigned from the symmetries of emptiness and the randomness of symmetry breaking.

The expansion rate and mass density of the universe

Ross claims that if the expansion rate of the universe, given by the Hubble parameter H, were larger, there would be no galaxy formation; if smaller, the universe would collapse prior to star formation. He also asserts that if the average mass density of the universe were larger, there would be too much deuterium from the big bang and stars would burn too rapidly. If it were smaller, there would be insufficient helium from the big bang and too few heavy elements would form.

In chapter 12, we saw that inflation results in the mass density of the universe being very close to the critical value ρ_c. This, in turn, implies that H also has a critical value. Only one of the two parameters is adjustable. Let's assume it's H.

Now, in the approximation of a linear expansion given by *Hubble's law* (see chapter 8), the age of the universe is given by $T = 1/H$. This is currently 13.8 billion years and is hardly fine-tuned for life. Life could just as well have evolved for $T = 12.8$ billion years or $T = 14.8$ billion years. In fact, suppose $T = 1.38$ billion years. Then we could not have life now, but it would come along ten billion years or so later. Or, suppose $T = 138$ billion years. The life will have already appeared 124 or so billion years earlier.

The cosmological constant

The cosmological constant is equivalent to an energy density of the vacuum and is the favorite candidate for the dark energy, which is responsible for the acceleration of the universe's expansion—constituting over 68 percent of the total mass/energy of the universe.

We saw in chapter 13 that calculations of the energy density of the vacuum that assume it equals the zero-point energy give answers that are 50–120 orders of magnitudes larger than the maximum value allowed by observations.

Physicists have not reached a consensus on the solution to the cosmological-constant problem. Some prominent figures, such as Steven Weinberg[15] and Leonard Susskind,[16] think the answer lies in multiple universes. Both refer to the fact that string theory, or its more advanced version called *M*-theory, offers a "landscape" of perhaps 10^{500} different possible universes. But we have no need for such speculation.

As I pointed out in chapter 13, the original energy-density calculations incorporated a fundamental error by summing all the states in a given volume. Since the entropy of a system is given by the number of accessible states of the system, the entropy calculated by summing over the volume will be greater than the entropy of a black hole of the same size, which depends on its area rather than its volume. But since we cannot see inside a black hole, the information that we have about what is inside is as small as it can be and so the entropy is as large as it can be.

Therefore, it was a mistake to calculate the number of states by summing over the volume. Correcting this by summing over the area, or, equivalently, setting the number of states equal to the entropy of a black hole equal to the size of the volume, we can naturally constrain the vacuum energy density. This calculation yields the result that an empty universe will have a vacuum energy density about equal to the critical density, just the value it appears to have.

For technical reasons, cosmologists are not ready to accept this solution to the cosmological-constant problem. Nevertheless, I think it is fair to conclude that the original calculation is simply wrong — as far wrong as any other calculation in the history of physics — and should be ignored. In any case, don't give up all your worldly goods and enter a monastery or convent because the cosmological constant is so small.

OTHER PARAMETERS

This takes care of the five parameters that are supposedly fine-tuned to such precision that even a tiny deviation would make life of any kind impossible. Note that only four are independent despite theist claims. Next let us move to those parameters for which proponents of fine-tuning can only use to claim that life would be very unlikely if the values of the parameters were slightly different.

The Hoyle prediction

In chapter 9, we covered the brilliant achievement of astronomer Fred Hoyle and his collaborators in showing how most of the elements of the periodic table are formed in stars during gravitational collapse after burning all their hydrogen fuel. In 1951, Hoyle predicted that the carbon nucleus would have an excited state at about 7.7 MeV above its ground state in order for enough carbon to be produced in stars to make life in the universe possible. This story is of great historical interest because it is the only case where anthropic reasoning has led to an empirically verified prediction. Shortly thereafter the excited state was found at 7.656 MeV.

However, calculations since have demonstrated that the same carbon would have been produced if the excited state were anyplace between 7.596 MeV and 7.716 MeV. Furthermore, sufficient carbon for life would have occurred for an excited state anywhere from just above the ground state to 7.933 MeV.[17] A state somewhere in such a large range is expected from standard nuclear theory. Furthermore, carbon is not the only element upon which life might be based.

Relative masses of the elementary particles

The masses of elementary particles affect many features of the universe, and a number of fine-tuning claims refer to their values. Let me begin with the mass difference between the neutron and the proton. If the difference in masses between the neutron and the proton were less than the sum of the masses of the electron and the neutrino (the neutrino mass in our universe is negligible for this purpose but may not be in some other universe), there would be no neutron decay. In the early universe, electrons and protons would combine to form neutrons and few, if any, protons would remain. If the mass difference were greater than the binding energies of nuclei, neutrons inside nuclei would decay, leaving no nuclei behind.

There is a range of 10 Mev or so for the mass difference to still be in the allowed region for the full periodic table to be formed. The actual mass difference is 1.29 Mev, so there is plenty of room for it to be larger. Since the neutron and proton masses are equal to a first approximation, and the difference results from a small electromagnetic correction, it is unlikely to be as high as 10 MeV.

Next let us bring in the mass of the electron, which also affects the neutron-decay story. A lower electron mass gives more room in parameter space for neutron decay, while a higher mass leaves less.

The ratio of the electron and proton masses helps determine the region of parameter space for which chemistry is unchanged from our universe. We can show that this region is quite substantial and no fine-tuning is evident here either.

Relative strengths of the forces and other physics parameters

The dimensionless relative force strengths are the next set of physical parameters whose fine-tuning is claimed for reasons I found wanting. The gravitational-strength parameter α_G is arbitrary. Thus, there is nothing about α_G to fine-tune. As we have seen, the so-called weakness of gravity compared to the electromagnetic force between elementary particles results from their low masses and not their relative inherent strengths.

Next let me consider the dimensionless strength of the weak interaction α_W. Ross claims it is fine-tuned to give the right amount of helium and heavy-element production in the big bang and in stars. The key is the ratio of neutrons to protons in the early universe when, as the universe cools, their production reactions drop out of equilibrium. A significant range of parameters is allowed.

The electromagnetic strength represented by the dimensionless parameter α, historically known as the fine-structure constant, has the famous value 1/137 at low energies. Ross tells us that there would be insufficient chemical bonding if it were different. But, as I proved in *The Fallacy of Fine-Tuning*, the many-electron Schrödinger equation, which governs most of chemistry, scales with α and the mass of the electron. Again, a very wide range allows for chemistry as we know it and, particularly, the chemistry of life.

There are many places where the value of α relative to other parameters comes in. I have attributed the weakness of gravity relative to electromagnetism in matter to the natural low masses of elementary particles. This can also be achieved with a higher value of α, but it's not likely to be orders of magnitude higher.

The relative values of α and the strong force parameter α_S also are important in several cases. When the two are allowed to vary, no fine-tuning is necessary to allow for both nuclear stability and the existence of free protons.

There are two other facts that most proponents of fine-tuning ignore: (1) the force parameters α, α_S, and α_W are not constant but vary with energy; (2) they are not independent. The force parameters are expected to be equal at some unification energy. Furthermore, the three are connected in the current standard model and are likely to remain connected in any model that succeeds it. They are unlikely to ever differ by many orders of magnitude.

Other parameters such as the decay rate of protons and the baryon excess in the early universe have quite a bit of room to vary before they result in life-threatening radiation.

Cosmic parameters

We have already disposed of the cosmic parameters that were deemed so crucial in making any livable universe possible: the mass density of the universe, the expansion rate, and the ratio of the number of protons and electrons are not only not fine-tuned, they are fixed by conventional physics and cosmology or, in the case of the Hubble parameter, practically any value would allow for life.

Similarly, the deuterium abundance has little relation to the existence of life. The amount needed for life is small, and a wide range of two orders of magnitude is allowed.

The Astronomer Royal of the United Kingdom, Martin Rees, and others have claimed that the lumpiness of matter, represented by a quantity Q, in the universe had to be fine-tuned within an order of magnitude to allow for galaxy formation. An order of magnitude is hardly the kind of fine-tuning the theists are claiming, which are more typically one part in fifty to one hundred orders. What's more, varying the nucleon mass along with Q allows, again, for more parameter space for life.

In chapter 14 we discussed the ΛCDM model that gives a precise fit to the anisotropies of the cosmic microwave background and is consistent with observations on galactic structure. That model contains only six adjustable parameters, none of which are included in the parameters Ross or other divine fine-tuners have made such a big show about. The density of matter is not a parameter but is assumed to equal the critical value. The expansion rate (Hubble parameter) is not an adjustable parameter but is calculated in the model. One parameter is the dark-energy density relative to the critical density. Rees's parameter Q is not included but is implicit in the calculation of galactic structure.

In short, the divine fine-tuners have to go back to the drawing board and run the ΛCDM model over a range of parameters and show that life of any form would be impossible unless the six parameters were exactly what they are of our universe.

Simulating universes

The aggregate properties of the universe as we know it today are determined by just three physics parameters: the electromagnetic strength α and the masses of the proton and electron, m_p and m_e. From these we can estimate quantities such as the maximum lifetime of stars, the minimum and maximum masses of planets, the minimum length of a planetary day, and the maximum length of a year for a habitable planet. Generating ten thousand universes in which the parameters are varied randomly on a logarithmic scale over a range of ten orders of magnitude, I find that 61 percent of the universes have stellar lifetimes over ten billion years, sufficient for some kind of life to evolve.

Collins has previously objected to my preliminary, twenty-year-old conclusion that long stellar lifetimes are not fine-tuned.[18] He argues that not all these universes are livable, that I have not accounted for life-inhibiting features. He refers to John Barrow and Frank Tipler, who in their classic (though containing numerous typographical and mathematical errors) *The Anthropic Cosmological Principle* estimated that $\alpha \leq 11.8\alpha_S$ for carbon to be stable.[19]

Since in my study I was varying all the parameters by ten orders of magnitude, I would not expect such a tight criterion to be satisfied very often. Nevertheless, I have checked and found the Barrow-Tipler limit to be satisfied 59 percent of the time. I have also studied what happens when the parameters are varied by just two orders of magnitude. Then 91 percent of the time we have $\alpha \leq 11.8\alpha_S$. Again I must stress that the fine-tuners are claiming far greater sensitivity than ranges of an order of magnitude.

Applying rather-tight limitations to all three parameters in order to produce life, 13 percent of all universes are capable of supporting some kind of life not too different from ours when I vary them by ten orders of magnitude. Varying by two orders of magnitude, which is more realistic since the parameters are not independent but related, I find that 92 percent of the universes have stellar lifetimes over ten billion years and 37 percent are capable of supporting some kind of life not too different from ours. Life very different from ours remains possible in a large fraction of the remaining universes, judging from the large stellar lifetimes for most.

I don't claim to have provided an explanation for the values of every parameter of physics and cosmology. I don't have to in order to refute the opposing claim that many are fine-tuned to incredible precision such as one in 120 orders of magnitude. Uncertainties of 1 percent, to 10 percent, or even an order of magnitude — as was the case for the lumpiness parameter Q — do not classify as fine-tuning.

SUMMARY OF THE CASE AGAINST FINE-TUNING

The following is a summary of the logical and scientific errors made by proponents of fine-tuning (not all make every error) that I have uncovered in my studies:

1. They make fine-tuning claims based on the parameters of our universe and our form of life, ignoring the possibility of other life-forms.
2. They claim fine-tuning for physics constants whose values are arbitrary, such as c, h, and G.
3. They assert fine-tuning for quantities whose values are precisely set by cosmological physics or have wide allowable ranges, such as the ratio of electrons to protons, the expansion rate of the universe, and the mass density of the universe. These are not even varied in the current standard cosmological model.
4. They assert that the relative strength of the electromagnetic and gravitational forces is fine-tuned, when in fact this quantity cannot be universally defined.
5. They assert that an excited state of the carbon nucleus had to be fine-tuned for stars to produce the carbon needed of life, when calculations show that a wide range of values for the energy level of that state will produce sufficient carbon.
6. They claim fine-tuning for the masses of elementary particles when the ranges of these masses are set by well-established physics and sufficiently constrained to give some form of life.
7. They assume the strengths of the various forces are constants that can independently change from universe to universe. In fact, they are dependent on one another and vary with energy, and their relative values and energy dependences are close to being pinned down by theory, in ranges that make some kind of life possible.

8. Most make a serious analytical mistake in taking all the parameters in the universe to be fixed while varying only one at a time. This fails to account to the fact that a change in one parameter can be compensated by a change in another, opening up more parameter space for a viable universe.

9. They misunderstand or misuse probability theory, ignoring the fact that events with "mind-boggling" low probabilities occur billions of times a day. The only way one can use a low probability to argue that something is unlikely is to compare it with the probabilities of all the alternatives. What is the probability of God? In *The Fallacy of Fine-Tuning*, I compared the calculations for the probability of God using sophisticated *Bayesian statistics* made by two physicists, one a believer and one a nonbeliever. The believer came up with 0.67, while the nonbeliever's result was 10^{-17}.[20]

10. They claim many parameters of Earth and the solar system are fine-tuned for life, failing to consider that with quadrillion planets in the visible universe in habitable regions of their stars, and the countless number beyond our horizon, a planet with the properties needed for life is likely to occur many times. Nevertheless, the universe is hardly life-friendly. If God wanted to fine-tune it for life, he could have made it a lot friendlier.

11. The fine-tuners are also wrong to reject the multiverse solution as "unscientific." It is not unscientific to speculate about invisible, unconfirmed phenomena that are predicted by existing models that, so far, agree with all the available data. The neutrino was predicted to exist in 1930 based on the well-established principle of energy conservation but was not detected until 1956, and even then indirectly. If the physics community and granting agencies had used the fine-tuners' criterion, my late colleague Fred Reines and his (also late) collaborator Clyde Cowan would not have been able to get the money and support to engage in the search that finally discovered the neutrino.

12. The current, highly successful ΛCDM cosmological model has only six parameters, none of which has been shown to be fine-tuned.

As my discussion illustrates, the explanations for apparent fine-tuning are technical and require adequate training to understand. A proper analysis finds there is no evidence that the universe is fine-tuned for life and all we have is yet another God-of-the-gaps argument that is doomed to failure by its implicit assumption that some phenomena exist that science will never be able to explain without introducing God into the explanation.

SOMETHING RATHER THAN NOTHING

Since cosmic creationists continue to fail in their attempts to use science to provide a credible case for a creator god, they frequently fall back on three questions that are increasingly more philosophical in the order presented below:

1. How can something come from nothing?
2. Where do the laws of physics come from?
3. Why is there something rather than nothing?

1. How can something come from nothing?

The first question can still be discussed in largely scientific terms. Let me restate it as follows: How can matter come from nonmatter?

The universe has mass, which is a measure of the amount of matter, and since mass and rest energy are equivalent, it would seem that the law of conservation of energy must have been broken to create the matter of the universe out of "nothing."

However, we have seen that when the negative potential energy of gravity is included, the total energy of the universe is zero, give or take quantum uncertainties. So the law of conservation of energy was not broken for the universe to appear from an earlier state of zero energy and zero matter.[21] Indeed, no law of physics was broken in bringing our universe into existence. And, of course, no law of physics was broken in bringing the multiverse into existence, since it always existed.

Now, I need to mention that there is considerable debate over whether one can define the total energy of the universe within the framework of general relativity. There are prominent physicists and cosmologists on both sides of the argument. I will not go into this question, which is highly technical.

However, I have a simple observation. As I have explained, according to Noether's theorem, energy conservation is required in any model that is to apply at all times. Einstein's general relativity is a model. If you are building a model based on general relativity you have two choices: (1) include energy and energy conservation in your model to help guarantee it is good at all times or, (2) exclude energy and energy conservation and accept the fact that the model isn't valid at all times.

2. Where do the laws of physics come from?

This leads to the second question: Where did the law of conservation of energy and, indeed, all the laws of physics come from? Robin Collins argues that any explanation for a universe without God only works if "the explanation does not merely transfer fine-tuning up one level to newly postulated laws, principles and parameters."[22]

I don't see how transferring fine-tuning up one level to God is any improvement. In fact, it is less informative since we have no idea how God did his fine-tuning whereas we have some idea from cosmological science how the universe may have come from nothing. In the first case we can do nothing useful with that assumption. In the second case we can make certain predictions that are empirically testable. For example, we can predict that the energy density of the universe will always remain at it critical value as measurements continue to improve.

In any case, we have seen that the "law" of conservation of energy follows from the fact that there is no special moment in time. It was not handed down by God. It's more necessary than God.

As described in chapter 11, physicists in the twentieth century discovered a set of principles I call metalaws that are required to be present in all physics models. In order to describe the universe objectively, physicists must formulate their models so that they describe observations in ways that are independent of the point of view of particular observers, what I call point-of-view invariance. This gives the model builders no choice but to include the great conservation principles or metalaws of energy, linear momentum, angular momentum, and electric charge. Point-of-view invariance leads to all of classical physics, including Newton's laws of mechanics and gravity, Maxwell's equations of electromagnetism, thermodynamics, fluid mechanics, and Einstein's theory of special relativity. Much of general relativity and quantum mechanics (if not all), including the Heisenberg uncertainty principle, also follow.[23]

When I argue that some parameter is in a range consistent with known physics, apologists come back with, "Where did physics come from?" My answer: the physics came from physicists formulating models to describe observations that must include metalaws that constitute the basic laws of physics. The metalaws do not set all the parameters of physics. Many occur by accident. However, as we have seen, the values of the parameters in the models that successfully describe all observations in our universe are within the ranges set by the metalaws.

3. Why is there something rather than nothing?

The third question is largely philosophical because it deals more with the meaning of words than with actual physics. In a book published in 2012 called *A Universe from Nothing: Why There Is Something Rather Than Nothing*,[24] cosmologist Lawrence Krauss describes how our universe could have arisen naturally from a preexisting structureless void he calls "nothing." I don't think anything he says disagrees with this book, which was written independently.

In a *New York Times* review of Krauss's book, philosopher David Albert asks "why the world should have consisted of the particular elementary stuff it does, as opposed to something else, or to nothing at all."[25] Krauss admits he does not know but suggests it may arise randomly, in which case some universe like ours would have arisen without a prescribed cause.

Clearly, no consensus exists on how to define "nothing." It may be impossible. To define "nothing" you have to give it some defining property, but, then, if it has a property it's not nothing!

Albert was not satisfied that Krauss answered the fundamental question: Why there is something rather than nothing, that is, why is there "being" rather than "nonbeing"? Again, I have a simple retort to Albert: Why should nonbeing, no matter how defined, be the default state of existence rather than being? Why is some creative act needed to convert nonbeing to being? Perhaps such an act is needed to convert being into nonbeing.

If nonbeing is the natural state, then why is there God? Once theologians assert that there is a God as opposed to nonbeing, they can't turn around and demand a cosmologist explain why there is a universe as opposed to nothing. They claim God is a necessary entity. Why can't a godless multiverse be a necessary entity?

As my late friend and colleague mathematician Norm Leavitt once said, "What is there? Everything. So what isn't there? Nothing."

But we can do even better than this standoff and make an argument for something being a more natural state than nothing. We can provide a plausible reason based on our knowledge of existing physics.

It is commonly thought that a complex physical system can only come about by the deliberate act of an intelligent designer who must necessarily be even more complex. The chain of design then leads back to God as Aristotle's Prime Mover and Aquinas's First Cause Uncaused, the maximally complex creator of all that is.

We even do not have to rely on sophisticated scientific arguments to see from common experience alone that Aristotle and Aquinas, as well as everyone since who has used the design argument, had it backward. In nature, complexity arises out of simplicity. Consider the phase transitions observed in familiar matter. In the absence of external heat, water vapor will naturally condense into liquid water, which then will freeze into solid ice. With each transition, we move from a state of higher symmetry to one of lower symmetry — from simplicity to complexity. Complexity is broken symmetry, and the transition from simple to complex occurs spontaneously. Simplicity begets complexity, not the other way around. The particular crystal structure that results from the liquid-water-to-ice transition is unpredictable, that is, random. No two snowflakes, which result from a direct vapor-to-solid transition, are alike.

Physical systems move naturally from simple to complex without the need for design. Indeed, the fact that specific events, such as atomic transitions, are random can be taken as strong evidence against any design, intelligent or stupid.

And so how do we get something from nothing? Since no thing is more symmetric than nothing, we would expect nothing to naturally undergo a phase transition to something. As Nobel laureate Frank Wilczek put it in a *Scientific American* article back in 1980, "Nothing is unstable."[26]

It also should be remarked that although the universe is something rather than nothing, it is pretty close to being nothing in the sense that it is only 30 percent matter, which makes it 70 percent nothing.

I have presented the material in this section, in one form or another, in several books. However, if you want to read an excellent elucidation of these ideas at a layperson level by a professional science writer, see chapter 7 of *The Never-Ending Days of Being Dead* by Marcus Chown.[27]

THE GRAND ACCIDENT

Stephen Hawking and Leonard Mlodinow addressed the something-from-nothing question in their 2010 book *The Grand Design*, mentioned in the previous chapter. They concluded, "Spontaneous creation is the reason there is something rather than nothing, why the universe exists, why we exist. It is not necessary to invoke God to light the blue touch paper and set the universe going."[28]

The book should have been called *The Grand Accident*, because that's what "spontaneous" refers to—an uncaused accident. The authors agree there was no design, grand or otherwise. I suspect the publisher picked the title to sell more books.

Hawking and Mlodinow provide a picture of spontaneous creation that follows from Feynman's sum-over-histories formulation of quantum mechanics, which was mentioned in chapter 15.

As we have seen, M-theory offers the possibility of 10^{500} universes with different properties. Hawking and Mlodinow propose that these are "alternate histories" to which Feynman's model can be applied. By summing over all the histories we get the probability for the universe that we observe.

In this view, the universe is taken as it is at the present time and the most probable path back to the origin is calculated. As the authors put it, "The universe appeared spontaneously, starting off in every possible way."[29] The 10^{500} universes converged on the one that has the necessary structure to produce life as we know it. With such a huge number of possibilities, many would be expected to allow for some kind of life different from ours as well as for our kind.

Hawking and Mlodinow also emphasize, "The multiverse is not a notion invented to account for the miracle of fine-tuning, as often charged by apologists." Rather, "it is a consequence of the no-boundary condition as well as many other theories of modern cosmology."[30] And once you have one accidental universe, you will have many.

THE GOD HYPOTHESIS

So, where does that leave the hypothesis of God? I often hear the argument that God is not a scientific hypothesis since he is a spirit that can never be observed. Furthermore, religion and science are two separate realms and one should not be applied to describe the other. However, if God has any role in the universe, the results of some of his actions should be observable even if he is not, and science can be called upon to attempt to test if those observations have a supernatural cause.

In the spirit of Ockham's razor, we must recognize that currently God is an additional hypothesis not required by the data. If he were, he would be included in the set of premises that constitute scientific theories. He is not. While we certainly do not know everything, we know of no observed fact that requires the existence of God. Indeed, many observed facts that are inconsistent with the God hypothesis serve to prove beyond a reasonable doubt that a God who plays an active role in the universe and in human lives does not exist.[31]

On winter mornings I can look out my office window to an empty field covered with fresh snow. Occasionally I will see footprints of wild animals—foxes, coyotes, or whatever. I rarely see the animals themselves, but I know they exist by the fact that they left footprints. God has left no footprints on the snows of time.

Of course there are many purported facts that are claimed as evidence for God or gods. These are generally referred to as miracles, that is, events inconsistent with mundane and scientific explanation but with moral significance for those who observe them. But every time a claim is thoroughly investigated, its miraculous nature evaporates. Teardrops from a statue of Mary are found to be cooking oil from the rectory kitchen. The woman who got up and walked at the evangelical revival was placed in a wheelchair when she walked into the hall. The miraculous imprint on Jesus' shroud was painted by a thirteenth-century con man.[32]

Lourdes, France, is perhaps the most famous site of purported miracles. Visiting there, you will find on display many crutches that have been discarded. In a quotation usually attributed to the French author Anatole France (1844–1924) but actually expressed by one of his friends, "One single wooden leg would have been much more convincing."[33]

Believing scientists and science-savvy Christian theologians have made heroic attempts to find a model of an active theist God, as opposed to an inactive, deist god, that is consistent with both science and Christianity. Most use quantum mechanics and chaos theory to find a place for God to act without the product of his actions showing up as miracles in our scientific instruments. But, as I showed in *Quantum Gods*,[34] and Robert Price and Edwin Suiminen have more recently confirmed in *Evolving out of Eden*,[35] it doesn't work. God simply isn't hidden in electron clouds.

A SUMMARY

This book has traced the history of humanity's view of the cosmos from the distant past to the present. Summarizing, we have seen that plausible scenarios exist for a natural origin of our universe — many worked out fully mathematically and published in peer-reviewed scientific journals. We now have a surprisingly simple model of cosmology that combines with the standard model of elementary particles to provide not only a description of the physical world that is fully consistent with all observations but also, in many cases, successful quantitative predictions of exquisite precision. Of course, neither should be taken as the "final word."

While more speculative but still based on our best knowledge, cosmologists have inferred that our universe is but one in an unbounded, eternal multiverse that contains an unlimited number of other universes. Although one has not yet been observed, the possibility exists that another universe may have left a detectable imprint on ours. While the multiverse hypothesis is hardly confirmed, it has sufficient scientific backing to take it seriously and consider the philosophical and theological consequences.

Theologians and apologists who are willing to accept for the purpose of argument the possibility of many universes insist that their existence still requires some prior, divine cause. If, as I have argued, the multiverse follows as a consequence of the known laws of physics, they ask: Where did those laws come from other than a supernatural lawgiver?

Most physicists and cosmologists simply accept the laws of physics as brute facts and prefer not to involve themselves in theological debate. I have tried to show that the multiverse scenario can be described without recourse to any special dynamical principles or laws that must be assumed as brute facts.

To understand this requires a drastic revision of the common notion of natural laws as "ordinances" handed down by some great legislature in the sky. As we have seen, what we call the basic principles and laws of physics, such as the conservation laws, are simply statements that automatically appear in the mathematical models of physics when those models are designed to be independent of the point of view of any particular observer. It follows that the models contain certain symmetries, some of which can be broken spontaneously to give the tiny amount of asymmetry, one part in one hundred thousand, that leads to complexity in structures, such as galaxies, stars, and planets.

Few people, even scientists, have come to grips with the fact that the universe is mostly particles moving around randomly and that the wonder that so enamors us is just a minor statistical fluctuation.

The beautiful galactic structures that we spend so much time admiring constitute just 0.5 percent of the mass of the universe and a billionth of the total number of elementary particles therein. Most of the universe is composed of particles in largely random motion. That is, our universe looks very much as it should look if it came from a perfectly symmetric, unstructured state that we can, for want of a better name, call "nothing."

The widespread assertion that our universe is fine-tuned for life is greatly overblown and not required by known physics. Our existence on Earth is a simple matter of natural selection. With every type of planet possible in the multiverse, we naturally evolved on one with the properties needed for intelligent life.

In short, nothing in our observations of the universe requires the existence of God. Furthermore, the absence of evidence that should be there for the actions of God rules out beyond a reasonable doubt the kind of God worshipped by most of humanity.

The universe visible from Earth contains an estimated 150 billion galaxies, each galaxy containing on the order of a hundred billion stars. It is known to be 13.8 billion years old with an uncertainty of less than one hundred million years. The most distant object we can in principle see is now forty-six billion light-years away, taking into account the expansion of the universe during the time the photons carrying its image traveled to our telescopes. This marks a horizon beyond which we cannot see because light has not had the time to reach us in the age of the universe.

As discussed in chapter 14, well-established inflationary cosmology implies that on the other side of our horizon lies a region containing *at least* twenty-three orders of magnitude as many galaxies as those inside, which arose from the same primordial seed that produced our universe. It is likely to be many orders of magnitude greater. Our visible universe can be likened to a grain of sand in the Sahara desert.

And that's just our personal universe. Besides that, a strong case can now be made that we live in an eternal multiverse containing an unlimited number of other universes. However, for the purposes of the discussion in this section I will ignore the possibility of other universes and just stick with the one that indisputably exists—our own.

Let us consider two alternative possibilities that are so far consistent with our best knowledge:

1. Intelligent life exists on only a single planet, Earth.
2. Life is rare, and intelligent life even rarer. But the universe is so vast that there still are countless numbers of intelligent beings in our universe.

1. We are alone

Believers are told that they are the special creation of a single divinity that created everything that is. While making sure every photon and electron in the universe behaves properly without us noticing, their God is also listening to their every thought and helping them to do the right thing—as he defines it.

This involves God controlling momentous events such as telling a president of the United States to go to war, as George W. Bush said God did,[36] or redirecting a tennis ball off a Christian lady's racquet so it wins her the point as she shouts, "Thank you, Jesus!"

Of course, a God of limitless power could do all that. By why would he have waited until just 150,000 years ago to create humans? And why would he have confined us to a tiny speck of dust in a vast ocean of space, with no chance, at least with our present physical makeup, of ever traveling much beyond the environs of Earth? If he desired the worship of humans so badly, then you would think he should have made it available to him for all times and at all places.

As we have seen, apologists make an illogical argument that they think is the clincher for the existence of God. They claim that life in the universe depends very sensitively on the value of a large number of physical parameters. According to this view, since the specific values of the parameters needed for life cannot possibly have resulted from chance (they can't prove that), they must have been "fine-tuned" by God in order to produce us.

Surely any God worthy of the name would not have been so incompetent as to build a vast, out-of-tune universe and then have to delicately twiddle all these knobs so that a single planet is capable of producing human beings. It would have made a lot more sense for him to have enabled us to live anywhere in the universe, even outer space. But the fact is—he did not.

So far, we know of no form of life anywhere in the universe besides Earth. At best we might one day find primitive life on Mars or someplace else in the solar system; any other intelligent life has to be a great distance away. Since 1979, the SETI program has listened for possible radio transmissions from alien civilizations with no success. Let's face it: the universe is not life-friendly. But it's not life-forbidding either, or else we wouldn't be here.

2. We are not alone

New estimates based on the brightness variations of thousands of stars measured by the Kepler Space Observatory suggest that within our horizon there may be as many as 5×10^{21} planets capable of sustaining biological life of some form.[37] We have seen that beyond our horizon exists a far vaster region of space. It is estimated that this region contains a *minimum* of twenty-three orders of magnitude as many galaxies as those inside our horizon but is likely to be much bigger.[38] This gives at least 10^{44} possible habitable planets in our universe.

This is not pure speculation. It is based on observations and the theory of cosmic inflation that is now empirically well established. So even if the probability of intelligent life for an otherwise-habitable planet is miniscule, say one part in a trillion-trillion (10^{-24}), this leaves 10^{20} planets with some form of intelligent life.

Of course, fundamentalists still believe literally in the cosmology of the Bible and think scientists are a bunch of frauds, so they have no self-contradiction here. Science is just wrong. There is just one universe created six thousand years ago with no evolution or climate change, which are just "hoaxes."

Moderate Protestants and Catholics, on the other hand, have yet one more conflict between science and religion among the many they must reconcile in order to both accept the findings of science and still hold onto some semblance of Christian faith.

The Judeo-Christian-Islamic God is a mighty God when viewed from the perspective of the desert tribes in the Middle East that conceived him. But that God is not mighty enough from the perspective of modern science.

Religion claims to teach us humility. But it really trades in a false pride, telling people they are the children of God, that they are the center of the universe and the reason for its existence — that they'll live forever if they just follow instructions. But it's an unearned pride. So, when science shows that we occupy but a tiny mote in space and time, the religious recoil from this lesson in humility.

Still, the very fact that in the short period of a few thousand years humans have been able learn so much about the universe by just looking up in the sky and at the world around them, and reasoning over what they saw, testifies that we are unique among the millions of species on this planet. We cannot yet compare ourselves with whatever intelligent life-forms might be out there. But we are special, at least on Earth and in the solar system. Even though magical thinking and hubris may still destroy us, we can hope that our unique abilities will lead us to a better future.

NOTES

PREFACE

1. For an excellent recent biography of Copernicus, see Dava Sobel, *A More Perfect Heaven: How Copernicus Revolutionized the Cosmos* (New York: Walker, 2011).

2. J. Richard Gott III et al., "A Map of the Universe," *Astrophysical Journal* 624, no. 2 (2008): 463.

3. H. M. Collins, "Introduction: Stages in the Empirical Program of Relativism," *Social Studies of Science* 11, no. 1 (1981): 3–10.

CHAPTER 1. FROM MYTH TO SCIENCE

1. John David North, *Cosmos: An Illustrated History of Astronomy and Cosmology* (Chicago: University of Chicago Press, 2008).

2. Stefano Liberati, "Greek Cosmology," from Spring 2003 course at the University of Maryland, http://www.physics.umd.edu/courses/Phys117/Liberati/Support/cosmology.pdf (accessed October 29, 2012).

3. A comprehensive review of cosmology in the ancient Near East can be found in Edward T. Babinski, "The Cosmology of the Bible," in *The Christian Delusion: Why Faith Fails*, edited by John W. Loftus (Amherst, NY: Prometheus Books, 2010), pp. 109–47.

4. Jean Bottéro, *Religion in Ancient Mesopotamia* (Chicago: University of Chicago Press, 2001), pp. 77–90.

5. Helge Kragh, *Conceptions of Cosmos from Myths to the Accelerating Universe: A History of Cosmology* (Oxford; New York: Oxford University Press, 2007), p. 9.

6. George L. Robinson, *Leaders of Israel: A Brief History of the Hebrews from the Earliest Times to the Downfall of Jerusalem, A.D. 70* (New York: International Committee of Young Men's Christian Associations, 1906).

7. Steve Wells notes 462 contradictions in the Old and New Testaments in Steve Wells, *The Skeptics Annotated Bible* (Moscow, ID: SAB Books, 2012).

8. George Smith, *The Chaldean Account of Genesis, Containing the Description of*

the Creation, the Fall of Man, the Deluge, the Tower of Babel, the Times of the Patriarchs, and Nimrod: Babylonian Fables, and Legends of the Gods; From the Cuneiform Inscriptions (New York: Scribner, Armstrong, 1876).

9. Gary D. Thompson, "The Development, Heyday, and Demise of Panbabylonism," 2012, http://members.westnet.com.au/Gary-David-Thompson/page9e .html (accessed October 28, 2012).

10. William Lane Craig, "Philosophical and Scientific Pointers to Creatio Ex Nihilo," *Journal of the American Scientific Affiliation* 32, no. 1 (1980): 5–13; Dinesh D'Souza, *What's So Great about Christianity* (Washington, DC: Regnery, 2007).

11. Tim Callahan, "Out of Nothing? The Genesis Creation Myth Is Not Unique," *Skeptic* 17, no. 3 (2012): 20–22.

12. Richard Alleyne, "God Is Not the Creator, Claims Academic," *Telegraph* (UK), October 8, 2009.

13. Callahan, "Out of Nothing?"

14. E. A. Wallis Budge, *Egyptian Ideas of the Future Life; Egyptian Religion* (New York: University Books, 1959), p. 38.

15. Ibid., p. 37.

16. Ibid., p. 42.

17. David Adams Leeming, *Creation Myths of the World: An Encyclopedia*, 2nd ed. (Santa Barbara, CA: ABC-CLIO, 2010), p. 3.

18. Kragh, *Conceptions of Cosmos*, p. 13.

19. Geoffrey S. Kirk, John E. Raven, and Malcolm Schofield, *The Presocratic Philosophers: A Critical History with a Selection of Texts*, 2nd ed. (Cambridge; New York: Cambridge University Press, 1983).

20. David C. Lindberg, *The Beginnings of Western Science: The European Scientific Tradition in Philosophical, Religious, and Institutional Context, Prehistory to A.D. 1450*, 2nd ed. (Chicago: University of Chicago Press, 2007), p. 27.

21. Arthur Fairbanks, ed. and trans., *The First Philosophers of Greece* (London: K. Paul, Trench, Trübner, 1898), pp. 8–16.

22. Mary Ackworth Orr, *Dante and the Early Astronomers* (Port Washington, NY: Kennikat, 1969), p. 53; originally published in 1914 (London, Edinburgh: Gall and Inglis).

23. Victor J. Stenger, *God and the Atom: From Democritus to the Higgs Boson* (Amherst, NY: Prometheus Books, 2013).

24. Orr, *Dante and the Early Astronomers*, pp. 55–56.

25. Epicurus, *The Art of Happiness*, trans. George K. Strodach (New York: Penguin Books, 2012).

26. A. E. Stallings and Richard Jenkyns, *Lucretius: The Nature of Things* (London; New York: Penguin, 2007), pp. 67–68.

27. Kragh, *Conceptions of Cosmos*, p. 18.

28. Stallings and Jenkyns, *Lucretius*, pp. 18–19.

29. John David North, *Cosmos: An Illustrated History of Astronomy and Cosmology* (Chicago: University of Chicago Press, 2008), p. 70.

30. Orr, *Dante and the Early Astronomers*, p. 61.

31. Ibid., p. 63.

32. Ibid., pp. 62–63.

33. Ibid., pp. 63–64.

34. Ibid., pp. 68–70.

35. Plato, *Timaeus* 31 a, b.

36. Lindberg, *Beginnings of Western Science*, p. 39.

37. Ibid., pp. 42–43.

38. From NASA's website at http://starchild.gsfc.nasa.gov/docs/StarChild/universe_level2/cosmology.html (accessed August 8, 2013).

39. Kragh, *Conceptions of Cosmos*, pp. 21–23.

40. Ibid., p. 25.

41. Otto Neugebauer, *Notes on Hipparchus* (Locust Valley, NY: J. J. Augustin, 1956).

42. Orr, *Dante and the Early Astronomers*, p. 101.

43. As quoted in Kragh, *Conceptions of Cosmos*, p. 29.

44. From *Webster's New World College Dictionary* (Cleveland, OH: Wiley Publishing, 2010), available at YourDictionary.com, http://www.yourdictionary.com/ptolemaic-system (accessed December 25, 2013).

45. See lists in Orr, *Dante and the Early Astronomers*, p. 116.

46. Ibid., p. 125.

CHAPTER 2. TOWARD THE NEW COSMOS

1. Helge Kragh, *Conceptions of Cosmos from Myths to the Accelerating Universe: a History of Cosmology* (Oxford; New York: Oxford University Press, 2007), p. 34.

2. Andrew Dickson White, *A History of the Warfare of Science with Theology in Christendom: Two Volumes in One*, Great Minds Series (Amherst, NY: Prometheus Books, 1993), p. 97. Many Christian authors have tried to undermine White by claiming that he said the Church officially opposed the idea that Earth is spherical. He did not, as you can read for yourself.

3. Ibid., p. 35.

4. Ibid.

5. Ibid., p. 37.

6. Ibid., pp. 32–33.

7. Ibid., p. 38.

8. Jim al-Khalili, *The House of Wisdom: How Arabic Science Saved Ancient Knowledge and Gave Us the Renaissance* (New York: Penguin, 2011).

9. Ibid., p. 132.

10. Ibid., p. 182.

11. Ibid., p. 212.

12. Ibid., pp. 213–21.

13. Bernard Lewis, *The Assassins: A Radical Sect in Islam* (New York: Basic Books, 2003).

14. Dava Sobel, *A More Perfect Heaven: How Copernicus Revolutionized the Cosmos* (New York: Walker, 2011).

15. Ibid., p. 7.

16. Ibid., p. 18.

17. Ibid., p. 20.

18. Giese lies at rest next to Copernicus in Frauenberg (Frombork) Cathedral.

19. Sobel, *More Perfect Heaven*, p. 174.

20. Ibid., p. 178.

21. Ibid., p. 186.

22. Amir R. Alexander, *Infinitesimal: How a Dangerous Mathematical Theory Shaped the Modern World* (New York: Scientific American/Farrar, Strauss, and Giroux, 2014).

23. Mary-Jane Rubenstein, *Worlds without End: The Many Lives of the Multiverse* (New York: Columbia University Press, 2014), pp. 88–105.

24. Owen Gingerich, *The Eye of Heaven: Ptolemy, Copernicus, Kepler* (New York: American Institute of Physics, 1993), p. 181.

25. Victor J. Stenger, *God and the Atom: From Democritus to the Higgs Boson* (Amherst, NY: Prometheus Books, 2013).

26. Galileo Galilei, *Sidereus nuncius, Or, the Sidereal Messenger,* trans. Albert Van Helden (Chicago: University of Chicago Press, 1989). First published in 1610.

27. As quoted in Helge Kragh, *Matter and Spirit in the Universe: Scientific and Religious Preludes to Modern Cosmology,* vol. 3, *History of Modern Physical Sciences* (London: Imperial College Press, 2004), p. 7.

28. I am following my usual convention of referring to the Judeo-Christian-Islamic supreme divinity as God (uppercase) and all other divinities as god (lowercase). I will also assume God is personal and male, as in tradition, while god is impersonal and asexual.

29. Gottfried Wilhelm von Leibniz and Samuel Clarke, *The Leibniz-Clarke Correspondence: Together with Extracts from Newton's* Principia *and* Opticks (Manchester, UK: Manchester University Press, 1956) p. 11.

CHAPTER 3. BEYOND UNAIDED HUMAN VISION

1. Immanuel Kant, *Universal Natural History and Theory of the Heavens* (Ann Arbor: University of Michigan Press, 1969).

2. Ibid., p. 141.

3. Helge Kragh, *Matter and Spirit in the Universe: Scientific and Religious Preludes to Modern Cosmology*, vol. 3, *History of Modern Physical Sciences* (London: Imperial College, 2004), p. 12.

4. Ibid., p. 14; Michael Hoskin, *The Construction of the Heavens: William Herschel's Cosmology* (Cambridge; New York: Cambridge University Press, 2012).

5. C. M. Walmsley et al., "Herschel: The First Science Highlights," *Astronomy & Astrophysics* 518 (2010): L10.

6. Gottfried Wilhelm Leibniz, *Theodicy: Essays on the Goodness of God, the Freedom of Man, and the Origin of Evil* (La Salle, IL: Open Court, 1985).

7. Richard Dawkins, *The Selfish Gene* (New York: Oxford University Press, 1976).

CHAPTER 4. GLIMPSES OF THE UNIMAGINED

1. Jacques Laskar, "Stability of the Solar System," http://www.scholarpedia.org/article/Stability_of_the_solar_system#Laplace-Lagrange_stability_of_the_Solar_System (accessed December 4, 2012).

2. Ibid.

3. Pierre Simon Laplace, *Exposition du système du monde* (Paris: Impr. du Cercle-Social, An IV de la République Française, 1796).

4. Pierre Simon Laplace, *A Philosophical Essay on Probabilities*, trans. F. W. Truscott and F. L. Emory from 6th French ed. (New York: Dover Publications, 1952).

5. Ken Adler, *The Measure of All Things: The Seven-Year Odyssey and Hidden Error That Transformed the World* (New York: Free Press, 2003).

6. Ibid.

7. Frank Stacey, "Kelvin's Age of the Earth Paradox Revisited," *Journal of Geophysical Research* 105, no. B6 (2000): 13,155–58.

8. Marcia Bartusiak, *The Day We Found the Universe* (New York: Pantheon Books, 2009), chap. 2.

CHAPTER 5. HEAT, LIGHT, AND THE ATOM

1. Helge Kragh, *Matter and Spirit in the Universe: Scientific and Religious Preludes to Modern Cosmology* (London: Imperial College Press, 2004); Helge Kragh, *Entropic Creation: Religious Contexts of Thermodynamics and Cosmology* (Aldershot, Hampshire, England; Burlington, VT: Ashgate, 2008).

2. Hermann von Helmholtz, *Science and Culture: Popular and Philosophical Essays*, ed. David Cahan (Chicago: Chicago University Press, 1995), p. 30.

3. Rudolf Clausius, "On the Second Fundamental Theorem of the Mechanical Theory of Heat," *Philosophical Magazine* 35 (1868): 404–19.

4. William Thomson, *Popular Lectures and Addresses*, vol. 1 (London: Macmillan, 1891), pp. 356–57.

5. Kragh, *Matter and Spirit in the Universe*, pp. 46–47.

6. James Clerk Maxwell, *The Scientific Papers of James Clerk Maxwell*, ed. W. D. Niven (New York: Dover Publications, 1965), pt. 2, p. 376.

7. Helge Kragh, *Conceptions of Cosmos from Myths to the Accelerating Universe: A History of Cosmology* (Oxford; New York: Oxford University Press, 2007), p. 105.

8. Ibid., pp. 105–106.

9. Victor J. Stenger, *God and the Atom: From Democritus to the Higgs Boson* (Amherst, NY: Prometheus Books, 2013).

10. Ludwig Boltzmann, "On Certain Questions of the Theory of Gases," *Nature* 51 (1895): 483–85.

CHAPTER 6. THE SECOND PHYSICS REVOLUTION

1. Walter Isaacson, *Einstein: His Life and Universe* (New York: Simon & Schuster, 2007), p. 575.

2. You will often hear someone say that the rest energy $E = mc^2$ contained in a mass m is huge because $c = 3 \times 10^8$ meters per second is so large. Actually, the value of c is arbitrary and the numerical value in this case is large because you chose to use a big number for c. Rest energy and mass are identical except for units.

3. In 2011, CERN reported that neutrinos had been measured moving faster than light. The media headlined that Einstein had been proven wrong. It turned out that the result was caused by a loose electrical connection. But even if it had been true, Einstein's speed limit would not have been violated. The particles would just have been tachyons, allowed by relativity.

4. Abraham Pais, *"Subtle Is the Lord": The Science and the Life of Albert Einstein* (Oxford; New York: Oxford University Press, 1982), p. 179.

5. Clifford M. Will, *Was Einstein Right? Putting General Relativity to the Test* (New York: Basic Books, 1986), p. 93.

6. Clifford M. Will, "The Confrontation between General Relativity and Experiment," 2006, http://relativity.livingreviews.org/Articles/lrr-2006-3/ (accessed December 24, 2012).

7. Stephen W. Hawking, "Black Hole Explosions," *Nature* 248, no. 5443 (1974): 30–31.

8. Nina Byers, "E. Noether's Discovery of the Deep Connection between Symmetries and Conservation Laws," 1999, http://www.physics.ucla.edu/~cwp/articles/noether.asg/noether.html (accessed January 15, 2013).

9. A complex number c is a set of two real numbers a and b of the form $c = a + ib$, where $i = \sqrt{-1}$, that is, $i^2 = -1$.

10. P. A. M. Dirac, *The Principles of Quantum Mechanics* (Oxford: Clarendon, 1930).

11. Victor J. Stenger, *Has Science Found God? The Latest Results in the Search for Purpose in the Universe* (Amherst, NY: Prometheus Books, 2003), pp. 351–53; Luciano Burderi and Tiziana Di Salvo, "The Quantum Clock: A Critical Discussion on Space-Time," J arXiv preprint arXiv:1207.0207 (2012).

12. In quantum mechanics, the component of angular momentum along a given axis z is given as $J_z = j_z \hbar$ where j_z is an integer, including zero, or a half integer.

CHAPTER 7. ISLAND UNIVERSES

1. Marcia Bartusiak, *The Day We Found the Universe* (New York: Pantheon Books, 2009), chap. 6; Nina Byers and Gary A. Williams, *Out of the Shadows: Contributions of Twentieth-Century Women to Physics* (Cambridge, UK; New York: Cambridge University Press, 2006).

2. Vesto M. Slipher, "Spectrographic Observations of Nebulae," *Popular Astronomy* 23 (1915): 21–24.

3. Bartusiak, *Day We Found the Universe*, chap. 5.

4. Ibid., chap. 4.

5. Ibid., chap. 10.

6. "Scientists Gather for 1920 Conclave," *Washington Post*, April 25, 1920, p. 38.

7. Harlow Shapley and Herber D. Curtis, "The Scale of the Universe," *Bulletin of the National Research Council* 2, no. 11 (1921): 171–217.

8. "Universe Multiplied a Thousand Times by Harvard Astronomer's Calculations," *New York Times*, May 31, 1921, p. 1.

9. Bartusiak, *Day We Found the Universe*, p. 194.

10. Ibid., p. 204.

11. "Finds Spiral Nebula Are Stellar Systems," *New York Times*, November 23, 1924, p. 6.

12. Bartusiak, *Day We Found the Universe*, p. 214.

13. Edwin Hubble, "Cepheids in Spiral Nebulae," *Publications of the American Astronomical Society* 5 (1925): 261–64.

14. Bartusiak, *Day We Found the Universe*, p. 217.

15. Ibid., pp. 220–24.

CHAPTER 8. A DYNAMIC COSMOS

1. Albert Einstein, "Cosmological Considerations on the General Theory of Relativity" (translated), in *Cosmological Constants: Papers in Modern Cosmology*, edited by Jeremy Bernstein and Gerald Feinberg (New York: Columbia University Press, 1986), pp. 16–26. Note that reprints of many of the early cosmological papers can be found in this volume.

2. Willem De Sitter, "On Einstein's Theory of Gravitation and Its Astronomical Consequences," *Monthly Notices of the Royal Astronomical Society* 78 (1917): 3–28.

3. Pierre Kerzberg, *The Invented Universe: The Einstein-De Sitter Controversy (1916-17) and the Rise of Relativistic Cosmology* (Oxford: Clarendon, 1989).

4. Howard P. Robertson, "On the Foundations of Relativistic Cosmology," *Proceedings of the National Academy of Sciences* 15, no. 11 (1929): 822–29.

5. Alexander Friedmann, "Über Die Krümmung Des Raumes," *Zeitschrift für Physik* 10, no. 1 (1922): 377–286; English translation in Alexander Friedmann, "On the Curvature of Space," *General Relativity and Gravitation* 31, no. 12 (1999): 1991–2000.

6. Ari Belenkiy, "Alexander Friedmann and the Origins of Modern Cosmology," *Physics Today* 65, no. 10 (2012): 38–43.

7. Georges Lemaître, "Un Univers homogène de masse constante et de rayon croissant rendant compte de la vitesse radiale des nébuleuses extra-galactiques" ("A Homogeneous Universe of Constant Mass and Growing Radius Accounting for the Radial Velocity of Extragalactic Nebulae"), *Annales de la Société Scientifique de Bruxelles* 47 (1927): 49.

8. See my discussion of this meeting and the debate in Victor J. Stenger, *The Unconscious Quantum: Metaphysics in Modern Physics and Cosmology* (Amherst, NY: Prometheus Books, 1995), pp. 66–74.

9. As I discovered a few years ago when my wife and I were staying at the

Metropole Hotel in Brussels where the 1927 meeting was held, a photograph of the attendees is prominently displayed in the lobby. See http://www.dailymail .co.uk/sciencetech/article-2002163/1927-Solvay-Conference-Electrons-Photons-Is -greatest-meeting-minds-ever.html (accessed January 28, 2013).

10. Robert Smith, "Edwin P. Hubble and the Transformation of Cosmology," *Physics Today* 43, no. 4 (1990): 52–58.

11. Marcia Bartusiak, *The Day We Found the Universe* (New York: Pantheon Books, 2009), chap. 14.

12. Edwin Hubble, "A Relation between Distance and Radial Velocity among Extra-Galactic Nebulae," *Proceedings of the National Academy of Sciences* 15, no. 3 (1929): 168–73. A free copy is available at http://www.pnas.org/content/15/3/168 .full?ijkey=fdc5eaebb2fb9bfb2745c815793c187c8c7363fa&keytype2=tf_ipsecsha (accessed January 2, 2013).

13. Milton Humason, "The Large Radial Velocity of N.G.C. 7619," *Proceedings of the National Academy of Sciences* 15, no. 3 (1929): 167–68.

14. Astronomers still use the antiquated parsec as their unit of distance. It is defined as the distance that produces a parallax of one second of arc for observations made six months apart on Earth. I will generally stick to the popularly more familiar light-years except where that would add to confusion, where 1 parsec = 3.26 light-years.

15. Robert P. Kirshner, "Hubble's Diagram and Cosmic Expansion," *Proceedings of the National Academy of Sciences* 101, no. 1 (2004): 8–13.

16. Lemaître, "Univers Homogène."

17. Georges Lemaître, "Expansion of the Universe, a Homogeneous Universe of Constant Mass and Increasing Radius Accounting for the Radial Velocity of Extra-Galactic Nebulae," *Monthly Notices of the Royal Astronomical Society* 91 (1931): 490–501.

18. M. Livio, "Lost in Translation: Mystery of the Missing Text Solved," *Nature* 479 (2011): 171–73.

19. Bartusiak, *Day We Found the Universe*, p. 249.

20. Edwin Powell Hubble, *The Realm of the Nebulae* (New Haven, CT; London: Yale University Press, H. Milford, Oxford University Press, 1936).

21. Helge Kragh, *Conceptions of Cosmos from Myths to the Accelerating Universe: A History of Cosmology* (Oxford; New York: Oxford University Press, 2007), p. 149.

22. Arthur S. Eddington, "The End of the World: From the Standpoint of Mathematical Physics," *Nature* 127 (1931): 447–53.

23. Arthur S. Eddington, *The Expanding Universe* (Cambridge: Cambridge University Press, 1933), pp. 56–57.

24. Georges Lemaître, "The Beginning of the World from the Point of View of Quantum Theory," *Nature* 127 (1931): 706.

25. Georges Lemaître, *The Primeval Atom: An Essay on Cosmogony*, trans. H. Betty and Serge A. Korff (New York: Van Nostrand, 1950).

26. Jean-Pierre Luminet, "Editorial Note to: Georges Lemaître, The Beginning of the World from the Point of View of Quantum Theory," *General Relativity and Gravitation* 43 (2011): 2911–28.

27. J. L. Schellenberg, *Divine Hiddenness and Human Reason* (Ithaca, NY: Cornell University Press, 1993).

28. Walter Baade, "A Revision of the Extra-Galactic Distance Scale," *Transactions of the International Astronomical Union* 8 (1952): 397–98.

29. Pope Pius XII, "The Proofs for the Existence of God in the Light of Modern Natural Science," address by Pope Pius XII to the Pontifical Academy of Sciences, November 22, 1951, reprinted as "Modern Science and the Existence of God," *Catholic Mind* 49 (1972): 182–92. Available online at http://www.papalencyclicals .net/Pius12/P12EXIST.HTM (accessed January 27, 2013). The papal quotes in this section are all from this document.

30. Kragh, *Conceptions of Cosmos*, pp. 150–51.

31. Paul Dirac, "A New Basis for Cosmology," *Proceedings of the Royal Society A* 165 (1938): 199–208.

32. See discussion in Victor J. Stenger, *The Fallacy of Fine-Tuning: Why the Universe Is Not Designed for Us* (Amherst, NY: Prometheus Books, 2011), pp. 59–62.

33. Edward Arthur Milne, *Relativity, Gravitation and World-Structure* (Oxford: Clarendon, 1935); Edward Arthur Milne, *Kinematic Relativity: A Sequel to Relativity, Gravitation and World Structure* (Oxford: Clarendon, 1948).

34. For a detailed discussion on Milne's science, philosophy, and theology, and the response of the scientific, philosophical, and theological communities, see Helge Kragh, *Matter and Spirit in the Universe: Scientific and Religious Preludes to Modern Cosmology* (London: Imperial College Press, 2004), pp. 200–29.

35. Victor J. Stenger, *The Comprehensible Cosmos: Where Do the Laws of Physics Come From?* (Amherst, NY: Prometheus Books, 2006), pp. 39–45.

36. Kragh, *Matter and Spirit in the Universe*, pp. 169–72.

37. Hubble, *Realm of the Nebulae*, p. 199.

38. Kragh, *Matter and Spirit in the Universe*, pp. 212–19.

39. Edward Arthur Milne, *Modern Cosmology and the Christian Idea of God* (Oxford: Clarendon, 1952), p. 62.

40. Milne, *Relativity, Gravitation and World-Structure*, p. 266.

41. For a layperson's survey of the history of dark matter, see Jeremiah P. Ostriker and Simon Mitton, *Heart of Darkness: Unraveling the Mysteries of the Invisible Universe* (Princeton, NJ; Oxford: Princeton University Press, 2013).

42. Reprinted with an explanatory introduction in Karl G. Jansky, "Elec-

trical Disturbances of Extraterrestrial Origin," *Proceedings of the IEEE* 86, no. 7 (1988): 1510–15.

CHAPTER 9. NUCLEAR COSMOLOGY

1. Jeremiah P. Ostriker and Simon Mitton, *Heart of Darkness: Unraveling the Mysteries of the Invisible Universe* (Princeton, NJ; Oxford: Princeton University Press, 2013), pp. 147–48.

2. Helge Kragh, *Conceptions of Cosmos from Myths to the Accelerating Universe: A History of Cosmology* (Oxford; New York: Oxford University Press, 2007), pp. 155–57.

3. Edwin Hubble, "Problems of Nebular Research," *Scientific Monthly* 51 (1940): 391–408.

4. Kragh, *Conceptions of Cosmos*, p. 160.

5. For a nice explanation, see the Nobel Prize talk by John N. Bahcall, "How the Sun Shines," http://www.nobelprize.org/nobel_prizes/physics/articles/fusion/ (accessed February 24, 2013).

6. Carl Friedrich von Weizäcker, "Element Transformation inside Stars," *Physicalische Zeitschrift* 39 (1938): 633–46, translated by Richard H. Milburn in Kenneth R. Lang and Owen Gingerich, *A Source Book in Astronomy and Astrophysics, 1900–1975* (Cambridge, MA: Harvard University Press, 1979), pp. 309–18.

7. Kragh, *Conceptions of Cosmos*, pp. 162–63.

8. A nice, short biography of Gamow can be found in Alexander Vilenkin, *Many Worlds in One: The Search for Other Universes* (New York: Hill and Wang, 2006), pp. 29–31.

9. Ralph Alpher and Robert Herman, *Genesis of the Big Bang* (New York: Oxford University Press, 2001), p. 71.

10. Ralph A. Alpher, Hans Bethe, and George Gamow, "The Origin of the Chemical Elements," *Physical Review* 73, no. 7 (1948): 803–804.

11. Note that when we are talking about temperatures on the order of a million degrees, it is unnecessary to specify the units.

12. Ralph A. Alpher and Robert C. Herman, "Remarks on the Evolution of the Expanding Universe," *Physical Review* 75 (1949): 1089–95.

13. Kragh, *Conceptions of Cosmos*, pp. 180–84.

14. Fred Hoyle, "A New Model for the Expanding Universe," *Monthly Notices of the Royal Astronomical Society* 108 (1948): 372–82.

15. Hermann Bondi and Thomas Gold, "The Steady-State Theory of the Expanding Universe," *Monthly Notices of the Royal Astronomical Society* 108 (1948): 252–70.

16. James Jeans, *Astronomy and Cosmogony* (Cambridge: Cambridge University Press, 1928).

17. Fred Hoyle, Geoffrey R. Burbidge, and Jayant Vishnu Narlikar, *A Different Approach to Cosmology: From a Static Universe through the Big Bang towards Reality*, paperback ed. (Cambridge; New York: Cambridge University Press, 2005).

18. Helge Kragh, "Quasi-Steady-State and Related Cosmological Models: A Historical Review," http://arxiv.org/pdf/1201.3449v1.pdf (accessed May 5, 2013).

19. E. Margaret Burbidge, G. R. Burbidge, William A. Fowler, and F. Hoyle, "Synthesis of Elements in the Stars," *Reviews of Modern Physics* 29, no. 4 (1957): 547–650.

20. E. E. Salpeter, "Nuclear Reactions in Stars without Hydrogen," *Astrophysical Journal* 115 (1952): 326–28.

21. Fred Hoyle, "On Nuclear Reactions Occurring in Very Hot Stars. I. The Synthesis of Elements from Carbon to Nickel," *Astrophysical Journal Supplement* 1 (1954): 121–46.

22. Fred Hoyle, "A New Model for the Expanding Universe," *Monthly Notices of the Royal Astronomical Society* 108 (1948): 372–82.

23. Fred Hoyle, "The Universe: Past and Present Reflections," *Annual Reviews of Astronomy and Astrophysics* 20 (1982): 1–35.

24. Fred Hoyle, *Home Is Where the Wind Blows: Chapters from a Cosmologist's Life* (Mill Valley, CA: University Science Books, 1994), p. 413.

25. Maarten Schmidt, "3C 273: A Star-like Object with Large Red-Shift," *Nature* 197 (1963): 1040.

26. Jesse L. Greenstein and Thomas A. Matthews, "Red-Shift of the Unusual Radio Source: 3C 48," *Nature* 197 (1963): 1041–42.

27. Jesse L. Greenstein and Maarten Schmidt, "The Quasi-Stellar Radio Sources 3C 48 and 3C 273," *Astrophysical Journal* 140 (1964): 1–34.

28. Dmitri Pogosian, "Astronomy 122: Astronomy of Stars and Galaxies," 2013, http://www.ualberta.ca/~pogosyan/teaching/ASTRO_122/lect27/lecture 27.html (accessed February 21, 2013).

29. A. Hewish et al., "Observation of a Rapidly Pulsating Radio Source," *Nature* 217 (1968): 709–13.

CHAPTER 10. RELICS OF THE BIG BANG

1. Steven Weinberg, *The First Three Minutes: A Modern View of the Origin of the Universe*, updated ed. (New York: Basic Books, 1993).

2. Arno A. Penzias and Robert W. Wilson, "A Measurement of Excess Antenna Temperature At 4080 Mc/s," *Astrophysical Journal* 142 (1965): 41921.

3. R. H. Dicke, P. J. E. Peebles, P. G. Roll, and D. T. Wilkinson, "Cosmic Black-Body Radiation," *Astrophysical Journal* 142 (1965): 414–19.

4. Weinberg, *First Three Minutes*, pp. 68–69.

5. Ralph A. Alpher, James W. Follin Jr., and Robert C. Herman, "Physical Conditions in the Initial Stages of the Expanding Universe," *Physical Review* 92 (1953): 1347–61.

6. Weinberg, *First Three Minutes*, p. 124.

7. Ibid., pp. 122–25.

8. Dennis Overbye, "Essay; Remembering David Schramm, Gentle Giant of Cosmology," *New York Times*, February 10, 1998, online at http://www.nytimes.com/1998/02/10/science/essay-remembering-david-schramm-gentle-giant-of-cosmology.html?pagewanted=all&src=pm (accessed February 6, 2013).

9. Michael S. Turner, "David Norman Schramm 1945–1997," 2009, http://www.nasonline.org/publications/biographical-memoirs/memoir-pdfs/schramm-david.pdf (accessed February 6, 2012).

10. Edward W. Kolb and Michael S. Turner, *The Early Universe* (Reading, MA: Addison-Wesley, 1990).

11. D. Z. Freedman, D. N. Schramm, and D. L. Tubbs, "The Weak Neutral Current and Its Effects in Stellar Collapse," *Annual Reviews of Nuclear Science* 27 (1977): 167–207.

12. Gary Steigman, David N. Schramm, and James E. Gunn, "Cosmological Limits to the Number of Massive Leptons," *Physics Letters B* 66, no. 2 (1977): 202–204.

13. A terrific source for always up-to-date cosmological information is Ned Wright's "Cosmological Tutorial," http://www.astro.ucla.edu/~wright/cosmolog.htm (accessed February 14, 2013). See his "Big Bang Nucleosynthesis" link for more details on this section.

14. Edward L. Wright, "Big Bang Nucleosynthesis," 2012, http://www.astro.ucla.edu/~wright/BBNS.html (accessed February 7, 2013).

15. S. Burles, K. M. Nollett, and M. S. Turner, "Big-Bang Nucleosynthesis: Linking Inner Space and Outer Space," 1999, http://arxiv.org/pdf/astro-ph/9903300v1.pdf (accessed February 7, 2013).

16. This figure was modified from Ned Wright's "Cosmological Tutorial."

17. D. N. Spergel, R. Bean, O. Dore, M. R. Nolta, C. L. Bennett, J. Dunkley, G. Hinshaw, N. Jarosik, E. Komatsu, and L. Page, "Wilkinson Microwave Anisotropy Probe (WMAP) Three Year Results: Implications for Cosmology," *Astrophysical Journal Suppl.* 170 (2007): 377, preprint at http://arxiv.org/abs/astroph/0603449 (accessed February 16, 2013); Richard H. Cyburt, Brian D. Fields, and Keith A. Olive, "Primordial Nucleosynthesis in Light of WMAP," *Physics Letters B* 567

(2003): 227–34, preprint at http://arxiv.org/abs/astro-ph/0302431v2 (accessed February 16, 2013).

18. Keith Olive and Evan D. Skillman, "A Realistic Determination of the Error on the Primordial Helium Abundance: Steps toward Non-Parametric Nebular Helium Abundances," *Astrophysical Journal* 617 (2004): 29–49.

19. J. G. Matthews, T. Kajino, and T. Shima, "Big Bang Nucleosynthesis with a New Neutron Lifetime," *Physical Review D* 71 (2005): 021302.

20. Hugh Ross, "Big Bang Model Refined by Fire," in *Mere Creation: Science, Faith & Intelligent Design*, ed. William A. Dembski (Downers Grove, IL: Intervarsity, 1988), pp. 363–83.

21. Victor J. Stenger, *The Fallacy of Fine-Tuning: Why the Universe Is Not Designed for Us* (Amherst, NY: Prometheus Books, 2011), p. 205.

CHAPTER 11. PARTICLES AND THE COSMOS

1. Evgenii Lifshitz, "On the Gravitational Stability of the Expanding Universe," *Journal of Physics — USSR* 10 (1946): 116.

2. Edward R. Harrison, "Fluctuations at the Threshold of Classical Cosmology," *Physical Review D* 1, no. 10 (1970): 2726.

3. Yakov B. Zel'dovich, "A Hypothesis, Unifying the Structure and the Entropy of the Universe, 1972," *Mon. Not. R. Astron. Soc* 160 (1972): 1–3.

4. Darcy Wentworth Thompson, *On Growth and Form* (Cambridge: Cambridge University Press, 1992), first published in 1917, p. 28.

5. Rainer K. Sachs and Arthur M. Wolfe, "Perturbations of a Cosmological Model and Angular Variations of the Microwave Background," *Astrophysical Journal* 147 (1967): 73.

6. Joseph Silk, "Cosmic Black-Body Radiation and Galaxy Formation," *Astrophysical Journal* 151 (1968): 459–71; Rashid A. Sunyaev and Ya B. Zel'dovich, "Small-Scale Fluctuations of Relic Radiation," *Astrophysics and Space Science* 7, no. 1 (1970): 3–19; Philip J. E. Peebles and J. T. Yu, "Primeval Adiabatic Perturbation in an Expanding Universe," *Astrophysical Journal* 162 (1970): 815–36.

7. D. Hanson et al., "Detection of B-mode Polarization in the Cosmic Microwave Background with Data from the South Pole Telescope," *Physical Review Letters* 111 (2013): 141301.

8. P. A. R. Ade et al., "A Measurement of the Cosmic Microwave Background B-Mode Polarization Power Spectrum at Sub-Degree Scales with POLARBEAR," (2014): to be published.

9. Albert Bosma, "The Distribution and Kinematics of Neutral Hydrogen

in Spiral Galaxies of Various Morphological Types" (doctoral thesis, University Gronigen, Gronigen, Netherlands, 1978). Available at http://ned.ipac.caltech .edu/level5/March05/Bosma/frames.html (accessed September 2, 2013).

10. Spin is an angular momentum and is generally expressed in units of $\hbar = h/2\pi$, where h is Planck's constant.

11. David Kaiser, "Physics and Feynman's Diagrams," *American Scientist* 93 (2005): 156–65.

12. Although, in my 2000 book *Timeless Reality* I suggest how Feynman diagrams can be taken more literally than is usually assumed.

13. Sin-Itiro Tomonoga, "On a Relativistically Invariant Formulation of the Quantum Theory of Wave Fields," *Progress in Theoretical Physics* 1, no. 2 (1946): 27–42; Julian Schwinger, "On Quantum-Electrodynamics and the Magnetic Moment of the Electron," *Physical Review* 73, no. 4 (1948): 416–17; Richard Feynman, "Space-Time Approach to Quantum Electrodynamics," *Physical Review* 76, no. 6 (1949): 769–89; Freeman J. Dyson, "The Radiation Theories of Tomonaga, Schwinger, and Feynman," *Physical Review* 75, no. 3 (1949): 486–502.

14. Victor J. Stenger, *The Comprehensible Cosmos: Where Do the Laws of Physics Come From?* (Amherst, NY: Prometheus Books, 2006).

15. Peter W. Higgs, "Broken Symmetries and the Masses of Gauge Bosons," *Physical Review Letters* 13 (1964): 508; F. Englert and R. Brout, "Broken Symmetry and the Mass of Gauge Vector Bosons," *Physical Review Letters* 13 (1964): 321; G. G. Guralnik, C. R. Hagen, and T. W. Kibble, "Global Conservation Laws and Massless Particles," *Physical Review Letters* 13 (1964): 585.

16. ATLAS Collaboration, "Observation of a New Particle in the Search for the Standard Model Higgs Boson with the ATLAS Detector at the LHC," *Physics Letters B* 716, no. 1 (2012): 1–29; CMS Collaboration, "Observation of a New Boson at a Mass of 125 Gev with the CMS Experiment at the LHC," *Physics Letters B* 716, no. 1 (2012): 30–61.

17. I have discussed this problem in two books: Victor J. Stenger, *The Unconscious Quantum: Metaphysics in Modern Physics and Cosmology* (Amherst, NY: Prometheus Books, 1995); Victor J. Stenger, *Quantum Gods: Creation, Chaos, and the Search for Cosmic Consciousness* (Amherst, NY: Prometheus Books, 2009).

18. David Tong, "Is Quantum Reality Analog after All?" *Scientific American* (November 26, 2012).

19. Brian Greene, *The Hidden Reality: Parallel Universes and the Deep Laws of the Cosmos* (New York: Alfred A. Knopf, 2011), p. 33.

20. Meinhard Kuhlmann, "What Is Real?" *Scientific American* (August 2013): 40–47.

21. J. Beringer and Particle Data Group et al., "The Review of Particle

Physics," *Physical Review D* 88 (2012): 010001, available online at http://pdg.lbl .gov/ (accessed February 22, 2013).

22. Andrei Sakharov, "Violation of CP Invariance, C Asymmetry, and Baryon Asymmetry of the Universe," *Journal of Experimental and Theoretical Physics Letters* 5 (1967): 24–27. English translation available at http://www.jetpletters.ac.ru/ ps/1643/article_25089.pdf courtesy of American Institute of Physics (accessed February 28, 2013).

23. Howard Georgi and S. L. Glashow, "Unity of All Elementary-Particle Forces," *Physical Review Letters* 32 (1974): 438–41.

24. T. J. Goldman and D. A. Ross, "A New Estimate of the Proton Decay Lifetime," *Physics Letters B* 84, no. 2 (1979): 208–11.

25. Super-Kamiokande Collaboration, "Proton Lifetime Is Longer Than 10^{34} Years," 2011, http://www-sk.icrr.u-tokyo.ac.jp/whatsnew/new-20091125-e.html (accessed February 22, 2013).

26. K. Hirata et al., "Observation of a Neutrino Burst from the Supernova SN1987A," *Physical Review Letters* 58 (1987): 1490–93; IMB Collaboration, "Neutrinos from SN1987A in the IMB Detector," *Nuclear Instruments and Methods in Physics Research* A264 (1988): 28–31.

27. Brian Greene, *The Elegant Universe: Superstrings, Hidden Dimensions, and the Quest for the Ultimate Theory* (New York: W. W. Norton, 1999).

CHAPTER 12. INFLATION

1. Hugh Ross, "Big Bang Model Refined by Fire," in *Mere Creation: Science, Faith & Intelligent Design*, ed. William A. Dembski, pp. 363–83 (Downers Grove, IL: Intervarsity, 1998), pp. 372–75.

2. Dinesh D'Souza, *Life after Death: The Evidence* (Washington, DC: Regnery, 2009), p. 84.

3. Stephen Hawking, *A Brief History of Time: From the Big Bang to Black Holes* (Toronto; New York: Bantam Books, 1988), pp. 121–22.

4. William Lane Craig, "The Craig-Pigliucci Debate: Does God Exist?" http://www.leaderu.com/offices/billcraig/docs/craig-pigliucci2.html (accessed July 12, 2013).

5. Paul A. M. Dirac, "Quantised Singularities in the Electromagnetic Field," *Proceedings of the Royal Society of London*, ser. A, containing papers of a mathematical and physical character (1931): 60–72.

6. Gerard 't Hooft, "Magnetic Monopoles in Unified Gauge Theories," *Nuclear Physics B* 79, no. 2 (1974): 276–84.

7. Alexander M. Polyakov, "Particle Spectrum in Quantum Field Theory," *Soviet Journal of Experimental and Theoretical Physics Letters* 20 (1974): 194–95.

8. T. W. B. Kibble, "Topology of Cosmic Domains and Strings," *Journal of Physics A: Mathematical and General* 9, no. 8 (1976): 1387–98.

9. John P. Preskill, "Cosmological Production of Superheavy Magnetic Monopoles," *Physical Review Letters* 43, no. 19 (1979): 1365–68.

10. For a good, nonmathematical discussion, see Alan H. Guth. *The Inflationary Universe: The Quest for a New Theory of Cosmic Origins* (Reading, MA: Addison-Wesley, 1997), pp. 147–65.

11. For a recent review containing a complete list of references to both theory and experiment, see D. Milstead and E. J. Weinberg, "Magnetic Monopoles," 2011, http://pdg.web.cern.ch/pdg/2011/reviews/rpp2011-rev-mag-monopole-searches.pdf (accessed March 9, 2013).

12. MACRO Collaboration, "Final Results of Magnetic Monopole Searches with the Macro Experiment," *European Journal of Physics* C25 (2002): 511–22.

13. Stephen Hawking and Roger Penrose, "The Singularities of Gravitational Collapse," *Proceedings of the Royal Society of London*, ser. A, 314 (1970): 529–48.

14. William Lane Craig and James D. Sinclair, "The *Kalam* Cosmological Argument," in *The Blackwell Companion to Natural Theology*, ed. William Lane Craig and James Porter Moreland (Chichester, UK; Malden, MA: Wiley-Blackwell, 2009), pp. 101–201.

15. Stephen Hawking, *A Brief History of Time: From the Big Bang to Black Holes* (Toronto; New York: Bantam Books, 1988), p. 50.

16. Demos Kazanas, "Dynamics of the Universe and Spontaneous Symmetry Breaking," *Astrophysical Journal* 241 (1980): L59–L63.

17. Katsuhiko Sato, "First-Order Phase Transitions of a Vacuum and the Expansion of the Universe," *Monthly Notices of the Royal Astronomical Society* 195 (1981): 467–79.

18. Alan Guth, "Inflationary Universe: A Possible Solution to the Horizon and Flatness Problems," *Physical Review D* 23, no. 2 (1981): 347–56.

19. Guth, *Inflationary Universe*.

20. Hawking, *Brief History of Time*, p. 128.

21. Alan H. Guth and Erick J. Weinberg, "Could the Universe Have Recovered from a Slow First-Order Phase Transition?" *Nuclear Physics B* 212, no. 2 (1983): 321–64.

22. Guth, *Inflationary Universe*, p. 203.

23. Andrei D. Linde, "A New Inflationary Universe Scenario: A Possible Solution of the Horizon, Flatness, Homogeneity, Isotropy and Primordial Monopole Problems," *Physics Letters B* 108 (1982): 389.

24. Andreas Albrecht and Paul J. Steinhardt, "Cosmology for Grand Unified

Theories with Radiatively Induced Symmetry Breaking," *Physical Review Letters* 48, no. 17 (1982): 1220–23.

25. Guth, *Inflationary Universe*, p. 206.

26. Thanks to Mark Whittle for this analogy.

27. Victor J. Stenger, *The Comprehensible Cosmos: Where Do the Laws of Physics Come From?* (Amherst, NY: Prometheus Books, 2006), mathematical supplement G.

28. Alan Michael Dressler, *Voyage to the Great Attractor: Exploring Intergalactic Space* (New York: A. A. Knopf, 1994).

29. R. Brent Tully and J. Richard Fisher, "A New Method of Determining Distances to Galaxies," *Astronomy and Astrophysics* 54 (1977): 661–73.

30. R. Brent Tully and J. Richard Fisher, *Nearby Galaxies Atlas* (Cambridge; New York: Cambridge University Press, 1987).

31. Margaret J. Geller and John P. Huchra, "Mapping the Universe," *Science* 246, no. 4932 (1989): 897–903.

32. Hélene Courtois et al., "Cosmography of the Local Universe," http://irfu .cea.fr/cosmography (accessed December 18, 2013).

33. Guth, *Inflationary Universe*, pp. 222–43.

34. James M. Bardeen, Paul J. Steinhardt, and Michael S. Turner, "Spontaneous Creation of Almost Scale-Free Density Perturbations in an Inflationary Universe," *Physical Review D* 28, no. 4 (1983): 679.

35. This is a chemistry term. For example, if you ionize hydrogen atoms into protons and electrons, they will "recombine" into atoms.

36. Smoot tells his story in the bestseller George Smoot and Keay Davidson, *Wrinkles in Time* (New York: W. Morrow, 1993).

CHAPTER 13. FALLING UP

1. See, for example, Eric J. Lerner, *The Big Bang Never Happened* (New York: Times Books/Random House, 1991).

2. George Smoot and Keay Davidson, *Wrinkles in Time* (New York: W. Morrow, 1993), p. 241.

3. John C. Mather et al., "Measurement of the Cosmic Microwave Background Spectrum by the COBE FIRAS Instrument," *Astrophysical Journal* 420 (1994): 439–44; The figure is from the NASA site http://lambda.gsfc.nasa.gov/product/ cobe/firas_image.cfm (accessed December 24, 2013).

4. As quoted on the cover of Smoot and Davidson, *Wrinkles in Time*.

5. George F. Smoot et al., "Structure in the COBE Differential Microwave Radiometer First-Year Maps," *Astrophysical Journal* 396 (1992): L1–L5.

6. Alan H. Guth, *The Inflationary Universe: The Quest for a New Theory of Cosmic Origins* (Reading, MA: Addison-Wesley, 1997), p. 242.

7. E. L. Wright et al., "Interpretation of the Cosmic Microwave Background Radiation Anisotropy Detected by the Cobe Differential Microwave Radiometer," *Astrophysical Journal* 396 (1992): L13–L18.

8. Jeremiah P. Ostriker and Simon Mitton, *Heart of Darkness: Unraveling the Mysteries of the Invisible Universe* (Princeton, NJ; Oxford: Princeton University Press, 2013), p. 201.

9. NASA Science News, "Hubble Sees the Fireball from a 'Kilonova,'" http://science.nasa.gov/science-news/science-at-nasa/2013/03aug_kilonova/ (accessed August 4, 2013).

10. Victor J. Stenger, "The Production of Very High Energy Photons and Neutrinos from Cosmic Proton Sources," *Astrophysical Journal* 284 (1984): 810–16.

11. Trevor C. Weekes et al., "Observation of TeV Gamma Rays from the Crab Nebula Using the Atmospheric Cerenkov Imaging Technique," *Astrophysical Journal* 342 (1989): 379–95.

12. F. A. Aharonian, A. N. Timokhin, and A. V. Plyasheshnikov, "On the Origin of Highest Energy Gamma-Rays from Mkn 501," *Astronomy and Astrophysics* 384, no. 3 (2002): 834–47.

13. Victor J. Stenger, "Neutrino Oscillations in DUMAND," in *Neutrino Mass: Mini-Conference and Workshop Transparencies*, ed. V. Barger and D. Cline, pp. 174–78 (Madison: University of Wisconsin, 1980). A scanned copy of the paper can be found at http://www.colorado.edu/philosophy/vstenger/Telemark.pdf (accessed April 14, 2013).

14. Super-Kamiokande Collaboration, "Evidence for Oscillation of Atmospheric Neutrinos," *Physical Review Letters* 81 (1998): 1562–67.

15. Because they are not stationary states but constantly mix with one another, the three neutrinos in the standard model do not have definite masses but are rather mixtures of three other neutrino states v_1, v_2, and v_3 that are stationary states. Technically, these are the states that have definite masses.

16. R. Svoboda and K. Gordan, "Neutrinos in the Sun," 1998, http://apod.nasa.gov/apod/ap980605.html (accessed February 26, 2013).

17. For a pedagogical review of dark matter, see Guido D'Amico, Marc Kamionkowski, and Kris Sigurdson, "Dark Matter Astrophysics," in *Dark Matter and Dark Energy: A Challenge for Modern Cosmology*, ed. Sabiuno Matarrese, Monica Colpi, Vittorio Gorini, and Ugo Moschella (Dordrecht: Springer, 2011), pp. 241–72.

18. To read more about sterile neutrinos, see Eric Hand, "Hunt for the Sterile Neutrino Heats Up," *Nature* 464 (2010): 334–35; also, Patrick Huber and Jon Link, "White Paper on Sterile Neutrinos," http://cnp.phys.vt.edu/white_paper/white paper.pdf (accessed September 10, 2012).

19. Leonard Susskind, *Cosmic Landscape: String Theory and the Illusion of Intelligent Design* (New York: Little, Brown, 2005).

20. Lawrence M. Krauss and Michael S. Turner, "The Cosmological Constant Is Back," *General Relativity and Gravitation* 27, no. 11 (1995): 1137–44.

21. G. de Vaucouleurs, "The Extragalactic Distance Scale and the Hubble Constant," *Observatory* 102 (1982): 178–94.

22. Saul Perlmutter et al., "Measurements of Ω and Λ from 42 High-Redshift Supernovae," *Astrophysical Journal* 517, no. 2 (1999): 565.

23. Adam G. Riess et al., "Observational Evidence from Supernovae for an Accelerating Universe and a Cosmological Constant," *Astronomical Journal* 116, no. 3 (1998): 1009.

24. Steven Weinberg, "The Cosmological Constant Problem," *Reviews of Modern Physics* 61, no. 1 (1989): 1–23.

25. Victor J. Stenger, *The Fallacy of Fine-Tuning: Why the Universe Is Not Designed for Us* (Amherst, NY: Prometheus Books, 2011), pp. 213–31. Weinberg's derivation can also be found therein.

26. NASA Lambda, "Data Center for Cosmic Microwave Background (CMB) Research," http://lambda.gsfc.nasa.gov/links/experimental_sites.cfm (accessed April 21, 2013).

27. C. Barth Netterfield et al., "A Measurement of the Angular Power Spectrum of the Anisotropy in the Cosmic Microwave Background," *Astrophysical Journal* 474, no. 1 (1997): 47.

28. Ravi Subrahmanyan et al., "A Search for Arcminutescale Anisotropy in the Cosmic Microwave Background," *Monthly Notices of the Royal Astronomical Society* 263, no. 2 (1993): 416.

29. From S. Hancock et al., "Constraints on Cosmological Parameters from Recent Measurements of Cosmic Microwave Background Anisotropy," *Monthly Notices of the Royal Astronomical Society* 294, no. 1 (1998): L1–L6.

CHAPTER 14. MODELING THE UNIVERSE

1. A list of redshift surveys and links to their published results can be found at http://www.astro.ljmu.ac.uk/~ikb/research/galaxy-redshift-surveys.html (accessed May 29, 2013). The web page was created by Ivan K. Baldry.

2. Daniel J. Eisenstein et al., "Detection of the Baryon Acoustic Peak in the Large-Scale Correlation Function of SDSS Luminous Red Galaxies," *Astrophysical Journal* 633, no. 2 (2005): 560.

3. Ibid.

4. Andrew H. Jaffe et al., "Cosmology from Maxima-1, Boomerang, and COBEe DMR Cosmic Microwave Background Observations," *Physical Review Letters* 86, no. 16 (2001): 3475–79.

5. C. L. Bennett et al., "Nine-Year Wilkinson Microwave Anisotropy Probe (WMAP) Observations: Final Maps and Results," arXiv preprint arXiv:1212.5225 (2012).

6. N. Jarosik et al., "Seven-Year Wilkinson Microwave Anisotropy Probe (WMAP) Observations: Sky Maps, Systematic Errors, and Basic Results," *Astrophysical Journal Supplement Series* 192, no. 2 (2011): 14.

7. G. Hinshaw et al., "Five-Year Wilkinson Microwave Anisotropy Probe Observations: Data Processing, Sky Maps, and Basic Results," *Astrophysical Journal Supplement Series* 180, no. 2 (2009): 225.

8. Mark Whittle, *Cosmology: The History and Nature of Our Universe* (Chantily, VA: Teaching, 2008), lectures 15 and 16.

9. Mark Whittle, "Big Bang Acoustics," 2000, http://www.astro.virginia.edu/~dmw8f/BBA_web/index_frames.html (accessed May 12, 2013).

10. Uros Seljak and Matias Zaldarriaga, "A Line of Sight Approach to Cosmic Microwave Background Anisotropies," *Astrophysical Journal* 469 (1996): 437–44.

11. Planck Collaboration, P. A. R. Ade, et al., "Planck 2013 Results. I. Overview of Products and Scientific Results," arXiv preprint arXiv:1303.5062 (2013).

12. Anja von der Linden et al., "Robust Weak-Lensing Mass Calibration of Planck Galaxy Clusters," *Monthly Notices of the Royal Astronomical Society* (2014): to be published.

13. Jean-Christophe Hamilton, "What Have We Learned from Observational Cosmology?" (paper presented at the Philosophical Aspects of Modern Cosmology Meeting, Grenada, Spain, September 22–23, 2013).

14. See figure 1 in Planck Collaboration, "Planck 2013 Results. XXII. Constraints on Inflation," arXiv preprint arXiv:1303.5082 (2013).

15. Paul J. Steinhardt and Neil Turok, *Endless Universe: Beyond the Big Bang* (New York: Doubleday, 2007).

16. Justin Khoury et al., "The Ekpyrotic Universe: Colliding Branes and the Origin of the Hot Big Bang, 2001," *Physical Review D* 64 (2001): 123522.

17. P. A. R. Ade et al., "Detection of B-Mode Polarization at Degree Angular Scales by BICEP2," *Physical Review Letters* 112, no. 24 (2014): 241101.

18. Dennis Overbye, "Space Ripples Reveal Big Bang's Smoking Gun," *New York Times*, March 17, 2014.

19. See, for example, plans for the LSST (Large Synoptic survey Telescope), available at http://www.lsst.org/lsst/ (accessed July 27, 2013).

20. Richard Massey, Thomas Kitching, and Johan Richard, "The Dark Matter of Gravitational Lensing," *Reports on Progress in Physics* 73, no. 8 (2010): 086901.

21. Francesco Arneodo, "Dark Matter Searches" (paper presented at the 32th International Symposium on Physics in Collision [PIC 2012], Strbské Pleso, Slovakia, September 12–15, 2012).

22. D. S. Akerib et al., "First Results from the LUX Dark Matter Experiment at the Sanford Underground Research Facility," 2013, http://arxiv.org/pdf/1310.8214 (accessed November 7, 2013).

23. Oscar Adriani et al., "An Anomalous Positron Abundance in Cosmic Rays with Energies 1.5–100 GeV," *Nature* 458, no. 7238 (2009): 607–609.

24. Aous A. Abdo et al., "Measurement of the Cosmic Ray e^+e^- Spectrum from 20 GeV to 1 TeV with the Fermi Large Area Telescope," *Physical Review Letters* 102, no. 18 (2009): 181101.

25. Tansu Daylan et al., "The Characterization of the Gamma-Ray Signal from the Central Milky Way: A Compelling Case for Annihilating Dark Matter," arXiv preprint arXiv (2014): 1402.6703.

26. Arneodo, "Dark Matter Searches."

27. M. Aguilar et al., "First Result from the Alpha Magnetic Spectrometer on the International Space Station: Precision Measurement of the Positron Fraction in Primary Cosmic Rays of 0.5–350 GeV," *Physical Review Letters* 110, no. 14 (2013): 141102.

28. Zurab G. Berezhiani and Rabindra N. Mohapatra, "Reconciling Present Neutrino Puzzles: Sterile Neutrinos as Mirror Neutrinos," *Physical Review D* 52, no. 11 (1995): 6607–11.

29. E. Bulbul et al., "Detection of an Unidentified Emission Line in the Stacked X-Ray Spectrum of Galaxy Clusters," *Astrophysical Journal* (in press, 2014); A. Boyarsky et al., "An Unidentified Line in X-Ray Spectra of the Andromeda Galaxy and Perseus Galaxy Cluster," 2014, http://arxiv.org/pdf/1402.4119v1.pdf (accessed March 13, 2014).

30. Richard A. Battye and Adam Moss, "Evidence for Massive Neutrinos from Cosmic Microwave Background and Lensing Observations," *Physical Review Letters* 112, no. 5 (2014): 051303; Mark Wyman et al., "Neutrinos Help Reconcile Planck Measurements with the Local Universe," *Physical Review Letters* 112, no. 5 (2014): 051302.

31. V. Berezinsky, "High Energy Neutrino Astronomy," *Nuclear Physics B-Proceedings Supplements* 19 (1991): 375–87; C. Spiering, "High-Energy Neutrino Astronomy: A Glimpse of the Promised Land J Arxiv Preprint," arXiv:1402.2096 (2014).

32. IceCube Collaboration, "First Observation of PeV-Energy Neutrinos with IceCube," http://arxiv.org/abs/1304.5356v2 (2013).

33. Floyd W. Stecker, "IceCube Observed PeV Neutrinos from AGN Cores," *Physical Review D* 88 (2013): 047301.

CHAPTER 15: THE ETERNAL MULTIVERSE

1. Stephen Hawking and Roger Penrose, "The Singularities of Gravitational Collapse," *Proceedings of the Royal Society of London*, ser. A, 314 (1970): 529–48.

2. William Lane Craig, *The Kalam Cosmological Argument* (London: Macmillan, 1979).

3. Stephen Hawking, *A Brief History of Time: From the Big Bang to Black Holes* (Toronto; New York: Bantam Books, 1988), p. 50.

4. Arvind Borde, Alan H. Guth, and Alexander Vilenkin, "Inflationary Spacetimes Are Not Past-Complete," *Physical Review Letters* 90 (2003): 151301.

5. Alexander Vilenkin, e-mail communication with author, May 21, 2010.

6. Anthony N. Aguirre and Steven Gratton, "Steady-State Eternal Inflation," *Physical Review D* 65, no. 8 (2002): 083507; "Inflation without a Beginning: A Null Boundary Proposal," *Physical Review D* 67 (2003): 083515 .

7. Sean M. Carrol, and Jennifer Chen, "Spontaneous Inflation and the Origin of the Arrow of Time," http://arxiv.org/abs/hep-th/0410270 (2004) (accessed May 22, 2010).

8. Sean Carroll and William Lane Craig, "Greer-Heard Forum: God and Cosmology," 2014, http://www.greerheard.com (accessed March 23, 2014).

9. William Lane Craig and James D. Sinclair, "The *Kalam* Cosmological Argument," in *The Blackwell Companion to Natural Theology*, ed. William Lane Craig and James Porter Moreland, pp. 101–201 (Chichester, UK; Malden, MA: Wiley-Blackwell, 2009).

10. James A. Lindsay, *Dot, Dot, Dot: Infinity Plus God Equals Folly* (Fareham, Hampshire, UK: Onus Books, 2013).

11. David Hilbert, "On the Infinite," in *Philosophy of Mathematics*, ed. Paul Benacerraf and Hillary Putnam (Englewood Cliffs, NJ: Prentice-Hall, 1964), pp. 139–41. Originally presented (in German) on June 24, 1925, before a congress of the Westphalian Mathematical Society, in honor of Karl Weierstrauss.

12. Lee Smolin, *Three Roads to Quantum Gravity* (New York: Basic Books, 2002).

13. Erik Verlinde, "On the Origin of Gravity and the Laws of Newton," *Journal of High Energy Physics*, no. 4 (2011): 1–27; Thanu Padmanabhan, "Grappling with Gravity," *Scientific American India* (January 2011): 31–35.

14. John Archibald Wheeler and Kenneth William Ford, *Geons, Black Holes, and Quantum Foam: A Life in Physics* (New York: Norton, 1998).

15. Robert J. Nemiroff et al., "Bounds on Spectral Dispersion from Fermi-Detected Gamma Ray Bursts," *Physical Review Letters* 108, no. 23 (2012): 231103.

16. Victor J. Stenger, *The Comprehensible Cosmos: Where Do the Laws of Physics Come From?* (Amherst, NY: Prometheus Books, 2006), pp. 312–19; "A Scenario for the Natural Origin of the Universe," *Philo* 9, no. 2 (2006): 93–102.

17. David Atkatz and Heinz Pagels, "Origin of the Universe as a Quantum Tunneling Event," *Physical Review Letters D* 25 (1982): 2065–73.

18. David Atkatz, "Quantum Cosmology for Pedestrians," *American Journal of Physics* 62, no. 7 (1994): 619–27.

19. Alexander Vilenkin, "Creation of Universes from Nothing," *Physics Letters B* 117, no. 1 (1982): 25–28.

20. James B. Hartle and Stephen W. Hawking, "Wave Function of the Universe," *Physical Review D* 28 (1983): 2960–75.

21. For those familiar with partial differential calculus, in Schrödinger wave mechanics the momentum component along the x-axis, p_x, is replaced by the differential that does not commute with x, from which the uncertainty principle can be derived. In Heisenberg's quantum mechanics, the observables are matrices. In Dirac's quantum mechanics, the observables are operators in linear vector space.

22. Bryce S. Dewitt, "Quantum Theory of Gravity. I. The Canonical Theory," *Physical Review* 160 (1967): 1113–48.

23. Victor J. Stenger, *Timeless Reality: Symmetry, Simplicity, and Multiple Universes* (Amherst, NY: Prometheus Books, 2000).

24. Christoph Schönborn, "Finding Design in Nature," *New York Times*, July 7, 2005.

25. William Lane Craig, "Opening Speech," in *Is Faith in God Reasonable?* ed. Paul Gould and Corey Miller (New York: Routledge, 2014).

26. Edward Robert Harrison, *Masks of the Universe* (New York London: Macmillan, Collier Macmillan, 1985), p. 252.

27. See, for example, Bernard Carr, ed., *Universe or Multiverse?* (Cambridge: Cambridge University Press, 2007); John Gribbin, *In Search of the Multiverse: Parallel Worlds, Hidden Dimensions, and the Ultimate Quest for the Frontiers of Reality* (Hoboken, NJ: Wiley, 2010); Steven L. Manly, *Visions of the Multiverse* (Pompton Plains, NJ: New Page Books, 2011).

28. Alexander Vilenkin, "Birth of Inflationary Universes," *Physical Review D* 27, no. 12 (1983): 2848–55. See also, *Many Worlds in One: The Search for Other Universes* (New York: Hill and Wang, 2006).

29. Andrei D. Linde, "Eternally Existing Self-Reproducing Chaotic Inflationary Universe," *Physics Letters B* 175, no. 4 (1986): 395–400.

30. Andrei Linde, "The Self-Reproducing Inflationary Universe," *Scientific American* 271, no. 5 (1994): 48–55.

31. Aguirre and Gratton, "Steady-State Eternal Inflation"; Aguirre and Gratton, "Inflation without a Beginning."

32. Craig, "Opening Speech."

33. Sean M. Carroll, *From Eternity to Here: The Quest for the Ultimate Theory of Time* (New York: Dutton, 2010), p. 355.

34. Planck Collaboration, P. A. R. Ade, et al., "Planck 2013 Results. XXIII. Isotropy and Statistics of the CMB," arXiv preprint arXiv:1303.5083 (2013).

35. Albert Einstein, Max Born, and Hedwig Born, *The Born-Einstein Letters: Correspondence between Albert Einstein and Max and Hedwig Born from 1916 to 1955, with Commentaries by Max Born* (London: Macmillan, 1971).

36. David Bohm and B. J. Hiley, *The Undivided Universe: An Ontological Interpretation of Quantum Theory* (London; New York: Routledge, 1993).

37. See my discussion in in Victor J. Stenger, *The Unconscious Quantum: Metaphysics in Modern Physics and Cosmology* (Amherst, NY: Prometheus Books, 1995).

38. Hugh Everett III, "'Relative State' Formulation of Quantum Mechanics," *Reviews of Modern Physics* 29 (1957): 434–62.

39. Bryce S. DeWitt, Hugh Everett, and Neill Graham, *The Many-Worlds Interpretation of Quantum Mechanics: A Fundamental Exposition* (Princeton, NJ: Princeton University Press, 1973); David Deutsch, *The Fabric of Reality: The Science of Parallel Universes – and Its Implications* (New York: Allen Lane, 1997).

40. Victor J. Stenger, *Timeless Reality: Symmetry, Simplicity, and Multiple Universes* (Amherst, NY: Prometheus Books, 2000).

41. Leonard Susskind, *Cosmic Landscape: String Theory and the Illusion of Intelligent Design* (New York: Little, Brown, 2005), pp. 316–24; Raphael Bousso and Leonard Susskind, "Multiverse Interpretation of Quantum Mechanics," *Physical Review D* 85, no. 4 (2012): 045007; Anthony Aguirre and Max Tegmark, "Born in an Infinite Universe: A Cosmological Interpretation of Quantum Mechanics," *Physical Review D* 84, no. 10 (2011): 105002.

42. Don N. Page and William K. Wootters, "Evolution without Evolution: Dynamics Described by Stationary Observables," *Physical Review D* 27, no. 12 (1983): 2885.

43. William K. Wootters, "'Time' Replaced by Quantum Correlations," *International Journal of Theoretical Physics* 23, no. 8 (1984): 701–11.

44. Pierre Martinetti, "Emergence of Time in Quantum Gravity: Is Time Necessarily Flowing?" http://arxiv.org/pdf/1203.4995 (2012); J. Butterfield and C. J. Isham, "On the Emergence of Time in Quantum Gravity," in *The Arguments of Time*, ed. J. Butterfield (Oxford: Oxford University Press, 1999), pp. 111–68.

45. Stenger, *Timeless Reality*.

46. Wootters, "'Time' Replaced by Quantum Correlations."

47. Max Tegmark, *Our Mathematical Universe: My Quest for the Ultimate Nature of Reality* (New York: Alfred A. Knopf, 2014).

48. Tegmark, "Mathematical Universe."

49. Stephen Hawking and Leonard Mlodinow, *The Grand Design* (New York: Bantam Books, 2010), pp. 42–43.

50. Ibid., p. 46.

51. William Lane Craig, "Hawking and Mlodinow: Philosophical Undertakers," 2013, http://www.reasonablefaith.org/hawking-and-mlodinow -philosophical-undertakers (accessed October 7, 2013).

52. Kirk D. Hagen, "Eternal Progression in a Multiverse: An Explorative Mormon Cosmology," *Dialogue: A Journal of Mormon Thought* 39, no. 2 (2006): 1–45.

53. Hyrum Leslie Andrus, *God, Man, and the Universe* (Salt Lake City, UT: Bookcraft, 1968), p. 144.

54. Ibid., p. 146.

55. B. H. Roberts, *New Witness for God* (Salt Lake City, UT: George Q. Canon & Sons, 1898), p. 448.

56. Andrew Crumey, "Parallel Worlds," 2013, http://www.aeonmagazine .com/world-views/can-the-multiverse-explain-the-course-of-history/?utm_source =Aeon+newsletter&utm_campaign=9265966a75-Weekly_9_July_20137_26_2013 &utm_medium=email&utm_term=0_411a82e59d-9265966a75-68603881 (accessed September 6, 2013).

57. Philip K. Dick, *The Man in the High Castle, a Novel* (New York: Putnam, 1962).

58. Philip Roth, *The Plot against America* (Boston: Houghton Mifflin, 2004).

59. Robert Cowley, ed., *The Collected What If?* (New York: Putnam, 1998).

60. As quoted in Crumey, "Parallel Worlds," http://www.aeonmagazine .com/world-views/can-the-multiverse-explain-the-course-of-history/?utm _source=Aeon+newsletter&utm_campaign=9265966a75-Weekly_9_July_20137 _26_2013&utm_medium=email&utm_term=0_411a82e59d-9265966a75-68603881 (accessed September 6, 2013).

61. For a discussion of the philosophy and history of the multiverse from a philosopher's viewpoint, see Mary-Jane. Rubenstein, *Worlds without End: The Many Lives of the Multiverse* (New York: Columbia University Press, 2014).

CHAPTER 16. LIFE AND GOD

1. Exoplanet Team, *The Extrasolar Planets Encyclopedia*, 2013, http://exo planet.eu (accessed November 6, 2013).

2. NASA Ames Research Center, "Kepler Discoveries," 2013, http://kepler .nasa.gov/Mission/discoveries/ (accessed November 6, 2013).

3. Erik A. Petigura, Andrew W. Howard, and Geoffrey W. Marcy, "Prevalence of Earth-Size Planets Orbiting Sun-like Stars," *Proceedings of the National Academy of Sciences* 110, no. 48 (2013): 19175–76.

4. Seth Shostak, "The Numbers Are Astronomical," 2013, http://www

.huffingtonpost.com/seth-shostak/the-numbers-are-astronomi_b_4214484.html (accessed November 6, 2013).

5. Alan H. Guth, *The Inflationary Universe: The Quest for a New Theory of Cosmic Origins* (Reading, MA: Addison-Wesley, 1997), p. 186.

6. John D. Barrow and Frank J. Tipler, *The Anthropic Cosmological Principle* (Oxford; New York: Oxford University Press, 1986).

7. William Lane Craig, "The Craig-Pigliucci Debate: Does God Exist?" http://www.leaderu.com/offices/billcraig/docs/craig-pigliucci2.html (accessed February 13, 2010).

8. Victor J. Stenger, "The Universe Shows No Evidence for Design," in *Debating Christian Theism*, ed. J. P. Moreland, Chad Meister, and Khaldoun A. Sweis (Oxford; New York: Oxford University Press, 2013), pp. 47–58.

9. Robin Collins, "The Fine-Tuning Evidence Is Convincing," in *Debating Christian Theism*, ed. J. P. Moreland, Chad Meister, and Khaldoun A. Sweis (Oxford; New York: Oxford University Press, 2013), pp. 35–46.

10. Victor J. Stenger, *The Fallacy of Fine-Tuning: Why the Universe Is Not Designed for Us* (Amherst, NY: Prometheus Books, 2011).

11. Hugh Ross, "Big Bang Model Refined by Fire," in *Mere Creation: Science, Faith & Intelligent Design*, ed. William A. Dembski (Downers Grove, IL: Intervarsity, 1998), pp. 363–83.

12. Collins, "Fine-Tuning Evidence Is Convincing," p. 38.

13. Rich Deem, "Evidence for the Fine Tuning of the Universe," 2013, http://www.godandscience.org/apologetics/designun.html (accessed July 12, 2013).

14. Robin Collins, "The Teleological Argument: An Exploration of the Fine-Tuning of the Universe," in *The Blackwell Companion to Natural Theology*, ed. William Lane Craig and James Porter Moreland (Chichester, UK; Malden, MA: Wiley-Blackwell, 2009), p. 43.

15. Steven Weinberg, "Living in the Multiverse," in *Universe or Multiverse?* ed. B. Carr (Cambridge: Cambridge University Press, 2007), chap. 2.

16. Leonard Susskind, *Cosmic Landscape: String Theory and the Illusion of Intelligent Design* (New York: Little, Brown, 2005).

17. M. Livio, D. Hollowell, A. Weiss, and J. W. Truran, "The Anthropic Significance of the Existence of an Excited State of C12," *Nature* 340 (1989): 281–84; see also figure 9.1, p. 170, in Stenger, *Fallacy of Fine-Tuning*.

18. Robin Collins, "Teleological Argument."

19. Barrow and Tipler, *Anthropic Cosmological Principle*, p. 326.

20. Stenger, *Fallacy of Fine-Tuning*, pp. 247–52.

21. Experts will point out that we technically cannot define a total energy of the universe for all geometries, although we can for a flat universe. In any case, we

can argue that the average energy density is zero, counting gravity as negative, which allows the same conclusion: the universe came from a state of zero energy.

22. Collins, "Fine-Tuning Evidence Is Convincing," p. 35.

23. Victor J. Stenger, *The Comprehensible Cosmos: Where Do the Laws of Physics Come From?* (Amherst, NY: Prometheus Books, 2006); Victor J. Stenger, "Where Do the Laws of Physics Come From?" 2007, http://philsci-archive.pitt.edu/3662/ (accessed August 21, 2013).

24. Lawrence M. Krauss, *A Universe from Nothing: Why There Is Something Rather Than Nothing* (New York: Free Press, 2012).

25. David Albert, "On the Origin of Everything," *New York Times*, March 25, 2012.

26. Frank Wilczek, "The Cosmic Asymmetry between Matter and Antimatter," *Scientific American* 243, no. 6 (1980): 82–90.

27. Marcus Chown, *The Never-Ending Days of Being Dead: Dispatches from the Frontline of Science* (London: Faber and Faber, 2007).

28. Stephen Hawking and Leonard Mlodinow, *The Grand Design* (New York: Bantam Books, 2010), p. 180.

29. Ibid., p. 136.

30. Ibid., p. 164.

31. Victor J. Stenger, *God: The Failed Hypothesis. How Science Shows That God Does Not Exist* (Amherst, NY: Prometheus Books, 2007).

32. For a grand tour of the full range of purported miracles and the personal investigations of the world's top paranormal investigator, see Joe Nickell, *The Science of Miracles: Investigating the Incredible* (Amherst, NY: Prometheus Books, 2013).

33. Jerry Coyne, "Why Evolution Is True," October 21, 2010, http://why evolutionistrue.wordpress.com/2010/10/21/115-years-of-debate-about-evidence -for-god/ (accessed August 19, 2013).

34. Victor J. Stenger, *Quantum Gods: Creation, Chaos, and the Search for Cosmic Consciousness* (Amherst, NY: Prometheus Books, 2009).

35. Robert M. Price and Edwin A. Suiminen, *Evolving out of Eden* (Valley, WA: Tellectual, 2013).

36. Ewin MacAskill, "George Bush: 'God Told Me to End the Tyranny in Iraq,'" 2005, http://www.theguardian.com/world/2005/oct/07/iraq.usa (accessed November 28, 2013).

37. Erik A. Petigura, Andrew W. Howard, and Geoffrey W. Marcy, "Prevalence of Earth-Size Planets Orbiting Sun-like Stars," *Proceedings of the National Academy of Sciences* 110, no. 48 (2013): 19175–76; Seth Shostak, "The Numbers Are Astronomical,"2013, http://www.huffingtonpost.com/seth-shostak/the -numbers-are-astronomi_b_4214484.html (accessed November 6, 2013).

38. Alan H. Guth, *The Inflationary Universe: the Quest for a New Theory of Cosmic Origins* (Reading, MA: Addison-Wesley, 1997), p. 186.

BIBLIOGRAPHY

Abdo, Aous A., et al. "Measurement of the Cosmic Ray e⁺e⁻Spectrum from 20 GeV to 1 TeV with the Fermi Large Area Telescope." *Physical Review Letters* 102, no. 18 (2009): 181101.

Ade, P. A. R., et al. "Planck 2013 Results. XVI. Cosmological Parameters." arXiv: 1303.5076. 2013. To be published.

Ade, P. A. R., et al. "Detection of B-Mode Polarization at Degree Angular Scales by BICEP2," *Physical Review Letters* 112, no. 24 (2014): 241101.

Adriani, Oscar, et al. "An Anomalous Positron Abundance in Cosmic Rays with Energies 1.5–100 GeV." *Nature* 458, no. 7238 (2009): 607–609.

Aguilar, M., et al. "The Alpha Magnetic Spectrometer (AMS) on the International Space Station: Part I–Results from the Test Flight on the Space Shuttle." *Physics Reports* 366, no. 6 (2002): 331–405.

Aguilar, M., et al. "First Result from the Alpha Magnetic Spectrometer on the International Space Station: Precision Measurement of the Positron Fraction in Primary Cosmic Rays Of 0.5–350 GeV." *Physical Review Letters* 110, no. 14 (2013): 141102.

Aguirre, Anthony, and Max Tegmark. "Born in an Infinite Universe: A Cosmological Interpretation of Quantum Mechanics." *Physical Review D* 84, no. 10 (2011): 105002.

Aguirre, Anthony, and Steven Gratton. "Inflation without a Beginning: A Null Boundary Proposal." *Physical Review* D67 (2003): 083515.

——. "Steady-State Eternal Inflation." *Physical Review D* 65, no. 8 (2002): 083507.

Aharonian, F. A., A. N. Timokhin, and A. V. Plyasheshnikov. "On the Origin of Highest Energy Gamma-Rays from Mkn 501." *Astronomy and Astrophysics* 384, no. 3 (2002): 834–47.

Akerib, D. S., et al. "First Results from the LUX Dark Matter Experiment at the Stanford Underground Research Facility." *Physical Review Letters* 112 (2014): 091303.

Alexander, Amir R. *Infinitesimal: How a Dangerous Mathematical Theory Shaped the Modern World.* New York: Scientific American/Farrar, Strauss and Giroux, 2014.

Al-Khalili, Jim. *The House of Wisdom: How Arabic Science Saved Ancient Knowledge and Gave Us the Renaissance.* New York: Penguin, 2011.

Albert, David. "On the Origin of Everything." *New York Times*, March 25, 2012.

Albrecht, Andreas and Paul J. Steinhardt. "Cosmology for Grand Unified Theories with Radiatively Induced Symmetry Breaking." *Physical Review Letters* 48, no. 17 (1982): 1220–23.

Alleyne, Richard. "God Is Not the Creator, Claims Academic." *Telegraph (U.K.)*, October 8, 2009.

Alpher, Ralph, and Robert Herman. *Genesis of the Big Bang*. New York: Oxford University Press, 2001.

——. "Remarks on the Evolution of the Expanding Universe." *Physical Review* 75 (1949): 1089–95.

Alpher, Ralph A., Hans Bethe, and George Gamow. "The Origin of the Chemical Elements." *Physical Review* 73, no. 7 (1948): 803–804.

Alpher, Ralph A., James W. Follin Jr., and Robert C. Herman. "Physical Conditions in the Initial Stages of the Expanding Universe." *Physical Review* 92 (1953): 1347–61.

Ambrosio, M., et al. "Search for Nucleon Decays Induced by GUT Magnetic Monopoles with the MACRO Experiment." *European Physical Journal C-Particles and Fields* 26, no. 2 (2002): 163–72.

Andrus, Hyrum Leslie. *God, Man, and the Universe*. Salt Lake City, UT: Bookcraft, 1968.

Arneodo, Francesco. "Dark Matter Searches." *Proceedings of the 32th International Symposium on Physics in Collision (PIC 2012)*. Strbské Pleso, Slovakia. 2012.

Atkatz, David. "Quantum Cosmology for Pedestrians." *American Journal of Physics* 62, no. 7 (1994): 619–627.

Atkatz, David, and Heinz Pagels. "Origin of the Universe as a Quantum Tunneling Event." *Physical Review Letters* D25 (1982): 2065–73.

ATLAS Collaboration. "Observation of a New Particle in the Search for the Standard Model Higgs Boson with the ATLAS Detector at the LHC." *Physics Letters B* 716, no. 1 (2012): 1–29.

Baade, Walter. "A Revision of the Extra-Galactic Distance Scale." *Transactions of the International Astronomical Union* 8 (1952): 397–98.

Babinski, Edward T. "The Cosmology of the Bible." In *The Christian Delusion: Why Faith Fails*, edited by John W. Loftus, 109–47. Amherst, NY: Prometheus Books, 2010.

Bahcall, John N. "How the Sun Shines." 2000. http://www.nobelprize.org/nobel_prizes/physics/articles/fusion/. Accessed February 24, 2013.

Banks, T. "Heretics of the False Vacuum: Gravitational Effects on and of Vacuum Decay II." arXiv preprint hep-th/0211160. 2002.

Barbour, Ian G. *Religion and Science: Historical and Contemporary Issues*. San Francisco: HarperSanFrancisco, 1997.

Barbour, Julian B. *Absolute or Relative Motion? A Study from Machian Point of View of the Discovery and the Structure of Dynamical Theories.* Cambridge; New York: Cambridge University Press, 1989.

Bardeen, James M. "Gauge-Invariant Cosmological Perturbations." *Physical Review D* 22, no. 8 (1980): 1882.

Bardeen, James M., Paul J. Steinhardt, and Michael S. Turner. "Spontaneous Creation of Almost Scale-Free Density Perturbations in an Inflationary Universe." *Physical Review D* 28 no. 4 (1983): 679.

Bartholomew, David J. *God, Chance, and Purpose: Can God Have It Both Ways?* Cambridge; New York: Cambridge University Press. 2008.

Bartusiak, Marcia. *Harlow Shapley 1885–1972: A Biographical Memoir.* New York: Pantheon Books, 2009.

——. *The Day We Found the Universe.* New York: Pantheon Books, 2009.

Bekenstein, Jacob D. "Is a Tabletop Search for Planck Scale Signals Feasible?" *Physical Review D* 86, no 12 (2012): 124040.

Belenkiy, Ari. "Alexander Friedmann and the Origins of Modern Cosmology." *Physics Today* 65, no. 10 (2012): 38–43.

Bennett, C. L., et al. "Nine-Year Wilkinson Microwave Anisotropy Probe (WMAP) Observations: Final Maps and Results." *Astrophysical Journal Supplement Series* 208, no. 2 (2013): 20.

Berezinsky, Veniamin Sergeevich. "High Energy Neutrino Astronomy." *Nuclear Physics B-Proceedings Supplements* 19 (1991): 375–87.

Beringer, J., et al. (Particle Data Group). "The Review of Particle Physics." *Physical Review D* 88 (2012): 010001.

Bernstein, Jeremy, and Gerald Feinberg. *Cosmological Constants: Papers in Modern Cosmology.* New York: Columbia University Press, 1986.

Birx, H. James. *Encyclopedia of Time: Science, Philosophy, Theology, & Culture.* Thousand Oaks, CA: Sage, 2009.

Boesgaard, Ann Merchant, and Gary Steigman. "Big Bang Nucleosynthesis: Theories and Observations." *Annual Reviews of Astronomy and Astrophysics* 23 (1985): 319–90.

Bohm, David, and B. J. Hiley. *The Undivided Universe: An Ontological Interpretation of Quantum Theory.* London; New York: Routledge, 1993.

Bok, Bart Jan, and Mildred Shapley Matthews. *Harlow Shapley, 1885–1972: A Biographical Memoir.* Washington, DC: National Academy of Sciences of the United States, 1978.

Boltzmann, Ludwig. "On Certain Questions of the Theory of Gases." *Nature* 51 (1895): 483–85.

Bond, J. R., A. H. Jaffe, and L. Knox. "Radical Compression of Cosmic Microwave Background Data." *Astrophysical Journal* 533, no. 1 (2000): 19.

Bondi, Hermann and Thomas Gold. "The Steady-State Theory of the Expanding Universe." *Monthly Notices of the Royal Astronomical Society* 108 (1948): 252–70.

Borde, Arvind, Alan H. Guth, and Alexander Vilenkin. "Inflationary Spacetimes Are Not Past-Complete." *Physical Review Letters* 90 (2003): 151301.

Bosma, Albert. "The Distribution and Kinematics of Neutral Hydrogen in Spiral Galaxies of Various Morphological Types." PhD thesis, Groningen University, 1978.

Bottéro, Jean. *Religion in Ancient Mesopotamia*. Chicago: University of Chicago Press, 2001.

Bousso, Raphael, and Leonard Susskind. "Multiverse Interpretation of Quantum Mechanics." *Physical Review D* 85, no. 4 (2012): 045007.

Brooke, John L. *The Refiner's Fire: The Making of Mormon Cosmology, 1644–1844*. Cambridge; New York: Cambridge University Press, 1994.

Budge, E. A. Wallis. *Egyptian Ideas of the Future Life: Egyptian Religion*. New York: University Books, 1959.

Bulbul, E., et al. "Detection of an Unidentified Emission Line in the Stacked X-Ray Spectrum of Galaxy Clusters." *Astrophysical Journal* (2014); to be published.

Burbidge, E. Margaret, et al. "Synthesis of Elements in the Stars." *Reviews of Modern Physics* 29, no. 4 (1957): 547–650.

Burderi, Luciano, and Tiziana Di Salvo. "The Quantum Clock: A Critical Discussion on Space-Time." arXiv preprint arXiv:1207.0207. 2012.

Burles, Scott, et al. "Sharpening the Predictions of Big-Bang Nucleosynthesis." *Physical Review Letters* 82, no. 21 (1999): 4176.

Butterfield, J., and C. J. Isham. "On the Emergence of Time in Quantum Gravity." In *The Arguments of Time*, edited by J. Butterfield, 111–68. Oxford: Oxford University Press, 1999.

Byers, Nina. "E. Noether's Discovery of the Deep Connection between Symmetries and Conservation Laws." *Proceedings of the Israel Mathematical Conference* 12 (1999): 67–82.

Byers, Nina, and Gary A. Williams. *Out of the Shadows: Contributions of Twentieth-Century Women to Physics*. Cambridge; New York: Cambridge University Press, 2006.

Callahan, Tim. "Out of Nothing? The Genesis Creation Myth Is Not Unique." *Skeptic* 17, no 3 (2012): 20–22.

Carr, Bernard, ed. *Universe or Multiverse?* Cambridge: Cambridge University Press, 2007.

Carroll, Sean M., William H. Press, and Edwin L. Turner. "The Cosmological Constant." *Annual Reviews of Astronomy and Astrophysics* 30 (1992): 499–542.

Carroll, Sean M. *From Eternity to Here: The Quest for the Ultimate Theory of Time*. New York: Dutton, 2010.

Carroll, Sean M., and Jennifer Chen. "Spontaneous Inflation and the Origin of the Arrow of Time." arXiv preprint hep-th/0410270. 2004.

Carroll, Sean M., and William Lane Craig. "Greer-Heard Forum: God and Cosmology." http://www.greerheard.com. Accessed March 23, 2014.

Catholic News Agency. "Believing in Aliens Not Opposed to Christianity, Vatican's Top Astronomer Says." 2008. http://www.catholicnewsagency.com/news/believing_in_aliens_not_opposed_to_christianity_vaticans_top_astronomer_says/. Accessed August 15, 2013.

Chown, Marcus. *The Never-Ending Days of Being Dead: Dispatches from the Frontline of Science*. London: Faber and Faber, 2007.

Clausius, Rudolf. "On the Second Fundamental Theorem of the Mechanical Theory of Heat." *Philosophical Magazine* 35 (1868): 404–19.

Coleman, Sidney, and Frank De Luccia. "Gravitational Effects on and of Vacuum Decay." *Physical Review D* 21, no. 12 (1980): 3305.

Coles, Peter, ed. *The Routledge Companion to the New Cosmology*. London: Routledge, 2001.

Collins, H. M. "Introduction: Stages in the Empirical Program of Relativism." *Social Studies of Science* 11, no. 1 (1981): 3–10.

Collins, Robin. "The Fine-Tuning Evidence Is Convincing." In *Debating Christian Theism*, edited by J. P. Moreland, Chad Meister, and Khaldoun A. Sweis, 35–46. Oxford; New York: Oxford University Press, 2013.

Courtois, Hélene M., et al. "Cosmography of the Local Universe." arXiv preprint arXiv:1306.0091. 2013.

Cowley, Robert, ed. *The Collected What If?* New York: Putnam, 1998.

Craig, William Lane. "The Craig-Pigliucci Debate: Does God Exist?" 1995. http://www.leaderu.com/offices/billcraig/docs/craig-pigliucci2.html. Accessed February 13, 2010.

——. "Five Arguments for God." *The Gospel Coalition*. http://thegospelcoalition.org/pdf-articles/Craig_Atheism.pdf. Accessed December 20, 2013.

——. "Hawking and Mlodinow: Philosophical Undertakers." http://www.reasonablefaith.org/hawking-and-mlodinow-philosophical-undertakers. Accessed October 7, 2013.

——. *The Kalam Cosmological Argument*. London: Macmillan, 1979.

——. "Opening Speech." In *Is Faith in God Reasonable?* edited by Paul Gould and Corey Miller. New York: Routledge, 2014.

——. "Philosophical and Scientific Pointers to Creatio ex Nihilo." *Journal of the American Scientific Affiliation* 32, no. 1 (1980): 5–13.

Craig, William Lane, and James D. Sinclair. "The *Kalam* Cosmological Argument." In *The Blackwell Companion to Natural Theology*, edited by William Lane Craig and James Porter Moreland, 101–201. Chichester, UK; Malden, MA: Wiley-Blackwell, 2009.

Craig, William Lane, and James Porter Moreland, eds. *The Blackwell Companion to Natural Theology.* Chichester, UK; Malden, MA: Wiley-Blackwell, 2009.

Crowe, Michael J. *Theories of the World From Antiquity to the Copernican Revolution.* Mineola, NY: Dover Publications, 2001.

Crumey, Andrew. "Parallel Worlds." 2013. http://www.aeonmagazine.com/world-views/can-the-multiverse-explain-the-course-of-history/?utm_source=Aeon+newsletter&utm_campaign=9265966a75-Weekly_9_July_20137_26_2013&utm_medium=email&utm_term=0_411a82e59d-9265966a75-68603881. Accessed September 6, 2013.

Cyburt, Richard H., Brian D. Fields, and Keith A. Olive. "Primordial Nucleosynthesis in Light of WMAP." *Physics Letters B* 567 (2003): 227–34.

D'Amico, Guido, Marc Kamionkowski, and Kris Sigurdson. "Dark Matter Astrophysics." In *Dark Matter and Dark Energy: A Challenge for Modern Cosmology,* edited by Sabiuno Matarrese et al., 241–72. Dordrecht: Springer, 2011.

Deem, Rich. "Evidence for the Fine Tuning of the Universe." 2013. http://www.godandscience.org/apologetics/designun.html. Accessed July 12, 2013.

De Sitter, Willem. "On Einstein's Theory of Gravitation and Its Astronomical Consequences." *Monthly Notices of the Royal Astronomical Society* 78 (1917): 3–28.

Deutsch, David. *The Fabric of Reality: The Science of Parallel Universes – and Its Implications.* New York: Allen Lane, 1997.

De Vaucouleurs, G. "The Extragalactic Distance Scale and the Hubble Constant." *Observatory* 102 (1982): 178–94.

Dewitt, Bryce S. "The Many-Universes Interpretation of Quantum Mechanics." In *The Many-worlds Interpretation of Quantum Mechanics,* edited by Bryce S. DeWitt and Neill Graham, 167–218. Princeton, NJ: Princeton University Press, 1973.

——. "Quantum Theory of Gravity. I. The Canonical Theory." *Physical Review* 160 (1967): 1113–48.

Dick, Philip K. *The Man in the High Castle, a Novel.* New York: Putnam, 1962.

Dicke, R. H., et al. "Cosmic Black-Body Radiation." *Astrophysical Journal* 142 (1965): 414–19.

Dirac, Paul. "A New Basis for Cosmology." *Proceedings of the Royal Society A* 165 (1938): 199–208.

——. *The Principles of Quantum Mechanics.* Oxford: Clarendon, 1930.

——. "Quantised Singularities in the Electromagnetic Field." *Proceedings of the Royal Society of London,* ser. A, containing papers of a mathematical and physical character (1931): 60–72.

Dressler, Alan Michael. *Voyage to the Great Attractor: Exploring Intergalactic Space*. New York: A. A. Knopf, 1994.

D'Souza, Dinesh. *Life after Death: The Evidence*. Washington, DC: Regnery, 2009.

———. *What's So Great about Christianity?* Washington, DC: Regnery, 2007.

Duncan, Comer, and Michael Bradie. "Physics and Philosophy of Space and Time," course at Bowling Green State University. 1998. http://physics.bgsu .edu/~gcd/ppst.html. Accessed December 20, 2013.

Dyson, Freeman J. "The Radiation Theories of Tomonaga, Schwinger, and Feynman." *Physical Review* 75, no. 3 (1949): 486–502.

Eddington, Arthur S. "The End of the World: From the Standpoint of Mathematical Physics." *Nature* 127 (1931): 447–53.

———. *The Expanding Universe*. Cambridge: Cambridge University Press, 1933.

Einstein, Albert. "Cosmological Considerations on the General Theory of Relativity." Translated. In *Cosmological Constants: Papers in Modern Cosmology*, edited by Jeremy Bernstein and Gerald Feinberg, 16–26. New York: Columbia University Press, 1986.

Einstein, Albert, Max Born, and Hedwig Born. *The Born-Einstein Letters: Correspondence between Albert Einstein and Max and Hedwig Born from 1916 to 1955, with Commentaries by Max Born*. London: Macmillan, 1971.

Eisenstein, Daniel J., et al. "Detection of the Baryon Acoustic Peak in the Large-Scale Correlation Function of SDSS Luminous Red Galaxies." *Astrophysical Journal* 633, no. 2 (2005): 560.

Englert, F., and R. Brout. "Broken Symmetry and the Mass of Gauge Vector Bosons." *Physical Review Letters* 13 (1964): 321.

Epicurus. *The Art of Happiness*. Translated with an introduction and commentary by George K. Strodach. New York: Penguin Books, 2012.

Everett, Hugh, III. "'Relative State' Formulation of Quantum Mechanics." *Reviews of Modern Physics* 29 (1957): 434–62.

Exoplanet Team. *The Extrasolar Planets Encyclopaedia*. 1995. exoplanet.eu. Accessed 2013.

Fairbanks, Arthur, ed. *The First Philosophers of Greece*. London: K. Paul, Trench, Trübner, 1898.

Ferngren, Gary B., Edward J. Larson, and Darrel W. Amundsen. *The History of Science and Religion in the Western Tradition: An Encyclopedia*. New York: Garland, 2000.

Feynman, Richard. "Space-Time Approach to Quantum Electrodynamics." *Physical Review* 76, no. 6 (1949): 769–89.

"Finds Spiral Nebula Are Stellar Systems." *New York Times*, November 23, 1924, p. 6.

Freedman, D. Z., D. N. Schramm, and D. L. Tubbs. "The Weak Neutral Current and Its Effects in Stellar Collapse." *Annual Reviews of Nuclear Science* 27 (1977): 167–207.

Friedmann, Alexander. "On the Curvature of Space." *General Relativity and Gravitation* 31, no. 12 (1999): 1991–2000.

———. "Über die Krümmung des Raumes." *Zeitschrift für Physik* 10, no. 1 (1922): 377–286.

Galilei, Galileo. *Sidereus nuncius, Or, The Sidereal Messenger*. Chicago: University of Chicago Press, 1989; first published in 1610.

Gamow, George. "The Origin of the Elements and the Separation of Galaxies." *Physical Review* 74 (1948): 505–506.

Ganga, Ken, et al. "Cross-Correlation between the 170 Ghz Survey Map and the COBE Differential Microwave Radiometer First-Year Maps." *Astrophysical Journal* 410 (1993): L57–60.

Garriga, Jaume, and Alexander Vilenkin. "Many Worlds in One." *Physical Review* D64 (2001): 043511.

Geller, Margaret J., and John P. Huchra. "Mapping the Universe." *Science* 246, no. 4932 (1989): 897–903.

Genz, Henning. *Nothingness: The Science of Empty Space*. Reading, MS: Perseus Books, 1999.

Georgi, Howard, and S. L. Glashow. "Unity of All Elementary-Particle Forces." *Physical Review Letters* 32 (1974): 438–41.

Gingerich, Owen. *The Eye of Heaven: Ptolemy, Copernicus, Kepler*. New York: American Institute of Physics, 1993.

Goldman, T. J., and D. A. Ross. "A New Estimate of the Proton Decay Lifetime." *Physics Letters B* 84, no. 2 (1979): 208–10.

Gonzalez, Guillermo, and Jay Wesley Richards. *The Privileged Planet: How Our Place in the Cosmos is Designed for Discovery*. Washington, DC; Lanham, MD: Regnery, 2004.

Gorbunov, D. S., and V. A. Rubakov. *Introduction to the Theory of the Early Universe: Hot Big Bang Theory*. Singapore; Hackensack, NJ: World Scientific, 2011.

Gott, J. Richard, III, et al. "A Map of the Universe." *Astrophysical Journal* 624, no. 2 (2008): 463.

Gott, Richard, III, et al. "Will the Universe Expand Forever?" *Scientific American*, March, 1976, p. 65.

Gould, Paul, and Corey Miller, eds. *Is Faith in God Reasonable? The Great Debate in Light of Contemporary Science, Philosophy, and Rhetoric*. New York: Routledge, 2014.

Grant, Edward. *Science and Religion, 400 B.C. to A.D. 1550: From Aristotle to Copernicus*. Westport, CT: Greenwood, 2004.

Greene, Brian. *The Hidden Reality: Parallel Universes and the Deep Laws of the Cosmos.* New York: Alfred A. Knopf, 2011.

Greenstein, Jesse L., and Maarten Schmidt. "The Quasi-Stellar Radio Sources 3C 48 and 3C 273." *Astrophysical Journal* 140 (1964): 1–34.

Greenstein, Jesse L., and Thomas A. Matthews. "Red-Shift of the Unusual Radio Source: 3C 48." *Nature* 197 (1963): 1041–42.

Gribbin, John. *In Search of the Multiverse: Parallel Worlds, Hidden Dimensions, and the Ultimate Quest for the Frontiers of Reality.* Hoboken, NJ: Wiley, 2010.

Guralnik, G. G., C. R. Hagen, and T. W. Kibble. "Global Conservation Laws and Massless Particles." *Physical Review Letters* 13 (1964): 585.

Guth, Alan H. "Inflationary Universe: A Possible Solution to the Horizon and Flatness Problems." *Physical Review D* 23, no. 2 (1981): 347–56.

——. *The Inflationary Universe: The Quest for a New Theory of Cosmic Origins.* Reading, MA: Addison-Wesley, 1997.

Guth, Alan H., and Erick J. Weinberg. "Could the Universe Have Recovered from a Slow First-Order Phase Transition?" *Nuclear Physics B* 212, no. 2 (1983): 321–64.

Guth, Alan H., and Paul J. Steinhardt. "The Inflationary Universe." *Scientific American* 250 (1984): 116–28.

Guth, Alan H., and S.-H. H. Tye. "Phase Transitions and Magnetic Monopole Production in the Very Early Universe." *Physical Review Letters* 44, no. 10 (1980): 631–35.

Guth, A. H. "Quantum Fluctuations in Cosmology and How They Lead to a Multiverse." *25th Solvay Conference on Physics, The Theory of the Quantum World* (2011).

Hagen, Kirk D. "Eternal Progression in a Multiverse: An Explorative Mormon Cosmology." *Dialogue: A Journal of Mormon Thought* 39, no. 2 (2006): 1–45.

Haines, Todd, et al. "Neutrinos from SN1987a in the IMB Detector." *Nuclear Instruments and Methods in Physics Research Section A: Accelerators, Spectrometers, Detectors and Associated Equipment* 264, no. 1 (1988): 28–31.

Halvorson, Hans, and Helge Kragh. "Theism and Physical Cosmology." In *Routledge Companion to Theism*, edited by Stewart Goetz, Victoria Harrison, and Charles Taliaferro. New York; Abingdon, UK: Routledge, 2009.

Hamilton, Jean-Christophe. "What Have We Learned from Observational Cosmology?" Philosophical Aspects of Modern Cosmology meeting, Grenada, Spain, September 22–23, 2013.

Hancock, S., et al. "Constraints on Cosmological Parameters from Recent Measurements of Cosmic Microwave Background Anisotropy." *Monthly Notices of the Royal Astronomical Society* 294, no. 1 (1998): L1–L6.

Hand, Eric. "Hunt for the Sterile Neutrino Heats Up." *Nature* 464 (2010): 334–35.

Hanson, D., et al. "Detection of B-Mode Polarization in the Cosmic Microwave Background with Data from the South Pole Telescope." *Physical Review Letters* 111 (2013): 141301.

Harrison, Edward R. "Fluctuations at the Threshold of Classical Cosmology." *Physical Review D* 1, no. 10 (1970): 2726.

——. *Masks of the Universe*. New York; London: Macmillan; Collier Macmillan, 1985.

Hartle, James B., and Stephen W. Hawking. "Wave Function of the Universe." *Physical Review* D28 (1983): 2960–75.

Hawking, Stephen. *A Brief History of Time: From the Big Bang to Black Holes*. Toronto; New York: Bantam Books, 1988.

——. "Quantum Cosmology." In *Three Hundred Years of Gravitation*, edited by S. W. Hawking and W. Israel, 631–51. Cambridge; New York: Cambridge University Press, 1987.

Hawking, Stephen, and Leonard Mlodinow. *The Grand Design*. New York: Bantam Books, 2010.

Hawking, Stephen, and Roger Penrose. "The Singularities of Gravitational Collapse." *Proceedings of the Royal Society of London*, ser. A, 314 (1970): 529–48.

Hawking, S. W., and W. Israel. *Three Hundred Years of Gravitation*. Cambridge; New York: Cambridge University Press, 1987.

Hewish, A., et al. "Observation of a Rapidly Pulsating Radio Source." *Nature* 217 (1968): 709–13.

Higgs, Peter W. "Broken Symmetries and the Masses of Gauge Bosons." *Physical Review Letters* 13 (1964): 508.

Hilbert, David. "On the Infinite." In *A Source Book in Mathematical Logic, 1879–1931*, edited by Jean Van Heijenoort, 370–71. Cambridge, MA: Harvard University Press, 1967.

——. "On the Infinite." In *Philosophy of Mathematics*, edited by Paul Benacerraf and Hillary Putnam, 139–41. Englewood Cliffs, NJ: Prentice-Hall, 1964. Originally presented (in German) on June 24, 1925, before a congress of the Westphalian Mathematical Society, in honor of Karl Weierstrauss.

Hinshaw, G., et al. "Five-Year Wilkinson Microwave Anisotropy Probe Observations: Data Processing, Sky Maps, and Basic Results." *Astrophysical Journal Supplement Series* 180, no. 2 (2009): 225.

Hirata, K., et al. "Observation of a Neutrino Burst from the Supernova SN1987A." *Physical Review Letters* 58 (1987): 1490–93.

Holt, Jim. *Why Does the World Exist? An Existential Detective Story*. New York: Liveright, 2012.

Holtzman, Jon A. "Microwave Background Anisotropies and Large-Scale Structure in Universes with Cold Dark Matter, Baryons, Radiation, and Massive and Massless Neutrinos." *Astrophysical Journal Supplement Series* 71 (1989): 1–24.

Hoskin, Michael A. *William Herschel and the Construction of the Heavens*. New York: Norton, 1964.

Hoyle, Fred. *Home Is Where the Wind Blows: Chapters from a Cosmologist's Life*. Mill Valley, CA: University Science Books, 1994.

——. "A New Model for the Expanding Universe." *Monthly Notices of the Royal Astronomical Society* 108 (1948): 372–82.

——. "On Nuclear Reactions Occurring in Very Hot Stars. I. The Synthesis of Elements from Carbon to Nickel." *Astrophysical Journal Supplement* 1 (1954): 121–46.

——. "The Universe: Past and Present Reflections." *Annual Reviews of Astronomy and Astrophysics* 20 (1982): 1–35.

——. "The Universe: Past and Present Reflections." *Engineering and Science*, (November 1981): 8–12.

Hoyle, Fred, Geoffrey R. Burbidge, and Jayant Vishnu Narlikar. *A Different Approach to Cosmology: From a Static Universe through the Big Bang Towards Reality*. Cambridge; New York: Cambridge University Press, 2005.

Hu, Wayne, and Martin White. "The Cosmic Symphony." 2004. http://background .uchicago.edu/~whu/SciAm/sym1.html. Accessed June 6, 2014.

Hubble, Edwin. "Cepheids in Spiral Nebulae." *Publications of the American Astronomical Society* 5 (1925): 261–64.

——. "Extra-Galactic Nebulae." *Astrophysical Journal* 64 (1926): 331–69.

——. "Problems of Nebular Research." *Scientific Monthly* 51 (1940): 391–408.

——. *The Realm of the Nebulae*. New Haven, CT; London: Yale University Press; H. Milford, Oxford University Press, 1936.

——. "A Relation between Distance and Radial Velocity among Extra-Galactic Nebulae." *Proceedings of the National Academy of Sciences* 15, no. 3 (1929): 168–73.

Huber, Patrick, and Jon Link. "White Paper on Sterile Neutrinos." 2012. http:// cnp.phys.vt.edu/white_paper/whitepaper.pdf. Accessed September 10, 2012.

Humason, Milton. "The Large Radial Velocity of N.G.C. 7619." *Proceedings of the National Academy of Sciences* 15, no. 3 (1929): 167–68.

IceCube Collaboration. "Evidence for High-Energy Extraterrestrial Neutrinos at the Icecube Detector." *Science* 342, no. 6161 (2013): 1242856.

Iocco, F., et al. "Primordial Nucleosynthesis: From Precision Cosmology to Fundamental Physics." *Physics Reports* 472 (2009): 1–76.

Isaacson, Walter. *Einstein: His Life and Universe*. New York: Simon & Schuster, 2007.

Jaffe, Andrew H., et al. "Cosmology from MAXIMA-1, BOOMERANG, and COBE DMR Cosmic Microwave Background Observations." *Physical Review Letters* 86, no. 16 (2001): 3475–79.

Jansky, Karl G. "Electrical Disturbances of Extraterrestrial Origin." *Proceedings of the IEEE* 86, no. 7 (1988): 1510–15.

Jarosik, N., et al. "Seven-Year Wilkinson Microwave Anisotropy Probe (WMAP) Observations: Sky Maps, Systematic Errors, and Basic Results." *Astrophysical Journal Supplement Series* 192, no. 2 (2011): 14.

Jastrow, Robert. *God and the Astronomers*. New York: Norton, 1992.

Jeans, James. *Astronomy and Cosmogony*. Cambridge: Cambridge University Press, 1928.

Kaiser, David. "Physics and Feynman's Diagrams." *American Scientist* 93 (2005): 156–65.

Kant, Immanuel. *Universal Natural History and Theory of the Heavens*. Ann Arbor: University of Michigan Press, 1969.

Kazanas, Demos. "Dynamics of the Universe and Spontaneous Symmetry Breaking." *Astrophysical Journal* 241 (1980): L59–63.

Kerzberg, Pierre. *The Invented Universe. The Einstein-de Sitter Controversy (1916–17) and the Rise of Relativistic Cosmology*. Oxford: Clarendon, 1989.

Kibble, T. W. B. "Topology of Cosmic Domains and Strings." *Journal of Physics A: Mathematical and General* 9, no. 8 (1976): 1387–98.

Kirk, Geoffrey S., John E. Raven, and Malcolm Schofield. *The Presocratic Philosophers: A Critical History with a Selection of Texts*. Cambridge; New York: Cambridge University Press, 1983.

Kirshner, Robert P. "Hubble's Diagram and Cosmic Expansion." *Proceedings of the National Academy of Sciences* 101, no. 1 (2004): 8–13.

Klauber, Robert D. "Mechanism for Vanishing Zero Point Energy." 2007. http://lanl.arxiv.org/abs/astro-ph/0309679v3. Accessed August 5, 2010.

Kolb, Edward W., and Michael S. Turner. *The Early Universe*. Reading, MA: Addison-Wesley, 1990.

Komatsu, E., et al. "WMAP Five-Year Observations: Cosmological Interpretation." *Astrophysical Journal Supplement Series* 180 (2009): 330–76.

Kragh, Helge. *Conceptions of Cosmos from Myths to the Accelerating Universe: A History of Cosmology*. Oxford; New York: Oxford University Press, 2007.

——. *Entropic Creation: Religious Contexts of Thermodynamics and Cosmology*. Aldershot, Hampshire, England; Burlington, VT: Ashgate, 2008.

——. *Matter and Spirit in the Universe: Scientific and Religious Preludes to Modern Cosmology*. London: Imperial College Press, 2004.

——. "Quasi-Steady-State and Related Cosmological Models: A Historical Review." arXiv preprint arXiv:1201.3449. 2012.

Krauss, Lawrence M. *A Universe from Nothing: Why There Is Something Rather Than Nothing*. New York: Free Press, 2012.

Krauss, Lawrence M., and Michael S. Turner. "The Cosmological Constant Is Back." *General Relativity and Gravitation* 27, no. 11 (1995): 1137–44.

Kuhlmann, Meinhard. "What Is Real?" *Scientific American*, August, 2013, pp. 40–47.

Kuhn, Thomas S. *The Structure of Scientific Revolutions*. Chicago: University of Chicago Press, 1962.

Lang, Kenneth R., and Owen Gingrich, eds. *A Source Book in Astronomy and Astrophysics, 1900–1975*. Cambridge, MA: Harvard University Press, 1979.

Lederman, Leon M., and Christopher T. Hill. *Beyond the God Particle*. Amherst, NY: Prometheus Books, 2013.

——. *Symmetry and the Beautiful Universe*. Amherst, NY: Prometheus Books, 2004.

Lederman, Leon M., and Dick Teresi. *The God Particle: If the Universe Is the Answer, What Is the Question?* Boston: Houghton Mifflin, 1993.

Leeming, David Adams. *Creation Myths of the World: An Encyclopedia*. Santa Barbara, CA: ABC-CLIO, 2010.

Leibniz, Gottfried Wilhelm. *Theodicy: Essays on the Goodness of God, the Freedom of Man, and the Origin of Evil*. La Salle, IL: Open Court, 1985.

Leibniz, Gottfried Wilhelm, and Samuel Clarke. *The Leibniz-Clarke Correspondence: Together with Extracts from Newton's* Principia *and* Opticks. Manchester, UK: Manchester University Press, 1956.

Lemaître, Georges. "The Beginning of the World from the Point of View of Quantum Theory." *Nature* 127 (1931): 706.

——. "Expansion of the Universe: A Homogeneous Universe of Constant Mass and Increasing Radius Accounting for the Radial Velocity of Extra-Galactic Nebulae." *Monthly Notices of the Royal Astronomical Society* 91 (1931): 490–501.

——. *The Primeval Atom: An Essay on Cosmogony*. New York: Van Nostrand, 1950.

——. "Un Univers homogène de masse constante et de rayon croissant rendant compte de la vitesse radiale des nébuleuses extra-galactiques" ("A Homogeneous Universe of Constant Mass and Growing Radius Accounting for the Radial Velocity of Extragalactic Nebulae"). *Annales de la Société Scientifique de Bruxelles* 47 (1927): 49.

Lerner, Eric J. *The Big Bang Never Happened*. New York: Times Books/Random House, 1991.

Leslie, John, ed. *Modern Physics and Cosmology*. Amherst, NY: Prometheus Books, 1998.

Lewis, Bernard. *The Assassins: A Radical Sect in Islam*. New York: Basic Books, 2003.

Liberati, Stefano. "Greek Cosmology: From Course at University of Maryland." 2003. http://www.physics.umd.edu/courses/Phys117/Liberati/Support/cosmology.pdf. Accessed October 29, 2012.

Lifshitz, Evgenii. "On the Gravitational Stability of the Expanding Universe." *Journal of Physics – USSR* 10 (1946): 116.

Lindberg, David C. *The Beginnings of Western Science: The European Scientific Tradition in Philosophical, Religious, and Institutional Context, Prehistory to A.D. 1450.* Chicago: University of Chicago Press, 2007.

Lindberg, David C., and Ronald L. Numbers. *God and Nature: Historical Essays on the Encounter between Christianity and Science.* Berkeley: University of California Press, 1986.

Linde, André. "Chaotic Inflation." *Physics Letters B* 129, no. 3 (1983): 177–81.

——. "Eternally Existing Self-Reproducing Chaotic Inflationary Universe." *Physics Letters B* 175, no. 4 (1986): 395–400.

——. "The Inflationary Universe." *Reports on Progress in Physics* 47 (1984): 925–86.

——. "A New Inflationary Universe Scenario: A Possible Solution of the Horizon, Flatness, Homogeneity, Isotropy and Primordial Monopole Problems." *Physics Letters B* 108 (1982): 389.

——. "Quantum Creation of the Inflationary Universe." *Physics Letters* 108B (1982): 389–92.

——. "The Self-Reproducing Inflationary Universe." *Scientific American* 271, no. 5 (1994): 48–55.

Lindsay, James A. *Dot, Dot, Dot: Infinity Plus God Equals Folly.* Fareham, Hampshire, UK: Onus Books, 2013.

Livio, M. "Lost in Translation: Mystery of the Missing Text Solved." *Nature* 479 (2011): 171–73.

Livio, M., et al. "The Anthropic Significance of the Existence of an Excited State of C12." *Nature* 340 (1989): 281–84.

Lucretius. *De rerum natura.* Cambridge, MA: Harvard University Press, 1992.

Luminet, Jean-Pierre. "Editorial Note to: Georges Lemaître, 'The Beginning of the World from the Point of View of Quantum Theory.'" *General Relativity and Gravitation* 43 (2011): 2911–28.

MacAskill, Ewin. "George Bush: 'God Told Me to End the Tyranny in Iraq.'" *Guardian*, 2005.

Manly, Steven L. *Visions of the Multiverse.* Pompton Plains, NJ: New Page Books, 2011.

Martinetti, Pierre. "Emergence of Time in Quantum Gravity: Is Time Necessarily Flowing?" J arXiv preprint arXiv:1203.4995. 2012.

Massey, Richard, Thomas Kitching, and Johan Richard. "The Dark Matter of Gravitational Lensing." *Reports on Progress in Physics* 73, no. 8 (2010): 086901.

Matarrese, Sabiuno, et al., eds. *Dark Matter and Dark Energy: A Challenge for Modern Cosmology.* Dordrecht: Springer, 2011.

Mather, John C., et al. "Measurement of the Cosmic Microwave Background Spectrum by the COBE FIRAS Instrument." *Astrophysical Journal* 420 (1994): 439–44.

Matthews, J. G., T. Kajino, and T. Shima. "Big Bang Nucleosynthesis with a New Neutron Lifetime." *Physical Review D* 71 (2005): 021302.

Maxwell, James Clerk. *The Scientific Papers of James Clerk Maxwell.* New York: Dover, 1965.

McLaughlin, P. J. *The Church and Modern Science.* New York: Philosophical Library, 1957.

Milgrom, Mordehai. "A Modification of the Newtonian Dynamics as a Possible Alternative to the Hidden Mass Hypothesis." *Astrophysical Journal* 270 (1983): 365–70.

Milne, Edward Arthur. *Kinematic Relativity; a Sequel to Relativity, Gravitation and World Structure.* Oxford: Clarendon, 1948.

———. *Modern Cosmology and the Christian Idea of God.* Oxford: Clarendon, 1952.

———. *Relativity, Gravitation and World-Structure.* Oxford: Clarendon, 1935.

Milstead, D., and E. J. Weinberg. "Magnetic Monopoles." 2011. http://pdg.web.cern.ch/pdg/2011/reviews/rpp2011-rev-mag-monopole-searches.pdf. Accessed March 9, 2013.

Moffat, J. W. "A New Nonsymmetric Gravitational Theory." *Physics Letters B* 355, no. 3 (1995): 447–52.

Moreva, Ekaterina, et al. "Time from Quantum Entanglement: An Experimental Illustration." arXiv preprint arXiv:1310.4691. 2013.

NASA Ames Research Center. "Kepler Discoveries." 2013. http://kepler.nasa.gov/Mission/discoveries/. Accessed November 6, 2013.

NASA/WMAP Science Team. "WMAP: Build a Universe." 2012. http://wmap.gsfc.nasa.gov/resources/camb_tool/index.html. Accessed May 11, 2013.

Nemiroff, Robert J., et al. "Bounds on Spectral Dispersion from Fermi-Detected Gamma Ray Bursts." *Physical Review Letters* 108, no. 23 (2012): 231103.

Netterfield, C. Barth, et al. "A Measurement of the Angular Power Spectrum of the Anisotropy in the Cosmic Microwave Background." *Astrophysical Journal* 474, no. 1 (1997): 47.

Neugebauer, Otto. *Notes on Hipparchus.* Locust Valley, NY: J. J. Augustin, 1956.

Nickell, Joe. *The Science of Miracles: Investigating the Incredible.* Amherst, NY: Prometheus Books, 2013.

Nolta, M. R., et al. "Five-Year Wilkinson Microwave Anisotropy Probe Observations: Angular Power Spectra." *Astrophysical Journal Supplement Series* 180, no. 2 (2009): 296.

North, John David. *Cosmos: An Illustrated History of Astronomy and Cosmology.* Chicago: University of Chicago Press, 2008.

Olive, Keith A., and Evan D. Skillman. "A Realistic Determination of the Error on the Primordial Helium Abundance: Steps Toward Nonparametric Nebular Helium Abundances." *Astrophysical Journal* 617, no. 1 (2004): 29.

Orr, Mary Ackworth. *Dante and the Early Astronomers*. Port Washington, NY: Kennikat, 1969; first published in 1914, London, Edinburgh: Gall and Inglis.

Ostriker, Jeremiah P., and Simon Mitton. *Heart of Darkness: Unraveling the Mysteries of the Invisible Universe*. Princeton, NJ; Oxford: Princeton University Press, 2013.

Overbye, Dennis. "Essay; Remembering David Schramm, Gentle Giant of Cosmology." *New York Times*, February 10, 1998.

——. "Space Ripples Reveal Big Bang's Smoking Gun." *New York Times*, March 17, 2014.

Padmanabhan, Thanu. "Grappling with Gravity." *Scientific American India*, January, 2011, pp. 31–35.

Page, Don. "Does God So Love the Multiverse?" In *Science and Religion in Dialogue*, edited by Melville Y. Stewart, 380–95. Chichester, UK: Wiley-Blackwell, 2010.

Page, Don N., and William K. Wootters. "Evolution without Evolution: Dynamics Described by Stationary Observables." *Physical Review D* 27, no. 12 (1983): 2885.

Pais, Abraham. *Subtle Is the Lord: The Science and the Life of Albert Einstein*. Oxford; New York: Oxford University Press, 1982.

Peebles, P. J. E. *The Large-Scale Structure of the Universe*. Princeton, NJ: Princeton University Press, 1980.

Peebles, Philip J. E., and J. T. Yu. "Primeval Adiabatic Perturbation in an Expanding Universe." *Astrophysical Journal* 162 (1970): 815–36.

Penrose, Roger. *The Emperor's New Mind: Concerning Computers, Minds, and the Laws of Physics*. Oxford; New York: Oxford University Press, 1989.

——. *The Road to Reality: A Complete Guide to the Laws of the Universe*. London: Jonathan Cape, 2004.

Penzias, Arno A., and Robert W. Wilson. "A Measurement of Excess Antenna Temperature at 4080 Mc/s." *Astrophysical Journal* 142 (1965): 419–21.

Perlmutter, Saul, et al. "Measurements of Ω and Λ from 42 High-Redshift Supernovae." *Astrophysical Journal* 517, no. 2 (1999): 565.

Petigura, Erik A., Andrew W. Howard, and Geoffrey W. Marcy. "Prevalence of Earth-Size Planets Orbiting Sun-like Stars." *Proceedings of the National Academy of Sciences* (2013): 19175–76.

Planck Collaboration, P. A. R. Ade, et al. "Planck 2013 Results. I. Overview of Products and Scientific Results." arXiv preprint arXiv:1303.5062. 2013.

——. "Planck 2013 Results. XXIII. Isotropy and Statistics of the CMB." arXiv preprint arXiv:1303.5083. 2013.

Pogosian, Dmitri. "Astronomy 122: Astronomy of Stars and Galaxies." 2013. http://www.ualberta.ca/~pogosyan/teaching/ASTRO_122/lect27/lecture27.html. Accessed February 21, 2013.

Polyakov, Alexander M. "Particle Spectrum in Quantum Field Theory." *Soviet Journal of Experimental and Theoretical Physics Letters* 20 (1974): 194–95.

Pope Pius XII. "The Proofs for the Existence of God in the Light of Modern Science." Address of Pope Pius XII to the Pontifical Academy of Sciences, November 22, 1951. http://www.papalencyclicals.net/Pius12/P12EXIST.HTM. Accessed January 27, 2013.

———. "The Proofs for the Existence of God in the Light of Modern Natural Science." *Catholic Mind* 49 (1972): 182–92.

Price, Huw. *Time's Arrow & Archimedes' Point: New Directions for the Physics of Time.* New York: Oxford University Press, 1996.

Price, Robert M., and Edwin A. Suiminen. *Evolving out of Eden.* Valley, WA: Tellectual, 2013.

Riess, Adam G., et al. "Observational Evidence from Supernovae for an Accelerating Universe and a Cosmological Constant." *Astronomical Journal* 116, no. 3 (1998): 1009.

Roberts, B. H. *New Witness for God.* Salt Lake City, UT: George Q. Canon & Sons, 1898.

Robertson, Howard P. "On the Foundations of Relativistic Cosmology." *Proceedings of the National Academy of Sciences* 15, no. 11 (1929): 822–29.

Robinson, George L. *Leaders of Israel: a Brief History of the Hebrews from the Earliest Times to the Downfall of Jerusalem, A.D. 70.* New York: International Committee of Young Men's Christian Associations, 1906.

Rogers, Eric M. *Physics for the Inquiring Mind; the Methods, Nature, and Philosophy of Physical Science.* Princeton, NJ: Princeton University Press, 1960.

Ross, Hugh. "Big Bang Model Refined By Fire." In *Mere Creation: Science, Faith & Intelligent Design,* edited by William A. Dembski, 363–83. Downers Grove, IL: Intervarsity, 1998.

———. *The Creator and the Cosmos: How the Greatest Scientific Discoveries of the Century Reveal God.* Colorado Springs, CO: NavPress, 1995.

Roth, Philip. *The Plot against America.* Boston: Houghton Mifflin, 2004.

Rundle, Bede. *Why There Is Something Rather Than Nothing.* Oxford; New York: Clarendon Oxford University Press, 2004.

Sachs, Rainer K., and Arthur M. Wolfe. "Perturbations of a Cosmological Model and Angular Variations of the Microwave Background." *Astrophysical Journal* 147 (1967): 73.

Sagan, Carl. *Cosmos.* New York: Random House, 1980.

Sakharov, Andrei. "Vacuum Quantum Fluctuations in Curved Space." *Doklady Akademii Nauk SSSR* 177, no. 1 (1967): 70–71.

———. "Violation of CP Invariance, C Asymmetry, and Baryon Asymmetry of the Universe." *Journal of Experimental and Theoretical Physics Letters* 5 (1967): 24–27.

Salpeter, E. E. "Nuclear Reactions in Stars without Hydrogen." *Astrophysical Journal* 115 (1952): 326–28.

Sato, Katsuhiko. "First-Order Phase Transitions of a Vacuum and the Expansion of the Universe." *Monthly Notices of the Royal Astronomical Society* 195 (1981): 467–79.

Schellenberg, J. L. *Divine Hiddenness and Human Reason*. Ithaca, NY: Cornell University Press, 1993.

Schmidt, Maarten. "3C 273: A Star-like Object with Large Red-Shift." *Nature* 197 (1963): 1040.

Schombert, James. "Cosmology: Course at the University of Oregon." 2012. http://abyss.uoregon.edu/~js/cosmo/. Accessed June 6, 2014.

Schönborn, Christoph. "Finding Design in Nature." *New York Times*, July 7, 2005.

Schwinger, Julian. "On Quantum-Electrodynamics and the Magnetic Moment of the Electron." *Physical Review* 73, no. 4 (1948): 416–17.

"Scientists Gather for 1920 Conclave." *Washington Post*, April 25, 1920, p. 38.

Sedley, D. N. *Creationism and Its Critics in Antiquity*. Berkeley: University of California Press, 2007.

Seljak, Uros, and Matias Zaldarriaga. "A Line of Sight Approach to Cosmic Microwave Background Anisotropies." *Astrophysical Journal* 469 (1996): 437–44.

Shapley, Harlow, and Herber D. Curtis. "The Scale of the Universe." *Bulletin of the National Research Council* 2, no. 11 (1921): 171–217.

Shostak, Seth. "The Numbers are Astronomical." 2013. http://www.huffingtonpost.com/seth-shostak/the-numbers-are-astronomi_b_4214484.html. Accessed November 6, 2013.

Silk, Joseph. *The Big Bang*. New York: W. H. Freeman, 2001.

——. "Cosmic Black-Body Radiation and Galaxy Formation." *Astrophysical Journal* 151 (1968): 459–71.

Silvestri, Alessandra, and Mark Trodden. "Approaches to Understanding Cosmic Acceleration." *Reports on Progress in Physics* 72 (2009): 096901.

Slipher, Vesto M. "Spectrographic Observations of Nebulae." *Popular Astronomy* 23 (1915): 21–24.

Smith, George. *The Chaldean Account of Genesis, Containing the Description of the Creation, the Fall of Man, the Deluge, the Tower of Babel, the Times of the Patriarchs, and Nimrod: Babylonian Fables, and Legends of the Gods; From the Cuneiform Inscriptions*. New York: Scribner, Armstrong, 1876.

Smith, Robert. "Edwin P. Hubble and the Transformation of Cosmology." *Physics Today* 43, no. 4 (1990): 52–58.

Smolin, Lee. *The Life of the Cosmos*. New York: Oxford University Press, 1997.

——. "A Perspective on the Landscape Problem." 2012. http://arxiv.org/pdf/1202.3373v1.pdf. Accessed September 5, 2012.

——. "The Status of Cosmological Natural Selection." 2006. http://arxiv.org/pdf/hep-th/0612185v1.pdf. Accessed September 5, 2012.

——. *Three Roads to Quantum Gravity*. New York: Basic Books. 2002.

Smoot, George, et al. "COBE Differential Microwave Radiometers-Instrument design and implementation." *Astrophysical Journal* 360 (1990): 685–95.

Smoot, George, and Keay Davidson. *Wrinkles in Time*. New York: W. Morrow, 1993.

Smoot, George F., et al. "Structure in the COBE Differential Microwave Radiometer First-Year Maps." *Astrophysical Journal* 396 (1992): L1–L5.

Sobel, Dava. *A More Perfect Heaven: How Copernicus Revolutionized the Cosmos*. New York: Walker, 2011.

Spergel, D. N., et al. "Wilkinson Microwave Anisotropy Probe (WMAP) Three Year Results: Implications for Cosmology." *Astrophysical Journal Suppl.* 170 (2007): 377.

Stacey, Frank. "Kelvin's Age of the Earth Paradox Revisited." *Journal of Geophysical Research* 105, no. B6 (2000): 155–58.

Stallings, A. E., and Richard Jenkyns. *Lucretius: The Nature of Things*. London; New York: Penguin, 2007.

Starobinsky, Alexei A. "A New Type of Isotropic Cosmological Models without Singularity." *Physics Letters B* 91, no. 1 (1980): 99–102.

Stecker, Floyd W. "IceCube Observed PeV Neutrinos from AGN Cores." J arXiv preprint arXiv:1305.7404. 2013.

Steigman, Gary, David N. Schramm, and James E. Gunn. "Cosmological Limits to the Number of Massive Leptons." *Physics Letters B* 66, no. 2 (1977): 202–204.

Steinhardt, Paul J., and Neil Turok. *Endless Universe: Beyond the Big Bang*. New York: Doubleday, 2007.

Stenger, Victor J. *The Comprehensible Cosmos: Where Do the Laws of Physics Come From?* Amherst, NY: Prometheus Books, 2006.

——. *The Fallacy of Fine-Tuning: Why the Universe Is Not Designed for Us*. Amherst, NY: Prometheus Books, 2011.

——. *God: the Failed Hypothesis. How Science Shows That God Does Not Exist*. Amherst, NY: Prometheus Books, 2007.

——. *God and the Atom: From Democritus to the Higgs Boson*. Amherst, NY: Prometheus Books, 2013.

——. *Has Science Found God? The Latest Results in the Search for Purpose in the Universe*. Amherst, NY: Prometheus Books, 2003.

——. "Neutrino Oscillations in DUMAND." In *Neutrino Mass: Mini-Conference and Workshop Transparencies*, edited by V. Barger, and D. Cline, 174–78. Madison: University of Wisconsin, 1980.

——. *Not By Design: the Origin of the Universe*. Amherst, NY: Prometheus Books, 1988.

———. "The Production of Very High Energy Photons and Neutrinos from Cosmic Proton Sources." *Astrophysical Journal* 284 (1984): 810–16.

———. *Quantum Gods: Creation, Chaos, and the Search for Cosmic Consciousness.* Amherst, NY: Prometheus Books, 2009.

———. "A Scenario for the Natural Origin of the Universe." *Philo* 9, no. 2 (2006): 93–102.

———. *Timeless Reality: Symmetry, Simplicity, and Multiple Universes.* Amherst, NY: Prometheus Books, 2000.

———. *The Unconscious Quantum: Metaphysics in Modern Physics and Cosmology.* Amherst, NY: Prometheus Books, 1995.

———. "The Universe Shows No Evidence for Design." In *Debating Christian Theism*, edited by J. P. Moreland, Chad Meister, and Khaldoun A. Sweis, 47–58. Oxford; New York: Oxford University Press, 2013.

Subrahmanyan, Ravi, et al. "A Search for Arcminute Scale Anisotropy in the Cosmic Microwave Background." *Monthly Notices of the Royal Astronomical Society* 263, no. 2 (1993): 416.

Sunyaev, Rashid A., and Ya B. Zel'dovich. "Small-Scale Fluctuations of Relic Radiation." *Astrophysics and Space Science* 7, no. 1 (1970): 3–19.

Super-Kamiokande Collaboration. "Evidence for Oscillation of Atmospheric Neutrinos." *Physical Review Letters* 81 (1998): 1562–67.

———. "Proton Lifetime Is Longer Than 1034 Years." 2011. http://www-sk.icrr.u -tokyo.ac.jp/whatsnew/new-20091125-e.html. Accessed February 22, 2013.

Susskind, Leonard. *Cosmic Landscape: String Theory and the Illusion of Intelligent Design.* New York: Little, Brown, 2005.

Svoboda, R., and K. Gordan. "Neutrinos in the Sun." 1998. http://apod.nasa.gov/apod/ap980605.html. Accessed January11, 2012.

Tegmark, Max. "The Mathematical Universe." *Foundations of Physics* 38, no. 2 (2008): 101–50.

———. *Our Mathematical Universe: My Quest for the Ultimate Nature of Reality.* New York: Alfred A. Knopf, 2014.

Tegmark, Max, and Martin J. Rees. "Why Is the CMB Fluctuation Level 10–5?" *Astrophysical Journal* 499 (1998): 526–32.

Tegmark, Max, et al. "Cosmological Parameters from SDSS and WMAP." *Physical Review D* 69, no. 10 (2004): 103501.

Tegmark, Max, Matias Zaldarriaga, and Andrew Hamilton. "Towards a Refined Cosmic Concordance Model: Joint 11-Parameter Constraints from the Cosmic Microwave Background and Large-Scale Structure." *Physical Review* D63 (2000): 043007.

Thompson, Darcy Wentworth. *On Growth and Form.* Cambridge: Cambridge University Press, 1992.

Thompson, Gary D. "The Development, Heyday, and Demise of Panbabylonism." 2012. http://members.westnet.com.au/Gary-David-Thompson/page9e.html. Accessed October 28, 2012.

't Hooft, Gerard. "Magnetic Monopoles in Unified Gauge Theories." *Nuclear Physics B* 79, no. 2 (1974): 276–84.

Tolman, Richard C. *Relativity, Thermodynamics and Cosmology*. Oxford: Clarendon, 1934.

Tomonoga, Sin-Itiro. "On a Relativistically Invariant Formulation of the Quantum Theory of Wave Fields." *Progress in Theoretical Physics* 1, no. 2 (1946): 27–42.

Tong, David. "Is Quantum Reality Analog after All?" *Scientific American*, November 26, 2012.

Tully, R. Brent, and J. Richard Fisher. *Nearby Galaxies Atlas*. Cambridge; New York: Cambridge University Press, 1987.

———. "A New Method of Determining Distances to Galaxies." *Astronomy and Astrophysics* 54 (1977): 661–73.

Turner, Michael S. "David Norman Schramm 1945–1997." 2009. http://www.nasonline.org/publications/biographical-memoirs/memoir-pdfs/schramm-david.pdf. Accessed February 6, 2012.

"Universe Multiplied a Thousand Times by Harvard Astronomer's Calculations." *New York Times*, May 31, 1921, p. 1.

Van Bebber, Mark. "What Does the Bible Say about Intelligent Life on Other Planets?" *Christian Answers*. 2014. http://christiananswers.net/q-eden/edn-c012.html?zoom_highlight=intelligent+life. Accessed February 17, 2014.

Verlinde, Erik. "On the Origin of Gravity and the Laws of Newton." *Journal of High Energy Physics* 2011, no. 4 (2011): 1–27.

Vilenkin, Alexander. "Birth of Inflationary Universes." *Physical Review D* 27, no. 12 (1983): 2848–55.

———. "Boundary Conditions in Quantum Cosmology." *Physical Review* D33 (1986): 3560–69.

———. "Creation of Universes from Nothing." *Physics Letters B* 117, no. 1 (1982): 25–28.

———. *Many Worlds In One: The Search for Other Universes*. New York: Hill and Wang, 2006.

———. "Predictions from Quantum Cosmology." *Physical Review Letters* 74 (1995): 846.

Von Helmholtz, Hermann. *Science and Culture. Popular and Philosophical Essays*. Chicago: Chicago University Press, 1995.

Von Weizäcker, Carl Friedrich. "Element Transformation inside Stars." *Physicalische Zeitschrift* 39 (1938): 633–46.

Walmsley, C. M., et al. "Herschel: The First Science Highlights." *Astronomy &Astrophysics* 518 (2010): L10.

Weekes, Trevor C., et al. "Observation of TeV Gamma Rays from the Crab Nebula Using the Atmospheric Cerenkov Imaging Technique." *Astrophysical Journal* 342 (1989): 379–95.

Weinberg, Steven. "The Cosmological Constant Problem." *Reviews of Modern Physics* 61, no. 1 (1989): 1–23.

———. *Dreams of a Final Theory.* New York: Pantheon Books, 1992.

———. *The First Three Minutes: A Modern View of the Origin of the Universe.* New York: Basic Books, 1993.

———. *Gravitation and Cosmology.* New York: Wiley, 1972.

———. "Living in the Multiverse." In *Universe Or Multiverse?* edited by B. Carr, chap. 2. Cambridge: Cambridge University Press, 2007.

Wells, Steve. *The Skeptics Annotated Bible.* Moscow, ID: SAB Books, 2012.

Wheeler, John A. "Superspace and the Nature of Quantum Geometrodynamics." In *Batelle Recontres,* edited by C. M. DeWitt and J. A. Wheeler. New York: Benjamin, 1968.

Wheeler, John Archibald, and Kenneth William Ford. *Geons, Black Holes, and Quantum Foam: A Life in Physics.* New York: Norton, 1998.

White, Andrew Dickson. *A History of the Warfare of Science with Theology in Christendom: Two Volumes in One.* Amherst, NY: Prometheus Books, 1993.

Whitle, Mark. "Big Bang Acoustics." 2000. http://www.astro.virginia.edu/~dmw8f/BBA_web/index_frames.html. Accessed May 12, 2013.

———. *Cosmology: The History and Nature of Our Universe.* Chantily, VA: Teaching, 2008.

Wilczek, Frank. "The Cosmic Asymmetry between Matter and Antimatter." *Scientific American* 243, no. 6 (1980): 82–90.

Will, Clifford M. "The Confrontation between General Relativity and Experiment." *Living Reviews in Relativity* 9, no. 3 (2006).

———. *Was Einstein Right? Putting General Relativity to the Test.* New York: Basic Books, 1986.

Wootters, William K. "'Time' Replaced by Quantum Correlations." *International Journal of Theoretical Physics* 23, no. 8 (1984): 701–11.

Wright, Edward L. "Big Bang Nucleosynthesis." 2012. http://www.astro.ucla.edu/~wright/BBNS.html. Accessed February 7, 2013.

———. "Ned Wright's Cosmological Tutorial." 2013. http://www.astro.ucla.edu/~wright/cosmolog.htm. Accessed February 14, 2013.

Wright, E. L., et al. "Interpretation of the Cosmic Microwave Background Radiation Anisotropy Detected by the COBE Differential Microwave Radiometer." *Astrophysical Journal* 396 (1992): L13–L18.

Wyman, Mark, et al. "Neutrinos Help Reconcile Planck Measurements with the Local Universe." *Physical Review Letters* 112, no. 5 (2014): 051302.

Zehavi, I., and A. Dekel. "Evidence for a Positive Cosmological Constant from Flows of Galaxies and Distant Supernovae." *Nature* 401 (1999): 252–54.

Zel'dovich, Yakov B. "A Hypothesis, Unifying the Structure and the Entropy of the Universe, 1972." *Monthly Notices of the Royal Astronomical Society* 160 (1972): 1–3.

Zuckerman, Miron, Jordan Silberman, and Judith A. Hall. "The Relation between Intelligence and Religiosity: A Meta-Analysis and Some Proposed Explanations." *Personality and Social Psychology Review* 17, no. 4 (2013): 325–54.

ABOUT THE AUTHOR

Victor J. Stenger is a retired elementary particle physicist and the author of twelve previous books, including the 2007 *New York Times* bestseller *God: The Failed Hypothesis. How Science Shows That God Does Not Exist.*

Dr. Stenger grew up in a Catholic working-class neighborhood in Bayonne, New Jersey. His father was a Lithuanian immigrant; his mother, the daughter of Hungarian immigrants. He attended public schools and received a bachelor of science degree in electrical engineering from Newark College of Engineering (now New Jersey Institute of Technology) in 1956. While at NCE, he was editor of the student newspaper and received several journalism awards.

Moving to Los Angeles on a Hughes Aircraft Company fellowship, Dr. Stenger received a master of science degree in physics from UCLA in 1959 and a doctorate in physics in 1963. He then took a position on the faculty of the University of Hawaii and retired to Colorado in 2000. His current positions include emeritus professor of physics at the University of Hawaii and adjunct professor of philosophy at the University of Colorado. Dr. Stenger has also held visiting positions on the faculties of the University of Heidelberg in Germany and Oxford University in England, and he has been a visiting researcher at Rutherford Laboratory in England; the National Nuclear Physics Laboratory in Frascati, Italy; and the University of Florence in Italy.

His research career spanned the period of great progress in elementary particle physics that ultimately led to the current standard model. He participated in experiments that helped establish the properties of strange particles, charmed quarks, gluons, and neutrinos. He also helped pioneer the emerging fields of very high-energy gamma ray and neutrino astronomy. In his last project before retiring, Dr. Stenger collaborated on the underground experiment in Japan that showed for the first time that the neutrino has mass. The Japanese leader of the project, Masatoshi Koshiba, shared the 2002 Nobel Prize for Physics for that work.

Victor J. Stenger has had a parallel career as an author of critically well-received popular-level books that interface between physics and cosmology and philosophy, religion, and pseudoscience.

Vic and his wife, Phylliss, have been happily married since 1962 and have two children and four grandchildren.

Dr. Stenger maintains a website where much of his writing can be found, at http://www.colorado.edu/philosophy/vstenger/.

INDEX